THE COMPLETE BOOK OF
CORVETTE

EVERY MODEL SINCE 1953

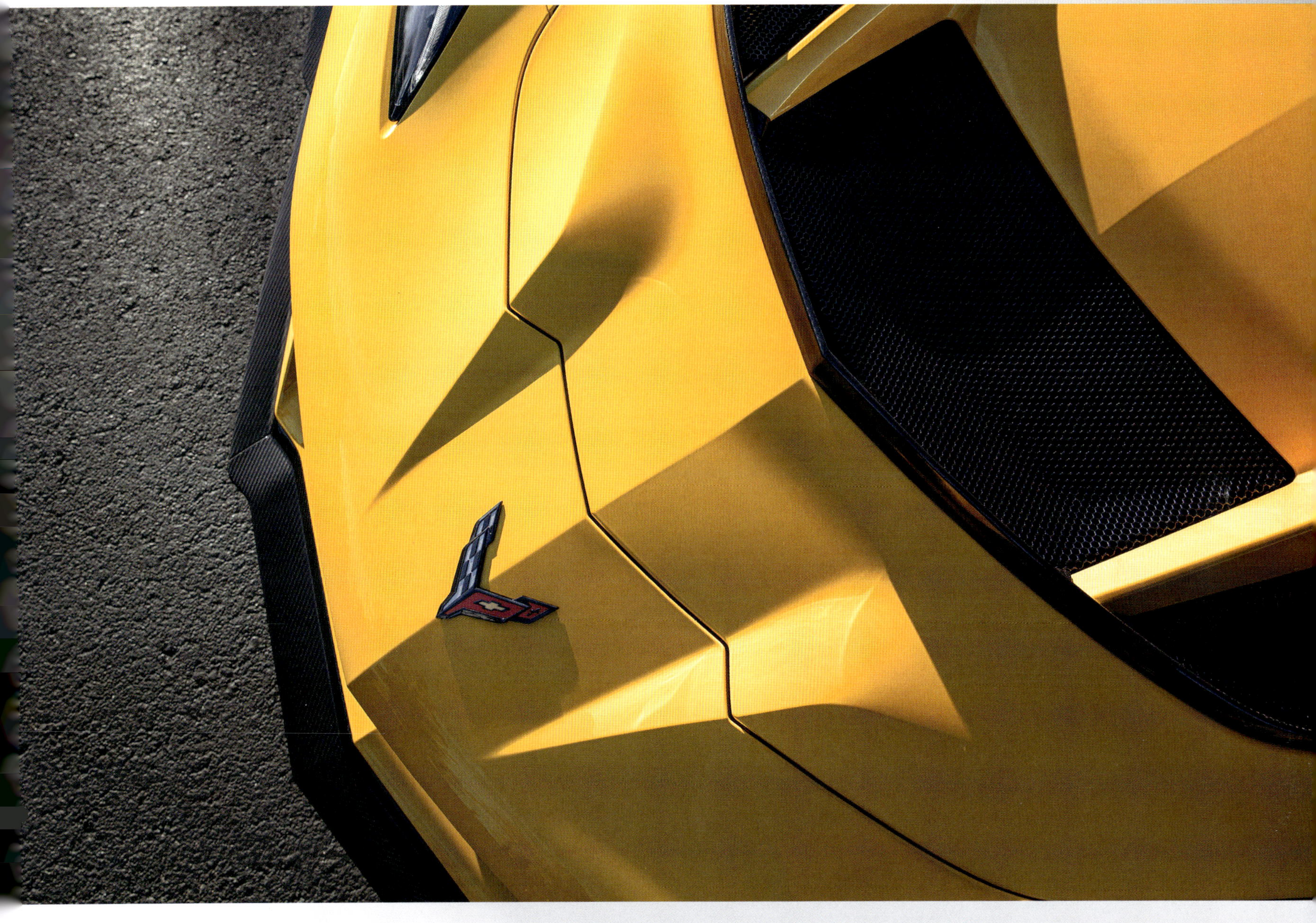

MIKE MUELLER

For my GM heroes Christo "The Great" Datini, Amazin' Larry Kinsel, and John "Just John" Kyros for sending this ol' cowboy into the sunset with a smile. And for right-hand girl Erin Welker, who taught me that it's only wise to be nicer to your editor. And for Jim and Tenna, the greatest Tigers fans in the history of Tigers fans.

Quarto.com

© 2025, 2020, 2014 Quarto Publishing Group USA Inc
Text © 2025, 2020, 2014 Mike Mueller
Photographs © 2025 General Motors, except where noted.

General Motors trademarks used under license to Quarto Publishing Group USA Inc.

First published in 2014 by Motorbooks, an imprint of The Quarto Group,
100 Cummings Center, Suite 265-D, Beverly, MA 01915, USA.
T (978) 282-9590 F (978) 283-2742

All rights reserved. No part of this book may be reproduced in any form without written permission of the copyright owners. All images in this book have been reproduced with the knowledge and prior consent of the artists concerned, and no responsibility is accepted by producer, publisher, or printer for any infringement of copyright or otherwise, arising from the contents of this publication. Every effort has been made to ensure that credits accurately comply with information supplied. We apologize for any inaccuracies that may have occurred and will resolve inaccurate or missing information in a subsequent reprinting of the book.

Motorbooks titles are also available at discount for retail, wholesale, promotional, and bulk purchase. For details, contact the Special Sales Manager by email at specialsales@quarto.com or by mail at The Quarto Group, Attn: Special Sales Manager, 100 Cummings Center, Suite 265-D, Beverly, MA 01915, USA.

10 9 8 7 6 5 4 3 2 1

ISBN: 978-0-7603-9447-2

The Library of Congress has cataloged the previous edition as follows:
Mueller, Mike, 1959-
    The complete book of Corvette : every model since 1953 / by Mike Mueller.—
        [Revised and updated]. pages cm
M's Corvette, from the first models in the early 50's up to the forthcoming 2014 Stingray. Complete Book of Corvette features over 400 archive photos from GM as well as many from the author's personal collection. Detailed spec sidebars and an appendix featuring every option available on every vehicle round out the book's historical content"—Provided by publisher.
    ISBN 978-0-7603-4574-0 (hardback)
  1. Corvette automobile—History. 2. Automobiles—United States. I. Title.
    TL215.C6M839725 2014
    629.222'2—dc23

2013047482

Editors: Jordan Wiklund and Zack Miller
Design: John Sticha and Rebecca Pagel
Layout: Rebecca Pagel and Cindy Samargia Laun

Printed in China

Front cover: Image courtesy of General Motors
Back cover: Image courtesy of General Motors
Back flap: Image courtesy of Mike Mueller
Frontis page: Image courtesy of General Motors
Title page: Image courtesy of General Motors

Contents

Acknowledgments 10
Introduction: The Stuff of Dreams 12

01 The Solid-Axle Years 16

1953 21
1954 24
1955 26
1956 28
1956 SR-2 31
1957 33
1957 SS racer 36
1958 40
1959 42
1959 Stingray racer 45
1960 47
CERV I 49
1961 51
Mako Shark I 53
1962 54

02 Enter the Sting Ray 58
1963 63
1963 Grand Sport 68
1964 71
CERV II 73
1965 74
Mako Shark II 77
1966 79
1967 82

03 Third Time's a Charm 88
1968 92
XP-882 94
1969 96
Manta Ray Show Car 101
1970 101
1971 106
XP-895 110
1972 111
Rotary Corvettes 112
1973 114
1974 117
1975 120
1976 122
1977 124
1978 125
1979 129
1980 130
1981 132
1982 136

04 Better Late than Never 140

1984	146
1985	149
1986	150
Corvette Indy	151
1987	154
1988	156
Corvette GTP	157
1989	158
Corvette Challenge racers	160
1990	161
CERV III	162
1991	165
1992	169
Sting Ray III	171
1993	172
1994	173
1995	175
1996	178

05 Long Live the King: ZR-1 186

1990	194
1991	197
1992	199
1993	200
1994	202
1995	203

06 50 Years Young 206
1997 210
1998 213
1999 216
C5-R racer 217
2000 219
2001 219
2002 222
2003 223
2004 226

07 Thoroughly Modern 228
2005 231
C6.R racer 232
2006 233
2007 236
2008 238
2009 241
2009 ZR1 243
2010 246
2011 248
2012 249
2013 253

08 World Beater 258
2014 262
2015 267
2015 Z06 269
C7.R racer 273
2016 275
2017 277
2018 279
2019 281

09 Makin' the Ol' Man Proud 284
2020 292
C8.R racer 296
2021 296
2022 300
2023 303
2023 Z06 304
2024 307
2024 E-Ray 309
2024 GT3.R racer 313
2025 314
2025 ZR1 318
Epilogue 322

Appendix 324
Index 348

ACKNOWLEDGMENTS

Acknowledgments

First and foremost, this book's author, Mike Mueller, has to thank his mother, Nancy, for making literally everything possible. Here she pauses during a 1994 drive in Champaign, Illinois, in Ray Quinlan's 1963 Sting Ray. Quinlan, at right in his unrestored 260 Cobra, was the National Corvette Museum's numero uno lifetime member, an honor he earned by donating his 1953 Corvette to the Bowling Green institution well before plans had even coalesced to build the building. Along with helping supply initial momentum to the NCM project, Quinlan also was guilty of opening early doors into the automotive journalism world for Mueller dating back to Mike's days at the University of Illinois. A World War II vet (a B-24 gunner in the Pacific theater), Ray died in September 2003. The devoted wife of now-retired machinist Jim Mueller for 56 years, Nancy succumbed to Alzheimer's in October 2019. Both are missed beyond words. *Mike Mueller*

AS DEVOTED FANS (all five of ya) may know, this is my fifth round detailing as much Corvette history as possible in a few hundred pages. After all, "Complete" usually means "complete," no? Originally contracted nearly two decades back, this epic's first edition ended with 2006's best Vette yet. The updated second edition closed on 2011's, then our third edition introduced you to 2014's world-beating seventh generation. But of course really big news followed in 2020 of the historic supercar that surely would've made Zora Duntov proud: the midengine C8, which we covered that year in our fourth edition.

Yet no matter how hefty the headlines, they always fade soon enough, mostly to make room for the latest prime notices. That being the unstoppable case, here we are, back again, to update ya'll with a fifth edition, this one predictably filling in the tale from 2021 to 2025.

Before you strap in for this E-ticket ride, I ask you to please take note of everyone who helped (and has helped) make this amazingly effective doorstop possible. Targeted foremost is my way-too-wise boss, Zack Miller, who continues (despite all that acumen) to send jobs this 65-year-old's way after some 32 years of trying to understand the Mueller definition of deadline. As he knows too well, I've never met one I couldn't miss. Kudos, too, go to the Z-man for somehow coaxing me back out of retirement for this. And I was napping so nicely . . .

At least Zack hasn't had to work as closely with me over the years as his assignment-editor extraordinaires: Lindsay Hitch, Jeffrey Zuehlke, Jordan Wiklund, and Jessi Schatz, the prime movers behind *Complete Book of Corvette*'s previous four production runs. Endless appreciation to each of you, wherever you are, including whatever mental state you're presently hiding in. And that brings us to Brooke Pelletier, who, with any luck, will escape her turn in the barrel while retaining a marble or two.

As for my biggest fan, always lovely Nancy Lou Mueller, I'll never forget everything you did during our 60 years together to truly make all this happen. Won't ever stop missing you, Ma. Same for my little brother Jim, whom we lost four years prior in 2014. As big a car guy as they come, Jimbo always was there to help, not to mention crank up the fun-o-meter to 11. Also lost my pa, Jim Mueller Sr., and another brother, Dave, since I last took to this soapbox. Of course, I wish yuze two guyz were here, too, for this, and not just because both also played such large roles over the years helping me get so many of these words/photos (plus so many others) to print. Think of you all every day.

I'd also like to thank my littlest bro, Ken. I'd like to but can't due to the fact he's never lifted a finger to help. But fortunately he can roll a mean game of Strat-O-Matic, football and/or baseball. Kenny, I only have two words for you: "Free Safety Blitz!" No, wait . . .

My sister, Kathy Young, and her husband, Frank, deserve beaucoup credit as well, most prominently

ACKNOWLEDGMENTS

Mueller with his first Corvette, an inspirational Christmas gift from his mother circa 1960. This 2-footpower vehicle was last seen, sans steering wheel, in great-grandfather Sam Hart's junk collection a few years later. *Nancy Mueller*

Mueller (left) first met the Corvette's revered "father," Zora Arkus-Duntov, during the National Corvette Museum's grand opening gala in September 1994. Yes, the two drinks were Duntov's. And they were stirred, not shaken. *John Heilig*

for all the good eats, endless Big Gulp Dr Peppers, and an always-warm bed awaiting me back home in Champaign, Illinois, after so many 500-mile days spent chasing down photographic subjects for this project and countless others around the Midwest. Hard to believe all those events are so many years ago now, right Kathy? But then you're still 29, no? Geez, I've got hangnails that old.

Speaking of photo ops, brother Jim's three Corvettes appear on following pages, as do many dozens of others that I had the pleasure of shooting—with a Hasselblad, a *real* camera. The list of gracious owners who mostly long ago also allowed me to capture their pride-and-joys (yes, on film) is extensive, a bit too long to fit here. So I hope you all will forgive me if I simply offer one blanket "thanks." Know that I know who you all are, and I still remember you quite fondly.

More gratitude goes way back to Larry Kinsel and Peggy Kelly at General Motors' Media Archives in Detroit, neither of whom ever turned me away whenever I came knocking. Again, how time flies. Larry actually tried to get away during the interim between edition number four and this one, but I found him. You can run but you can't hide, right my old friend?

Speaking further of ancient history, Matt Currin and Bob Tate also were of great help during my earliest photo searches in 2006, which began up Renaissance Center way with a hands-on review of 12,000 images—in one day. Man, did I have fun or what? Can't recall on which floor, but in my defense, there are a lot there, am I right?

I furthermore need to mention Mike Antonick, who allowed material from his annual *Corvette Black Book* to be included on the final pages here.

Readers interested in the latest update of this invaluable reference source can check out www.corvetteblackbook.com. Trust me, you want this. It contains far, far more than we've excerpted. Talk about big bang for the buck.

Now we find ourselves at the present, where there remains a long list of GM people who're still left wondering what the heck hit 'em. I don't believe I've ever known more people who did more, who invested more time, who endured more, who put up with me more, longer than this. Winston Churchill might be proud this time.

Along with Larry Kinsel, there's John Kyros and Rebecca Bushman, both of whom apparently know a little about archiving photos—or a lot. John, you've heard this before, please listen again: you will be remembered in my will. But not too soon, mehopes.

Another longtime comrade, ace archivist (and veteran wiffleballer) Christo "The Great" Datini, wasn't even supposed "to be here." But he still stepped up, stepped in, and probably stepped in it too many times while repeatedly saving my bacon. In your case, TGD, please have a cauliflower on me. And also say hey to Jim and Tenna for me while you're at it. You know they're *my* ma and pa now, right? Also, if you find her, tell Tenna there's more than enough Willie Horton to go around. Go Tigers, Skubal forever!

An additional archives master (and former Corvette team engineering officer) Bernie Vascotto seemingly came out of nowhere to help expand my horizons during my search for the truth. And, as usual, Corvette product manager Harlan Charles set me straight more than once. Thanks, Harlan, for the great input. Priceless, I say, priceless.

But my fortunes didn't end there. Ed Piatek, former Corvette chief engineer and present chief engineer of performance electric vehicles (EV), somehow managed to find much time to do one of many things he does well: tell detailed, invaluable stories like he didn't have better things to do. Cauliflower, heck, Ed, in your case the steak and Lowenbrau's mine to serve. Gratitude of course goes as well to present Corvette engineering chief Josh Holder, along with GM Communications guys Chad Lyons and Trevor Thompkins.

I'll wind this down by extending ample thanks to one more old friend, Tom Read, who a few years back traded a Powertrain Communications position for a similar role at GM's Arlington Assembly plant in Texas—about the same day I was moving from Arlington to Atlanta. No, Tom, despite all appearances, I'm not trying to get out of that lunch offer. At least not as far as you know.

Let's not forget Kyle Cheromcha at *The Drive* and Joe Skibinski at the Indianapolis Motor Speedway for coming through at the last second with fantastic photo support. Many thanks also go to all the folks at thedrive.com, as well as those at the National Corvette Museum, gmauthority.com, and corvetteblogger.com for all the great reporting. Keep up the fabulous work, y'all. If you're not familiar with any of these sites, you need to be.

Last but certainly not least (if I know what's good for me), I can't help but lay a big hug on my best friend ever, Erin Welker, who also has emerged as my trusted editor, able-bodied (damn straight!) assistant, and . . . well, she's now my new boss.

So, give it a rest, will ya, Zack?

—*Mike Mueller, August 2024*

INTRODUCTION

> Suffice it to say that "icon" is not a big enough word, either for Chevrolet's beloved two-seater or the main man behind it.

Introduction
The Stuff of Dreams

Yes, it was finally true after so many years of rumors: 2020's all-new C8 Corvette featured a mid-mounted engine, just as the breed's first chief engineer, Zora Arkus-Duntov, long imagined.

EIGHTY-SIX-YEAR-OLD Zora Arkus-Duntov died in April 1996 never having fulfilled one of his fondest dreams. No worries, though. Most everything else came up roses for the legendary figure long revered as the father of the Corvette, an honor he earned by resuscitating an ailing foundling and raising it up right, remaking it into America's Sports Car. One mustard stain on his resume? We all should be so lucky.

Consider all the meat in his portfolio. For starters, no other domestic factory performance car has ever managed to put Duntov's baby in a corner. Original T-bird? Apples and oranges. Shelby-American's Cobra? Too crude and way too few and far between. AMX? From an entirely different league. Viper and Ford GT? Strong rivals, sure, at least as far as brutal sex appeal was concerned, but where was the staying power? Not to mention the relative affordability and unmatched comfort/convenience.

Suffice it to say that "icon" is not a big enough word, either for Chevrolet's beloved two-seater or the main man behind it. The few pages here, additionally, aren't sufficient to tell Duntov's storied story, an intriguing tale chock-full of automotive immortality. Hence his enshrinement into the Specialty Equipment Market Association Hall of Fame (1973), the Automotive Hall of Fame (1991), the International Drag Racing Hall of Fame (1994), and the National Corvette Museum Hall of Fame (1998). Pulitzer-winning journalist George Will even proclaimed in 1996 that "if you do not mourn his passing, you are not a good American."

Yet there remained the aforementioned history Zora failed to make: He never managed to build his supreme sports car, a midengine Corvette. He continually championed the ideal, and the press

INTRODUCTION

While the platform beneath first-generation, solid-axle Corvettes went unchanged from 1953 to 1962, the body went through notable makeovers, including the addition of quad headlights in 1958. The so-called "boat-tail" back end arrived in 1961, previewing some of the upcoming second-generation Sting Ray body features that debuted two years later. Deleting the chrome trim around the last of the solid-axle breed's bodyside cove panels for 1962 meant that two-tone paint schemes were no longer available. The bright rocker moldings also were new that year.

predicted midengine breakthroughs more than once, but reality just wouldn't have it. At least not while Duntov was still with us. In July 2019 the dream finally came true in the form of the radically redesigned eighth-generation Corvette, a world-class supercar that carried its V-8 behind the seat. How proud might papa have been?

Dating back much further than Duntov's plans, the midengine ideal has long translated into earth-shaking performance, first in European racing prior to World War II, then on the street-legal supercar market that began taking serious shape around the globe during the 1960s. Maximizing horsepower alone doesn't necessarily adorn a car in a cape. But stuffing mucho muscle into a supremely balanced package, one with its heaviest component mounted amidships, goes a long way towards making, say, a Ferrari a Ferrari.

Making the 2020 Corvette more like its supercar competitors, meanwhile, represented an obvious milestone moment that surprised some while predictably leaving others asking what took so long? "Redesigning the Corvette from the ground up presented the team a historic opportunity, something Chevrolet designers have desired for over 60 years," said GM global design vice president Michael Simcoe in June 2019. Designers, certainly. But too bad thrifty execs of decades past repeatedly sidestepped such temptation, denying Duntov a shot at perhaps his most treasured contribution to Corvette history.

> With nearly 500 standard horses, the 2020 rendition was the strongest base Vette yet, capable (in top trim) of hitting 60 mph from rest in less than three seconds.

This time, however, market and technical pressures overruled cost concerns. As GM president Mark Reuss explained it, "the traditional front-engine vehicle reached its limits of performance, necessitating the new layout." The moment clearly had come to go truly global, or risk perhaps finally reaching the road's end, a fate also alleged more than once over the years.

"Our mission was to develop a new type of sports car combining the successful attributes of Corvette with the performance and driving experience of midengine supercars," announced chief engineer Tadge Juechter during the latest generation's nation-wide introduction. "We know [the 2020] Corvette can stand tall with the best the world has to offer," added GM global design vice president Michael Simcoe. "It's now [also] the best of America, a new arrival in the midengine sports car class."

With nearly 500 standard horses, the 2020 rendition was the strongest base Vette yet, capable (in top trim) of hitting 60 mph from rest in less than three seconds, another new Corvette record. A supercar? Damn straight.

And a great vehicle to carry the Corvette banner high into its next generation, as mentioned, its eighth. Code names for each came into vogue during the 1997 model's long haul to market after Chevy began using its in-house "C5" designation in public. Retroactive labeling then seemed only logical: original "solid-axle" 1953–62 examples made up the C1 group, the C2 family consisted of 1963–67 Sting Rays, the C3 era spanned from 1968 to 1982, and the C4 ran from 1984 to 1996 after skipping the 1983 model year due to developmental delays. Originally considered in 1988, the fifth gen was initially scheduled to debut in August 1992. But various pitfalls pushed that rollout back, too, resulting in a January 1997 intro. The C6 followed eight years later, the C7 came in 2014, and now we have the game-changing C8.

13

INTRODUCTION

The Corvette's real father wasn't Zora Arkus-Duntov, it was GM styling exec Harley Earl, a man who threw some serious weight around the corporation during his heyday. He also was responsible for various memorable postwar design trends, most notably the classic tailfin. Here Earl stands tall next to 1951's LeSabre, a legendary concept car that first demonstrated those fins, along with the wraparound windshield that GM production cars began popularizing in 1953.

In 2023 Chevrolet introduced the most powerful naturally aspired Corvette to date, the 670-horsepower Z06. Even more history followed the next year in the form of the legacy's first all-wheel-drive model, the E-Ray, which, as that name implied, implemented another major first: electric motivation for the front wheels. All told, this gas-electric hybrid produced 655 horses, enough muscle to make it the quickest Vette yet. At that time.

INTRODUCTION

Midengine flights of fantasy date back to 1960's CERV I, which, among other things, also served as a testbed for the innovative independent rear suspension introduced along with the first Sting Ray in 1963. At right is 1964's CERV II, which officially added the Chevrolet Experimental Research Vehicle acronym into Corvette lexicon, in turn retroactively creating CERV I identification for its forerunner. Like CERV II, 1990's CERV III (center) also featured its engine—a 650-hp LT5 V-8 fed by twin turbochargers—mounted amidships.

The Corvette's heart-and-soul dating back to 1955, Chevrolet's venerable small-block V-8 really did go where no engine has gone before. On November 29, 2011, technicians, VIP journalists, and key past/present team employees went to work in Wixom, Michigan, to help bolt together the 100 millionth example of this high-winding legend. And, of course, the only honorable choice for this milestone moment was the C7's supercharged 638-hp LS9, at the time the most powerful—small-block or otherwise—V-8 ever installed in a regular-production GM vehicle.

Just when Corvette customers thought they'd seen it all, along came the latest, greatest ZR1 in 2025, a world-beating supercar armed with a twin-turbo, 32-valve, DOHC V-8 that made 1,064 horsepower. GM claimed quarter-mile performance of less than 10 seconds for this record-shattering rocket ship. "Out of this world" didn't begin to tell the tale.

When this legacy originated on a rotating auto show stage in New York in January 1953, among wowed witnesses was Zora Arkus-Duntov himself. During a 1967 interview with *Hot Rod* magazine's Jim McFarland, he recalled his initial impressions: "Now there's potential. I thought it wasn't a good car yet, but if you're going to do something, this looks good."

He earlier had written Chevrolet chief engineer Ed Cole hoping for a job interview but received only a lukewarm response. Reinspired after encountering the first Corvette, Duntov queried Cole a second time with better results: He went to work at Chevrolet Engineering in May 1953. Once on the payroll, Duntov contacted Cole yet again, this time to detail predictions for the company's fiberglass-bodied roadster. Soon afterward he was assigned to the Corvette's engineering team, where he would stay until retirement in 1975.

> Seven decades after its humble birth, America's Sports Car remains an ageless wonder.

Duntov wasn't officially named the Corvette's first chief engineer until 1967. The second, Dave McLellan, reigned from 1975 to 1992. Cadillac man David Hill took over from there, retiring in January 2006. Filling Hill's shoes was Tom Wallace, formerly the lead engineer for GM's small and midsize trucks. After Wallace took early retirement in 2008, Tadge Juechter stepped up and remained in charge until he retired just as these keys were being punched in the summer of 2024.

As for the car's true patriarch, that claim actually belonged to GM exec and styling guru Harley Earl, who first envisioned production of a Bow-Tie-badged sports car late in 1951. It was his powerful support that provided Ed Cole the opportunity to put wheels to his dream. Going from initial sketches to Cole's auto show prototype required a scant 18 months.

Production began on a makeshift line in Flint, Michigan, in June 1953 before moving to Missouri in December. Chevrolet's aging St. Louis plant remained home until 1981, when Corvette assembly was relocated to its present facility in Bowling Green, Kentucky.

Seven decades after its humble birth, America's Sports Car remains an ageless wonder. Just as Zora planned.

CHAPTER ONE

01 The Solid-Axle Years

C1 1953–1962

C1 1953–1962

Chevrolet officials sure had high hopes when they ordered the renovation of a section of their aging St. Louis assembly plant in order to relocate Corvette production there for 1954. After building a mere 300 1953 models in Flint, the Bowtie boys planned to move as many as 10,000 Corvettes a year out of their Missouri facility. Silly them—nearly a third of the 3,640 '54 Corvettes they managed to build that first year in St. Louis sat unsold outside the plant in January 1955, leaving more than one General Motors exec poised to pull the plug right then and there.

- All solid-axle Corvettes were convertibles
- Only automatic-backed six-cylinder engines offered in 1953 and 1954
- V-8 power option introduced (1955)
- Standard V-8 power (1956)
- Three-speed manual transmission introduced (1955)
- Removable hardtop option introduced (1956)
- Optional two-tone paint introduced (1956)
- Ed Cole becomes Chevrolet general manager (1956)
- Optional fuel injection introduced (1957)
- Optional four-speed manual transmission introduced (1957)
- William Mitchell becomes General Motors design chief (1958)
- Annual production finally surpasses 10,000 (1960)
- Last Corvette enclosed trunk until 1998 (1962)

APPARENTLY AMERICA WASN'T READY for its own sports car. Making matters even worse was Ford's September 1954 introduction of its sporty, upscale two-seat Thunderbird, which outsold its Chevy counterpart by a whopping 23–1 margin (16,155 to 700) in 1955. Such sad numbers would speak for themselves today. But Detroit was a far different place 50 years ago. Bean counters had yet to take control, and the gallant hearts of a few strong-willed movers and shakers could actually outweigh the conservative, cost-conscious minds of the executive brain trust. Harley Earl alone wielded enough power atop GM's ivory tower to prevent anyone from tossing his baby out with the bathwater no matter how badly sales sagged. And rapidly rising Chief Engineer Ed Cole wasn't about to see his crew's hard work go to waste, at least not without a fight. It was left to Zora Duntov to help convince adventurous Yankees that they indeed wanted their own brand of sports car. Replacing the Corvette's original Stovebolt six-cylinder with Chevy's hot new V-8 in 1955 instantly helped turn the tide; reshaping the entire package into a timeless work of art the following year sealed the deal. A half-century later, we can only wonder why anyone at GM ever had a worry at all.

Major doubts in 1953 hinged on various factors, not the least of which was the untested market then awaiting Earl's innovative idea. While some American servicemen had brought the sports car bug back from World War II duty in Europe, their enthusiasm didn't exactly infect the masses. Small, sporty two-seaters were still few and far between when the first Corvette came to be. Of the 4.16 million automobiles registered in the United States in 1952, a mere one-quarter of 1 percent of that mix consisted of sports cars, most of them cute little MG TDs from jolly old England. But fortunately, there were those in Detroit who didn't believe that numbers told the whole story. Among them was Duntov.

"Considering the statistics, the American public does not want a sports car at all," began Zora's address before the Society of Automotive Engineers (SAE) in 1953. "But do the statistics

Above: All critics agreed that the new 1956 shape came off far more pleasing than the original Corvette design. The cove panels in the fenders allowed for optional two-tone paint schemes.

Opposite: Some Corvettes came with chrome headlight bezels in 1956, while some featured painted units.

17

CHAPTER ONE

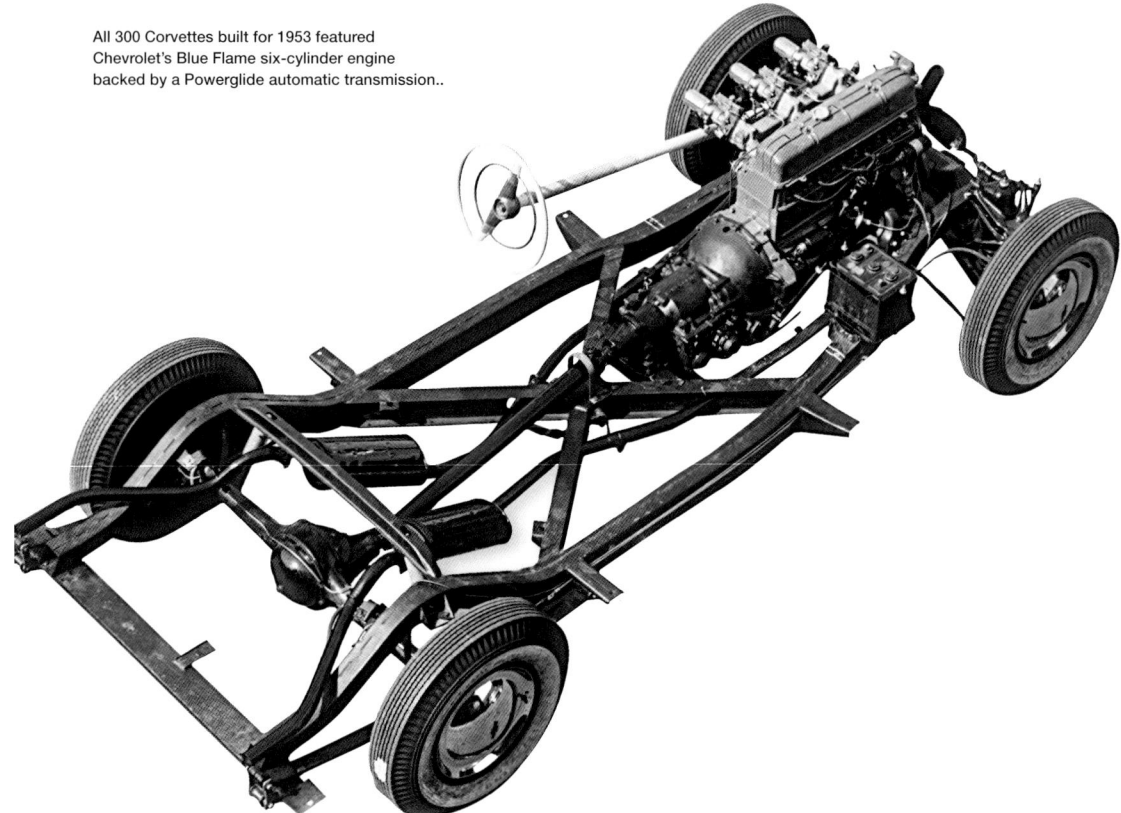

All 300 Corvettes built for 1953 featured Chevrolet's Blue Flame six-cylinder engine backed by a Powerglide automatic transmission..

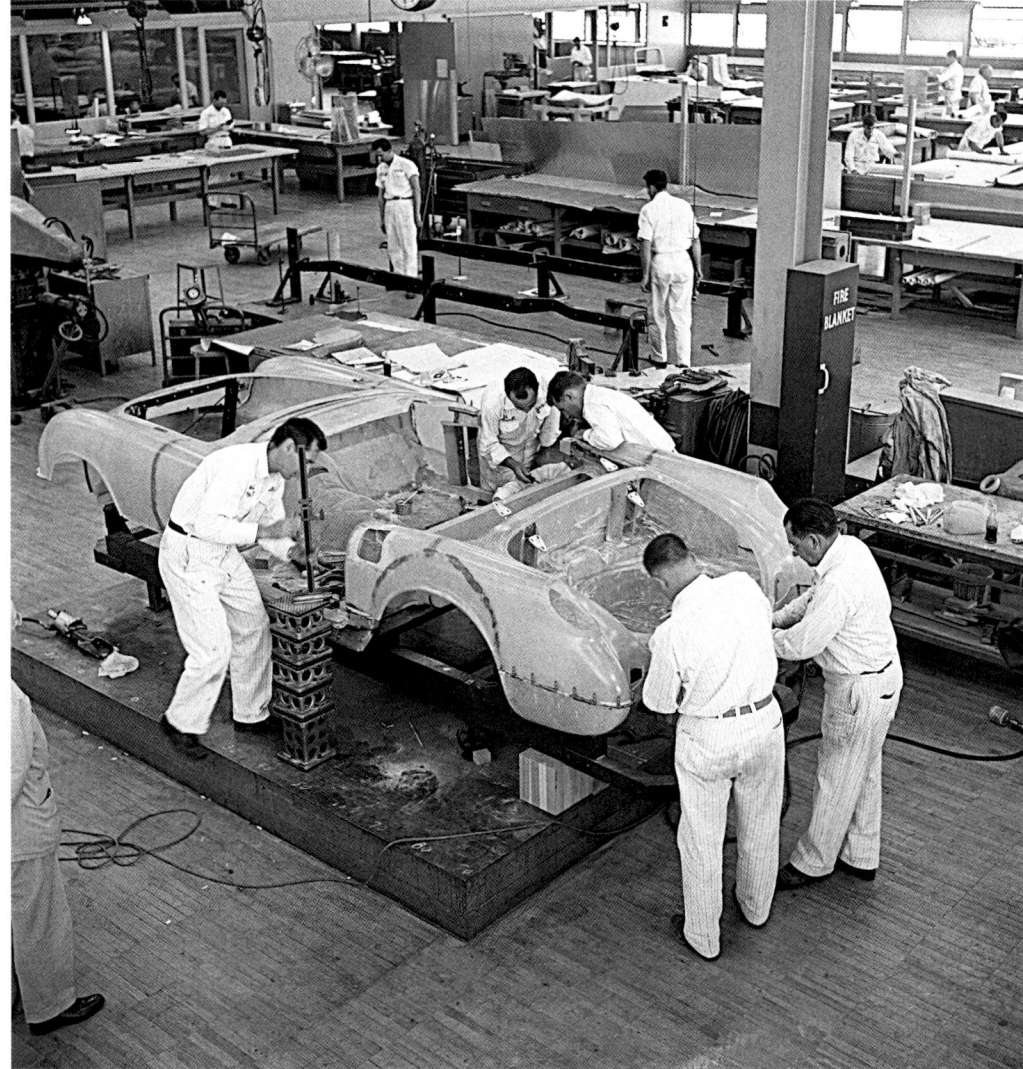

Below: Construction began on the prototype Corvette, designated EX-122, late in 1952; the goal was to show it off during General Motors' traveling Motorama show, which opened in New York in January 1953.

give a true picture? As far as the American market is concerned, it is still an unknown quantity, since [sic] an American sports car catering to American tastes, roads, ways of living, and national character has not yet been on the market."

Duntov wasn't the first to present a case for the production of a certified red-white-and-blue sports car. Legendary automotive journalist Ken Purdy had, four years before, foretold a second coming of sorts. In his 1949 *True* magazine feature entitled "The Two-Seater Comes Back," he wrote that all things that go around eventually make an encore:

"Before the Kaiser War, when Americans were serious about their motoring, the fast, high-performance two-seater automobile was as common as the 5-cent schooner of beer, and a lot more fun. But time passed, and inevitably the U.S. automobile began to change from an instrument of sport, like a pair of skis, into a device for economical mass transportation, and the two-seater was lost in the shuffle. Comes now a cloud on the horizon bigger than a man's hand which may portend a revival on this side of the water of the sports car—an automobile built for the sole purpose of going like a bat out of hell and never mind whether the girlfriend likes it or not."

"There is no reason why America should not be able to produce a good sports car," added *Argosy's* Ralph Stein in 1950. "We have engineers and designers with enough on the ball to create a crackerjack car, but, from observations, it looks very much as if they don't know what it takes. With a fast-growing band of sports car fans, however, the demand will gradually make itself felt."

That demand, however, was slow in coming after World War II, as most Americans concerned themselves with resuming, or starting, their family lives. Two-seat transportation was definitely out of the question once kids entered the picture. And with so few car buyers clamoring for fun machines as the 1950s dawned, Detroit wasn't in any particular hurry to start offering them. Then.

Fortunately, sports car fans did start banding together during the early 1950s, thanks in part to U.S. Air Force General Curtis LeMay, who hoped to boost military morale by arranging sports car races at various Strategic Air Command bases. Reportedly, LeMay also put a bug in good friend Harley Earl's ear concerning the prospects of GM building a true sports car. Furthering the cause was the emergence of the Sports Car Club of America (SCCA), which, during its first years immediately following the end of the war, showcased only foreign machines because domestic counterparts essentially were nowhere to be found.

Early American two-seat pioneers included Crosley's tiny Hot Shot and race car builder Frank Kurtis' Kurtis Kraft, both introduced in 1949 and both built in equally tiny numbers. An American–British hybrid, the Nash-Healey, followed in 1951 and remained in production until August 1954, with

C1 1953–1962

Chevrolet chief engineer Ed Cole (left) and division general manager Thomas Keating admire EX-122, the company's first experimental Corvette, during its Motorama debut in January 1953. Notice the unique trim and small pushbutton door latch release. Regular production Corvettes didn't get exterior door handles until 1956.

> "There is no reason why America should not be able to produce a good sports car," said *Argosy's* Ralph Stein in 1950.

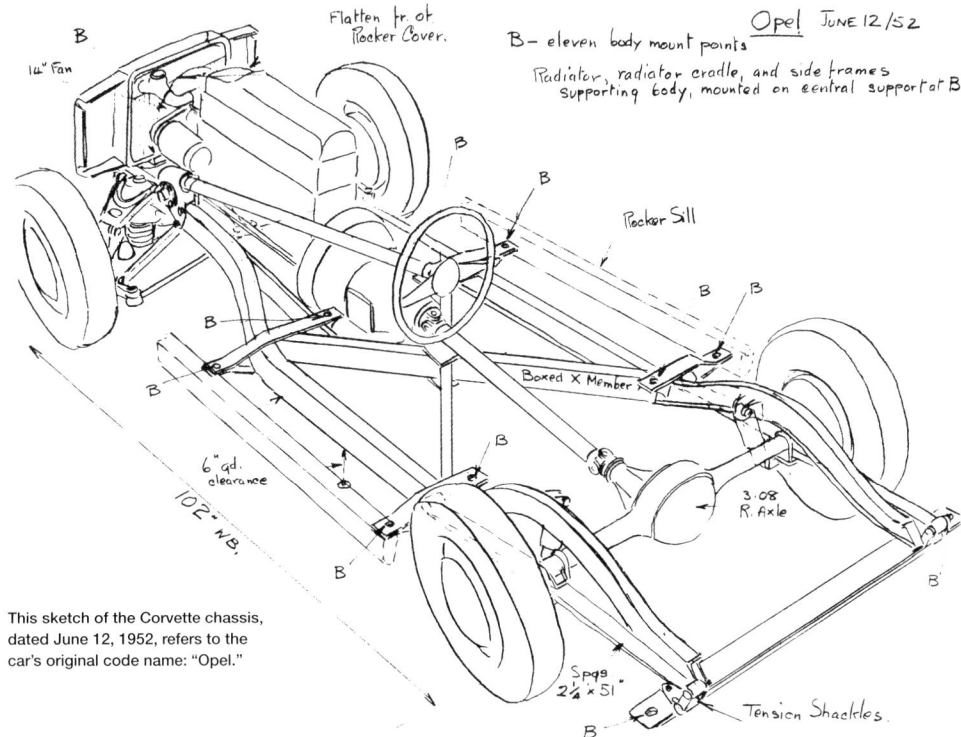

This sketch of the Corvette chassis, dated June 12, 1952, refers to the car's original code name: "Opel."

the total run barely surpassing 500. Also offered that latter year (again, in a highly limited fashion) was the all-American Kaiser Darrin, with its intriguing sliding doors that disappeared into the front fenders. Equally of note, the Kaiser Darrin's sleek body was molded from glass-reinforced plastic (GRP).

Discounting various experiments in the 1930s, GRP construction first came into vogue during World War II as a cheap, lightweight replacement for metals then in short supply. Making this stuff was simple enough: mix together hardening polyester resins with a woven mat made of fine glass fibers. The result was fiberglass, not to be confused with the trade name "Fiberglas" that Owens Corning has long used for its glass-fiber insulation.

It was only a matter of time before automakers discovered this high-tech material (later renamed "FRP," for fiberglass-reinforced plastic) after the war. Fiberglass body kits began appearing by the end of 1951, followed by America's first fiberglass-bodied regular production car, the Woodill Wildfire, in 1952. In February that year, *Life* magazine published an article entitled "Plastic Bodies for Autos," attracting even more attention, most prominently from Henry Kaiser, who then contracted fiberglass boat builder Glasspar to produce the body for his Kaiser Darrin.

Chevrolet people were attracted at the same time, and by the end of 1952, GM was exploring the feasibility of GRP body construction. According to research and developmental head Maurice Olley, whom Ed Cole had hired away from Rolls-Royce after becoming Chevy chief engineer in April 1952, fiberglass produced "a very usable body of light weight which will not rust [and] will take a paint finish. A fiberglass panel of body quality three times as thick as steel will weigh half as much and will have approximately equal stiffness."

Harley Earl, too, took note of GRP possibilities. He had begun considering a sporty two-seater late in 1951, then started actual design work after an experimental fiberglass-bodied car was brought to GM Styling in March 1952.

Earl assigned designer Robert McLean the task of establishing parameters for his dream machine. McLean's initial drawings depicted a low two-seater with a wide stance and a short 102-inch wheelbase. Earl's direct contributions to the design included his trendy wraparound windshield and clear headlight covers that fared the recessed headlamps into rounded fenders.

A full-size clay mockup of the design was prepared in April 1952, followed by a plaster model, which was shown to Ed Cole. Cole loved what he saw, and he joined Chevrolet General Manager Thomas Keating to present the model to GM President Harlow Curtice, with the goal to build an experimental prototype for the upcoming Motorama auto show, scheduled for New York's Waldorf-

CHAPTER ONE

The New York Motorama incorporated a rotating display, and as the car turned, its hood and trunk opened and closed hydraulically.

Works of art themselves, the various master models used in 1953 were carved from mahogany. Molds were pulled off these masters at GM's Parts Fabrication shop in Warren, Michigan, and "Parts Fab" workers then used these molds to hand-lay early fiberglass Corvette bodies.

Various tweaks helped boost output to 150 horses for the Corvette application. Behind the boosted Blue Flame six went Chevrolet's two-speed Powerglide automatic transmission, a not-so-popular choice that represented the easiest, least-expensive way to complete the driveline on such short notice.

Topping it all off was a hand-laid fiberglass shell made from plaster molds pulled directly from McLean's original clay model. But while plastic body construction would, of course, carry over from the show car into regular production, it wasn't the first choice. As Body Engineer Jim Premo later told the assembled SAE, "The body on the show model was made of reinforced plastic purely as an expedient to get the job built quickly. At the time of the Waldorf show, we were actually concentrating on a steel body utilizing kirksite tooling." Though kirksite dies were cheaper and could be fabricated more quickly than typical steel counterparts, they had much shorter working lives. Such considerations, however, were rendered moot once Premo's team proved they could work successfully in fiberglass. Besides, the auto show crowd loved the stuff.

"People seemed to be captivated by the idea of the fiberglass plastic body," continued Premo. "Furthermore, information being given to us by the reinforced plastic industry seemed to indicate the practicability of fabricating plastic body parts for automobiles on a large scale."

With the decision made to transform the Motorama prototype into a production reality, GM officials immediately began accepting bids for Corvette body construction following the car's debut in January 1953. The Molded Fiber Glass Company in Ashtabula, Ohio, won the contract, and then had to hustle to open a special plant dedicated to the process. Opened in April, the facility wasn't fully up and running until July 1954, leaving the MFG body people no choice but to subcontract some of the work to make up for slack in the meantime.

Once fabricated in Ohio, the various body parts were shipped to Michigan, where they were glued together and finished by hand. Completed bodies were then mated to their chassis on the Flint assembly line, which began rolling in June 1953. The rest is automotive history.

Astoria in January 1953. Curtice gave the project two thumbs way up, opening the door for Maurice Olley's team to create a suitable chassis. In June 1952, Olley drew up an initial design, code-named "Opel." Chevrolet officials, however, preferred the name "Corvette," taken from a lightly armored, certainly fast warship built during World War II.

Olley's foundation featured a rigid X-member frame. While the front suspension consisted of many stock passenger-car parts, the rear layout incorporated special leaf springs to help locate a modified live axle, because a modern Hotchkiss drive was used instead of the archaic torque tube found beneath regular Chevrolet vehicles. Braking and steering systems too came right off the standard parts shelf.

And with Chevy's all-new V-8 still two years away, Cole was forced to rely on the division's existing Blue Flame six-cylinder, which made 115 horsepower from 235 cubic inches in typical trim.

> For the most part the show car simply rolled right off its stage onto Main Street USA.

20

C1 1953–1962

Chevrolet wasted no time hauling this 1953 pre-production prototype around the country for advertising and promotional photos. Notice the different trim treatment compared to regular-production models.

1953

Model availability	convertible
Wheelbase	102 inches
Length	167.25 inches
Width	72.2 inches
Height	51.5 inches
Shipping weight	2,705 pounds
Tread (front/rear, in inches)	57/58.8
Tires	6.70x15 inches
Brakes	11-inch drums
Wheels	15x5 inches
Fuel tank	18 gallons
Front suspension	independent short and long wishbones, coil springs, and stabilizer bar
Rear suspension	solid axle with longitudinal leaf springs; tubular shock absorbers
Steering	worm and sector, 16:1 ratio
Engine	150-horsepower 235-ci OHV inline six-cylinder with three Carter one-barrel carburetors
Transmission	Powerglide automatic with floor shifter
Axle ratio	3.55:1

1953

Variety certainly wasn't the spice of life as far as Corvette buyers were concerned early on. All 300 Flint-built 1953 models were identically equipped with the Powerglide-backed Blue Flame six, and all were painted Polo White with contrasting red interiors. Although a heater and signal-seeking radio were listed as options, they were included in all 300 deals whether customers wanted them or not. Various minor trim differences set the production versions apart from their Motorama forerunner, but for the most part the show car simply rolled right off its stage onto Main Street USA.

Supplying the standard 150 horses was a nicely warmed-over version of Chevrolet's overhead-valve inline six-cylinder. Modifications included more compression (8:1 instead of the stock 7.5:1 squeeze), a more aggressive solid-lifter cam, a beefed-up valvetrain, and a special induction setup featuring three side-draft Carter one-barrel carburetors mounted on an aluminum intake manifold. Handling spent gases was a split exhaust manifold that dumped into dual pipes and mufflers.

CHAPTER ONE

According to *Motor Life's* Hank Gamble, "The Corvette is a beauty—and it goes!"

The first Corvette rolled off the Flint line on June 30, 1953.

The Corvette's fiberglass floorpan was comparatively light, but this stunt shot doesn't exactly tell the real story. This molding weighed 210 pounds.

The various body parts were bonded together at the Flint facility for final assembly.

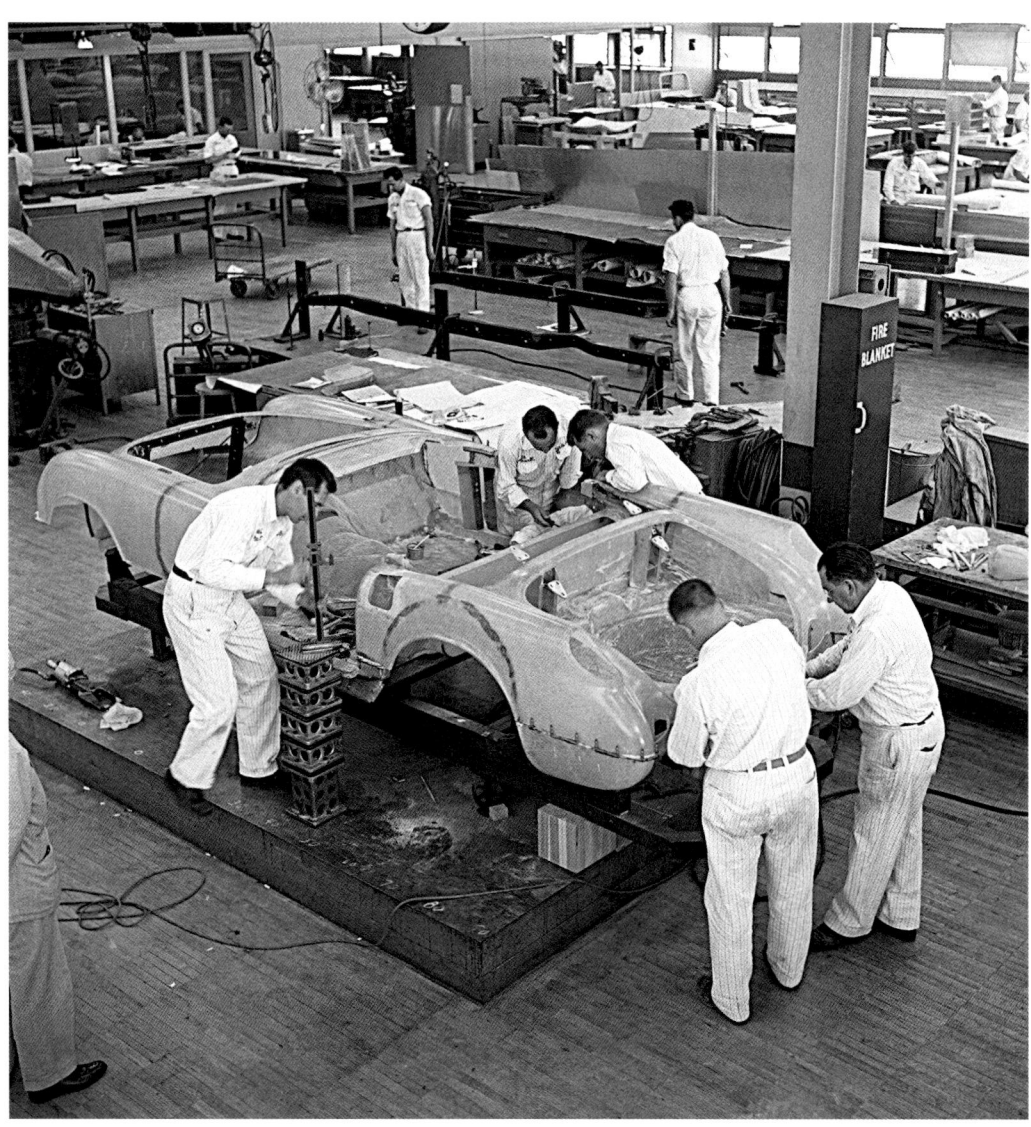

Following the approval of Harlow Curtice and Ed Cole in June 1952, Harley Earl got his Design Center staff scrambling to meet the January 1953 Motorama deadline. Assembling the handmade prototype required many hands in this early December 1952 photo.

22

Above: As relentless as the work schedule was at the proving ground, so was the "modeling" schedule to which Chevrolet's ad agency Campbell Ewald subjected the Corvette prototype. Chief photographer Myron Scott—who named the car—shot in image-enhancing locations around the United States.

Above left: Fed by three side-draft carburetors, the 1953 Corvette's Blue Flame six-cylinder produced 150 horsepower. Notice the small, bullet-type air intakes attached to each carburetor. Mike Mueller

Left: Along with red upholstery, all 1953 Corvettes were equipped with heaters and signal-seek radios, items listed as options in official paperwork that year. Mike Mueller

The sum of these parts equaled a top end of nearly 110 miles per hour, a sensational achievement for the day. Equally impressive at the time, the 1953 Corvette was able to run from rest to 60 miles per hour in only 11 seconds. According to *Motor Life's* Hank Gamble," The Corvette is a beauty—and it *goes!*"

Road & Track staffers felt that straight-line speed was the car's most "outstanding characteristic," but they didn't stop praising there. "The second most outstanding characteristic is its really good combination of riding and handling qualities. The Corvette corners flat like a genuine sports car."

Yet could a sports car be a sports car without a clutch and stick? Though it featured a sporty floor shifter, the original Corvette's Powerglide auto box left purists scoffing—and Chevy people defensive.

"The use of an automatic transmission has been criticized by those who believe that sports car enthusiasts want nothing but a four-speed crash shift," explained Maurice Olley. "The answer is that the typical sports car enthusiast, like the 'average man,' or the square root of minus one, is an imaginary quantity. Also, as the sports car appeals to a wider and wider section of the public, the center of gravity of this theoretical individual is shifting from the austerity of the pioneer toward the luxury of modern ideas. There is no need to apologize for the performance of this car with its automatic transmission."

Others outside Chevrolet confines weren't so sure: "That statement should get a rise from 100,000 *Road & Track* readers!" wrote *R&T*'s John Bond.

As it was, whether they bought Bond's magazine or not, few Americans got the chance to make their own conclusions concerning Chevrolet's first Corvette. Hoping to promote an exclusive image for their innovative two-seater, company officials initially limited availability of the first production run to VIP customers only. Most of the eager show-goers who saw the Motorama Corvette in January 1953 never even sniffed an opportunity to touch a regular-production counterpart.

"If you've got an itch to get behind the wheel of a Chevrolet Corvette, you might as well scratch it," wrote *Motor Trend's* Floyd Lawrence in the late summer of 1953. "Better are your chances of winning the Mille Miglia on a kiddie car." After

CHAPTER ONE

Above: Tachometer location drew early complaints. A lead-footed driver had to look down and to the right to find the rev counter at a time when keeping both eyes on the road was preferred. *Mike Mueller*

Right: In December 1953, Corvette assembly started up in St. Louis, Missouri, with the ambitious goal of producing 1,000 cars per month. As these models inch along the assembly line, they have removable side windows in place.

Below: Owners in 1953 noted a problem with body panel staining in back due to the original Corvette's short exhaust tips. These tips were lengthened early during the 1954 run to cure the malady. *Mike Mueller*

explaining that the very few cars then available were going to GM executives, Lawrence also quipped, "If [the] present distribution pattern continues, the hoped-for output of 300 units this year will scarcely take care of the top GM brass."

By the end of the year, only about 180 of the 300 1953 Corvettes built had found homes, as Chevrolet couldn't drum up enough very important people willing to come to its invitation-only party. Much momentum had already been lost by the time the VIP qualification was dropped in the summer of 1954. As Don MacDonald explained in *Motor Trend*, "The long gap between initial publicity and availability has cooled the desires of many buyers."

1954

St. Louis assembly plant manager William Mosher was told in March 1953 that Corvette production eventually would move to his facility, and it was there that the first 1954 model was completed late in December. Reportedly as many as 15 Corvettes were built that month in Missouri before the St. Louis line really started cooking in January. The plant was soon rolling out 50 two-seaters a day as Chevrolet geared up for its highly optimistic 10,000-cars-a-year production minimum. But the process soon stalled as demand remained lukewarm. In June, Chevy officials cut back production in St. Louis and halted fiberglass body panel construction entirely in Ohio.

Not even the addition of optional paint choices could turn the heads of potential customers, who stayed away in droves. While nearly 85 percent of the 1954 models were again painted Polo White, about 300 received Pennant Blue finishes, roughly

> Not even the addition of optional paint choices could turn the heads of potential customers, who stayed away in droves.

100 were sprayed Sportsman Red, and less than 10 rolled out in black. The rubberized canvas convertible-top color changed for 1954, from black to beige. A beige interior also was introduced. Additional updates were barely noticeable. Distinguishing between those 300 1953 Corvettes and the following 1954 models wasn't easy at a glance, with the most prominent clue coming in back, where the latter cars were treated to longer exhaust extensions to cure an exhaust staining problem that plagued the Flint-built cars.

Some early 1954 Corvettes received the short extensions before this change was made.

Another running change involved the installation of a different cam early in 1954, a move that upped the Blue Flame six's output to 155 horsepower. At some point that year, engineers also traded the triple carburetors' three bullet-shaped air inlets for a more functional arrangement of two round air cleaners perched in conventional, upright fashion.

1954

Model availability	convertible
Wheelbase	102 inches
Length	167 inches
Width	72.2 inches
Height	51.5 inches
Shipping weight	2,705 pounds
Tread (front/rear, in inches)	57/59
Tires	6.70x15 inches
Brakes	11-inch drums
Wheels	15x5 inches
Fuel tank	18 gallons
Front suspension	independent short and long wishbones, coil springs, and stabilizer bar
Rear suspension	solid axle with longitudinal leaf springs; tubular shock absorbers
Steering	worm and sector, 16:1 ratio
Engine	150-horsepower 235-ci OHV inline six-cylinder (early); 155-horsepower 235-ci OHV inline six-cylinder (midyear change), both with three Carter one-barrel carburetors
Transmission	Powerglide automatic with floor shifter
Axle ratio	3.55:1

Exterior color choices were introduced for the 1954 Corvette; Pennant Blue is shown here. In back are two later models, a 1978 Indy pace car replica and a 1984 Corvette. *Mike Mueller*

Midyear changes for 1954 included trading the Blue Flame six's bullet-style air inlets for two round air cleaners. A cam change early in the year also boosted output from 150 to 155 horsepower. *Mike Mueller*

CHAPTER ONE

At Flint Assembly and afterward at St. Louis, the Corvette essentially was a handmade automobile. Technicians glued and bonded body panels together, waited for them to dry and cure, and then sanded and smoothed them before painting.

With top up and side curtains in place, a fully trimmed 1954 body drops onto its rolling chassis. Hoods were installed after the body drop in Flint the previous year.

Below: On April 27, 1954, various Chicago-area dealers met at their sales zone's GM Training Center to become familiar with Chevrolet's Corvette. When they left that meeting with the cars they were to sell, their "convoy" was photographed touring down Lake Shore Drive in full formation.

Most 1954 Corvettes featured a hood latch mechanism controlled by only one interior release, while some early 1954 cars and all 1953 models used two releases, one for each latch.

Various baubles were added to the 1954 options list, although all 3,640 Corvettes built that year apparently were again equipped with every available feature. The Powerglide automatic was designated an option even though it was once more the only transmission choice, to the continued dismay of sports car purists and performance-seeking playboys.

> The really big news for 1955 involved a switch from the Blue Flame six to Chevrolet's new overhead-valve V-8.

1955

Notable upgrades for 1955 again began with expanded color choices, although some mystery still exists concerning the actual palette offered that year. Polo White was once more the popular choice on the outside, with about half of the 700 Corvettes built for 1955 receiving this familiar finish. Gypsy Red and the newly offered Harvest Gold adorned the bulk of the remaining orders. Pennant Blue briefly carried over from 1954, but reportedly was dropped in April 1955. Some sources say that Pennant Blue was then replaced by Corvette Copper, while others list Coppertone Bronze. By either name, this latter shade was all but unknown in 1955.

Interior colors included red, dark beige, light beige, green, and yellow, with the latter two appropriately reserved for the yellowish Harvest Gold exterior paint. Convertible-top shades were also expanded, as white and dark green (again for the Harvest Gold cars) joined beige. All these artistic touches, however, simply represented icing on the cake. The really big news for 1955 involved a switch from the Blue Flame six to Chevrolet's new overhead-valve V-8, the fire-breathing small block that overnight transformed the old, reliable Chevy into the "Hot One." This modern, high-winding powerplant displaced 265 cubic inches and made 180 horsepower in "Power Pack" trim beneath a 1955 Bel Air's hood. Its potential between fiberglass flanks was obvious.

A prototype V-8 Corvette was undergoing testing as early as May 1954 under the direction of performance consultant and three-time Indy

C1 1953–1962

As Harley Earl's design staff experimented with ideas for the 1956 model year, they developed the coves and removed the taillight fins. They experimented with routing the exhaust pipe through the rear quarter, as they had done with the Nomad show car, or out through a vertical trim piece.

500 winner Mauri Rose. Cole hired Rose in August 1952 to oversee the division's performance parts development projects, and his earliest challenges included developing the somewhat tricky triple-carb setup for the Blue Flame six. Dropping the new V-8 into the Stovebolt's place was comparatively simple, with only one minor frame modification required to allow clearance for the 265's fuel pump. "Installation of the compact V-8 in the Corvette is very neat," wrote *Motor Life*'s Ken Fermoyle.

"The engine fits so nicely, in fact, that one suspects that the possibility of using a V-8 was considered when the Corvette was designed."

With a lumpier cam and a Carter four-barrel carburetor topped by a low-restriction chrome air cleaner, the Corvette's 265 V-8 was rated at 195 hefty horsepower. Compression remained 8:1. On the outside, this born-again hot rod was identified by the large gold "V" added to the "Corvette" script on each fender.

An early Corvette prototype was later modified as a V-8 prototype. This car goes airborne during chassis tests at GM's Milford Proving Grounds in May 1954.

1955

Model availability	convertible
Wheelbase	102 inches
Length	167 inches
Width	72.24 inches
Height	48.5 inches
Shipping weight	2,695 pounds (six-cylinder), 2,665 pounds (V-8)
Tread (front/rear, in inches)	56.7/58.8
Tires	6.70x15 inches
Brakes	11-inch drums
Wheels	15x5 inches
Fuel tank	18 gallons
Front suspension	independent short and long wishbones, coil springs, and stabilizer bar
Rear suspension	solid axle with longitudinal leaf springs; tubular shock absorbers
Steering	worm and sector, 16:1 ratio
Engine (early)	155-horsepower OHV inline six-cylinder with three Carter one-barrel carburetors
Engine (midyear addition)	195-horsepower 265-ci V-8 with single four-barrel carburetor
Transmission (early)	Powerglide automatic
Transmission (late)	three-speed manual
Axle ratio	3.55:1

The big news for 1955 involved the introduction of Chevrolet's hot new overhead-valve V-8, which replaced the Blue Flame six-cylinder beneath fiberglass hoods early in the year. Announcing the 265-ci V-8's presence was a large gold V added to the Chevrolet fender script. *Mike Mueller*

CHAPTER ONE

Though Belgium-born Russian engineer Zora Arkus-Duntov wasn't even around when America's Sports Car was initially conceived, he has long been called the Corvette's "father" for what he did to raise Harley Earl's baby right. Sure, Chevrolet's introduction of its historic overhead-valve V-8 in 1955 went a long way toward helping the newborn Corvette onto its feet. But who knows what might have happened to this unproven flight of fancy had Duntov gone to work somewhere else in May 1953. The main driving force behind Chevy's dream machine almost from the get-go, he wasn't officially named Corvette chief engineer until 1967.

Below: Chevrolet's new small-block V-8 engine debuted in 1955 at 265 cubic inches and made 195 horsepower for Corvette applications. Bright shielding (visible behind the carburetor here) was added atop the distributor to prevent ignition voltage from interfering with radio reception—the Corvette's fiberglass body couldn't supply this shielding like a passenger-car's steel shell does.

While some (as few as seven, perhaps) very early 1955 Corvettes were equipped with the 155-horse Blue Flame six, the majority featured the 265 V-8. Enhancing the attraction further was a new three-speed manual transmission, offered exclusively behind the V-8. Estimates put three-speed Corvette production that year at about 75. As in previous years, all remaining listed options were included on all 1955 models.

As for more important numbers, according to *Road & Track*, the muscled-up V-8 Corvette could go from 0–60 in a reasonably scant 8.7 seconds. Quarter-mile performance was listed at 16.5 seconds, and the top end was about 120 miles per hour. "Loaded for bear," was *R&T*'s description of the 195-horse Corvette. "The V-8 engine makes this a far more interesting automobile and has upped performance to a point at least as good as anything in its price class," added Ken Fermoyle.

Such accolades aside, many witnesses remained hesitant to predict a brightened future for Chevrolet's struggling two-seater. "Whether addition of the V-8 engine will hypo Corvette sales remains to be seen," continued Fermoyle. "The blazing performance the Corvette now offers should attract more buyers." Meanwhile, *Road & Track*'s Euro-conscious critics continued to complain about the car's yeoman chassis, which, while good by American standards, couldn't quite compete with what lay beneath the world's best sporting machines.

Fortunately, Zora Duntov and crew weren't through tinkering. Not by a long shot.

1956

Boosting power was key to the Corvette's turnaround after 1955, but so too was Chevrolet's "Americanization" of the sports car ideal. Not only was America not quite ready for the original Corvette, the original Corvette wasn't quite ready for its country's countrymen. Like sports car purists, American sensibilities were at first offended by Chevrolet's unique open-air machine, which originally featured more downsides—from a Yankee perspective, that is—than its initial Powerglide-only restriction.

While surely intrigued by this new breed, many early customers simply couldn't get past the car's comparatively crude nature. Unlike Ford's first T-bird, which, along with standard V-8 power, featured conventional roll-up windows and a classy, convenient removable hardtop, Chevy's original Corvette protected its occupants from the weather only with a clumsy, hand-folded rag top and pesky side curtains. In reference to the situation, *Motor Trend*'s Don MacDonald explained that apparently Chevrolet's "conception of the Corvette market is that no owner will be caught in the rain without a spare Cadillac." Americans also were not thrilled about a feature familiar to many European sports car fanatics—opening the doors required a reach inside to fumble around for the interior door knob. Exterior door handles were not present.

So it was that both form and function were radically redone for 1956. Love at first glance was guaranteed by a restyled, modern-looking body that represented a marked improvement on what came before, as well as a faithfully updated rendition of the original image. While the toothy grille up front helped remind onlookers that the 1956 model was indeed a Corvette, the conventionally located headlights and recessed taillights more easily inspired thoughts of forward motion. Overall, the new look was sleek, clean, and truly American—no more Eurostyle headlamp stone shields here.

Equally American were newfound comfort and convenience qualities: fiberglass firsts for 1956 included exterior door handles, an adjustable seat for the passenger, and real side windows that actually rolled up and down. Even more newfangled features appeared on the options list. Power steering was introduced, as was hydraulic

The shapes and forms of the 1956 and 1957 Corvette remain the purest and simplest in many enthusiasts' eyes.

operation for the folding convertible top and a removable roof that made the latest, greatest Corvette nearly 100 percent weatherproof. About two out of three customers chose this $215 option in 1956.

Beneath all that beauty was a mildly revamped chassis, courtesy of Zora Duntov. In his mind, the first Corvette was a car with "two ends fighting each other"—this because considerable oversteer was designed in up front and almost as much roll understeer was inherent in back. Duntov changed this situation in 1956. As he wrote in *Auto Age*, "The target was to attain such handling characteristics that the driver of some ability could get really high performance safely."

This goal was met by adjusting geometry. Caster angle was increased in front by adding shims where the suspension-locating crossmember attached to the frame. Roll oversteer was minimized by changing the steering's main idler arm angle, again by shimming. In back, roll understeer was reduced by revamping the leaf spring hangers to decrease the slope of the springs.

More power also appeared in 1956, as the Corvette V-8 received a compression boost from 8:1 to 9.25:1. A second Carter four-barrel carburetor was added to help produce 225 healthy horses in top-performance trim. Standard power came from a single-carb 265, rated at 210 horsepower. The standard transmission in 1956 was the three-speed manual, with the two-speed Powerglide automatic now an option.

According to *Road & Track*, a three-speed 225-horse model could click off the quarter-mile in 15.8 seconds and run 0–60 in 7.3 ticks, numbers that inspired *Sports Car Illustrated*'s Roger Huntington to call the dual-carb 265 "one of the hottest production engines in the world." But there was more to come.

Duntov was confident he could squeeze 150 miles per hour out of the new '56 Corvette, and to that end he developed the legendary "Duntov cam,"

1956

Model availability	convertible, with optional removable hardtop
Wheelbase	102 inches
Length	168 inches
Width	70.46 inches
Height	51.9 inches (top up), 50.98 inches (with hardtop)
Shipping weight	2,825 pounds (with Powerglide), 2,730 pounds (with three-speed manual)
Tread (front/rear, in inches)	57/59
Tires	6.70x15 inches
Brakes	11-inch drums
Wheels	15x5 inches
Fuel tank	16.4 gallons
Front suspension	independent short and long wishbones, coil springs, and stabilizer bar
Rear suspension	solid axle with longitudinal leaf springs; tubular shock absorbers
Steering	recirculating ball, 16:1 ratio
Standard drivetrain	210-horsepower 265-ci V-8 with single four-barrel carburetor, three-speed manual transmission, 3.70:1 axle ratio
Optional engine	225-horsepower 265-ci V-8 with dual four-barrel carburetors
Optional engine	240-horsepower 265-ci V-8 with dual four-barrel carburetors and special high-lift camshaft (three-speed manual transmission only)
Optional transmission	Powerglide automatic (with 3.55:1 axle ratio)
Optional gear ratios	3.27:1, 4.11:1

CHAPTER ONE

In Harley Earl's mind, morphing the Corvette into a variety of body styles wasn't out of the question. Witness the Nomad wagon and Corvair fastback in 1955 and the two-door Corvette Impala hardtop (shown here) that debuted at the 1956 Motorama.

What a difference a total restyle can make. Chevrolet built only 300 1953 Corvettes (right); production for 1956 models (left) was 3,467. *Mike Mueller*

> Love at first glance was guaranteed by a restyled, modern-looking body that represented a marked improvement on what came before.

A beautiful new body, roll-up windows, and conventional door handles helped transform the Corvette into a real winner in 1956.

a potent solid-lifter bump stick that really brought the little 265 to life. Output for the Duntov-cam V-8, although not officially listed in 1956, was commonly put at 240 horsepower. Chevrolet paperwork recommended that this engine be used "for racing purposes only."

Duntov did just that in February 1956 during Daytona Beach's fabled Speed Week trials. On the Florida sands, he managed a two-way average speed of 150.533 miles per hour in the new Corvette, while comrade John Fitch set a two-way record of 145.543 miles per hour. A third Corvette driver, champion aerobatic pilot and stunning spokesmodel Betty Skelton, managed a 137.773-mile-per-hour two-way clocking.

John Fitch then led a four-car team south to Sebring the following month for the 12-hour endurance race there. While Chevrolet officials publicly tried to distance themselves early on from this all-out racing effort, the team obviously was fully factory backed. Chevy-subsidized engineering advancements beneath those blue-striped white-fiberglass skins were plenty, including special heavy-duty springs and shocks, and brakes that featured finned drums with wider shoes wearing sintered "cerametallix" linings.

The latter components represented real keys to survival on a course that could literally melt both car and driver. Two of the Sebring Corvette racers failed to survive the day in March 1956, as

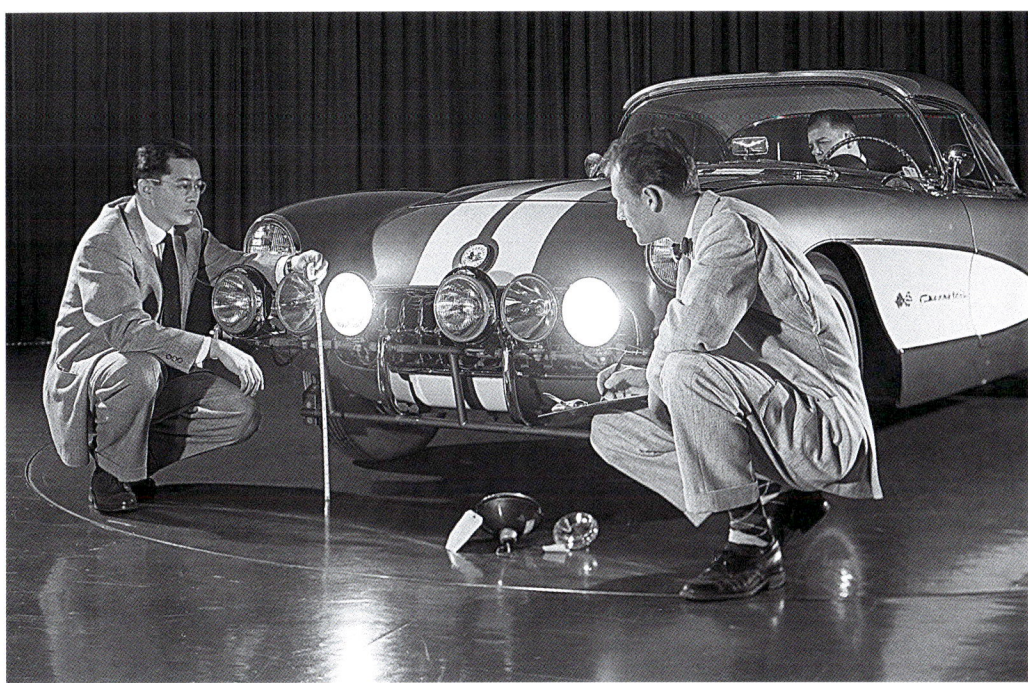

Though it never happened, evidence that Chevrolet people were considering taking the SS racer to LeMans in 1957 for France's famed 24-Hour endurance contest is demonstrated here in this December 25, 1956 photo—driving lights would've been required to race all through the night.

Brainy, beautiful, and tough, Betty Skelton was already an experienced, well-known pilot and test driver when GM put her to work in 1956. She also served as corporate spokesperson during the many Motorama auto show tours to follow.

> According to *Road & Track*, a three-speed 225-horse model could clock off the quarter-mile in 15.8 seconds and run 0–60 in 7.3 ticks.

did 35 rivals. But Fitch's machine, co-driven by Walt Hansgen, finished ninth overall and tops in its class, an achievement that was worth a fortune in publicity for Chevrolet. The now-famous "Real McCoy" advertisement appeared immediately after the race. The ad described the '56 Corvette as "a tough, road-gripping, torpedo on wheels with the stamina to last through the brutal 12 hours of Sebring." But Campbell-Ewald's copywriters also touted the two-seater's more civilized side, claiming it was "the most remarkable car made in America today." Why? "Because it is two cars wrapped up in one sleek skin. One is a luxury car with glove-soft upholstery, wind-up windows, a removable hardtop, ample luggage space, a velvety ride, and all the power assists you could want. The other is a sports car."

Critics who earlier questioned that latter aspect overnight began changing their minds. "Without qualification, General Motors is now building a sports car," wrote Karl Ludvigsen in *Sports Car Illustrated*.

It was also building a successful one. Corvette sales shot up by 400 percent in 1956, with the bulk of these fitted with the 225-horse 265. In one year's time, Chevrolet had taken what would soon become known as America's only sports car and transformed it into the world's only *American* sports car.

1956 SR-2

If prospective buyers still weren't sure just how hot Chevrolet's two-seater had become by 1956, they only needed to watch Zora Duntov, John Fitch, and Betty Skelton during their high-flying record runs down Daytona Beach that February. After Fitch's trip to Sebring the following month, it was time, as ads proclaimed, to "bring on the hay bales." The Corvette now truly was an honest-to-goodness racing machine.

Nonetheless, Jerry Earl remained unconvinced. So what if his dad worked for General Motors? Earl liked to race sports cars, and when it came time to hit the track in 1956, he initially chose a Ferrari to do all the hitting. Father, however, knew best, and there was no way GM Vice President Harley Earl was going to let any son of his drive an Italian exotic when he could be behind the wheel of Detroit iron—or Detroit fiberglass.

The elder Earl convinced his hot-blooded offspring to dump his Ferrari for a specially prepared Corvette racer featuring a custom fiberglass shell on a muscled-up chassis. Included in the deal were all the heavy-duty suspension tricks veteran driver John Fitch had put to the test at Sebring in March 1956. Inside Chevy Engineering, the beefed production parts supplied for Fitch's four-car Sebring team had become known by an "SR" designation, which may or may not have originally stood for either "sports racing" or "Sebring racer." In any case, the Fitch team's four Corvette racers were the first competition SRs, meaning that any variation to follow could've naturally become the second. Thus, the Corvette Harley Earl ordered for his son ended up wearing the moniker "SR-2."

Demonstrating just how nice it is to have friends in high places, Jerry Earl's custom racer—listed under shop order number 90090—took no more than four weeks to complete, leading some to believe that the first SR-2's body was simply dropped right onto an existing track-ready chassis from one of Fitch's Sebring racers. The body featured metallic blue paint, an extended snout, and bright aluminum bodyside cove panels. Both the louvers on the hood and the vents on each door were functional, the latter delivering all-important cooling air to the rear brakes. Up front, twin short screens replaced the windshield, while a small tailfin was added down the center of the deck lid in back.

Twin cutout exhausts, an oversized fuel tank, and a set of Halibrand mag wheels were also included. Inside, Earl's SR-2 initially looked more like a show car and less like a racer, a situation that wasn't entirely far from the truth. All that extra flash translated into extra weight, which in turn translated into a disappointing racing debut at Wisconsin's Road America in June 1956.

CHAPTER ONE

GM sent three Corvette racers to Daytona Beach in February 1956 to kick up a little sand and set a few speed records. The cars were driven by Zora Duntov, John Fitch, and Betty Skelton.

Campbell-Ewald, Chevy's ad agency and Betty Skelton's employer, wasted little time promoting the Corvette's racing exploits early in 1956. Indeed, this little two-seater was "the real McCoy."

Ad man Barney Clark recruited Dr. Dick Thompson to drive a Chevrolet-backed Corvette at Sebring in 1956. He used this photo of Dr. Thompson for a hard-hitting racing ad, and to prove Corvette was now in the racing game.

32

C1 1953–1962

Above: Three SR-2 Corvette racers were built in 1956—two competition machines and one show car. This SR-2 racer was built for Bill Mitchell. *Mike Mueller*

Left: At Sebring in 1957, the no. 4 car finished 12th overall, first in class. Behind it sits the no. 2 SR-2, updated since its introduction in 1956.

A return to the drawing board transformed a lightened SR-2 into a more competitive machine, but true success didn't come until Earl sold his custom Corvette late in 1957 to Jim Jeffords, who campaigned it for Chicago dealer Nickey Chevrolet. Wearing Nickey's distinctive "Purple People Eater" paint scheme, Jeffords' SR-2 sped off with an SCCA B/Production Championship in 1958.

By that time, Chevrolet's first SR-2 Corvette was sporting a reshaped, enlarged tailfin set off to the driver's side in order to incorporate both a headrest and an integral roll bar. Earl requested the "high fin" after seeing it on Chevrolet's second SR-2, built for GM Styling mogul Bill Mitchell.

Varying slightly here and there compared to its forerunner, Mitchell's red-and-white SR-2 made its high-profile public debut in February 1957 at Daytona, where Buck Baker drove it through the Flying Mile at 152.866 miles per hour, a figure topped only by a D-type Jaguar. SR-2 number two finished 16th at Sebring a month later and was eventually relegated to a GM storage basement in 1958.

Following Mitchell's lead, GM President Harlow Curtice requested a third SR-2 in 1956, only his was constructed entirely for show. Virtually a stock '56 Corvette underneath, Curtice's "low fin" SR-2 also featured metallic blue paint but rolled on Dayton wire wheels and was fitted with a removable stainless-steel hardtop. Curtice drove his SR-2 briefly and then sold it to a neighbor.

All three SR-2 Corvette racers have gone through various owners over the years, and each has experienced a modification or two. Originally powered by dual-carb small blocks, the trio today rely on fuel injection, with the two race cars now equipped with bored and stroked V-8s. Four-speeds are found in the SR-2s as well, even though, like the fuelie setup, this transmission type didn't appear on the Corvette options list until 1957.

1957

Big news for 1957 began with a bigger engine as Chevrolet's 265 small-block V-8 was bored out to 283 cubic inches. Maximum output predictably increased, with the top-shelf dual-carb option now producing a whopping 270 horsepower.

To better handle those ponies, engineers on April 9, 1957, released the Corvette's first optional four-speed, the Borg-Warner T-10. According to a Chevy press release, "The four forward speeds of the new transmission are synchronized to provide a swift and smooth response." Walt Woron of *Motor Trend* felt those words represented

33

CHAPTER ONE

The top-dog fuelie Corvette in 1957 featured a solid-lifter V-8 that produced 283 horsepower. A four-speed was the only transmission offered behind the 283/283. Two 283-horse injection options were listed: RPOs 579B and 579E. The latter was the race-ready air box package. Shown here is one of the 43 air box Corvette models built for 1957. *Mike Mueller*

an understatement: "When you can whip the stick around from one gear to any other the way you'd stir a can of paint, that's a gearbox that's synchronized."

Yet as much as the four-speed represented just what the doctor ordered to help maximize Corvette performance, an even hotter new option also debuted for 1957—Ramjet fuel injection (FI). Initially the work of engineer John Dolza, Ramjet injection was perfected with the help of Duntov, whom Cole assigned to the project in early 1955. Fuel injection was nothing new on the world stage, but it was a real sensation on the American market.

In *Road & Track*'s words, "The fuel injection engine is an absolute jewel, quiet and remarkably docile when driven gently around town, yet instantly transformable into a roaring brute when pushed hard." Built by Rochester Products, Ramjet injection delivered fuel more evenly in a much more efficient manner than those dual four-barrels, and it did so instantly. The FI equipment also eliminated flooding and fuel starvation common to the carburetors of the day whenever hard turns sent the gas supply in the bowl centrifuging off sideways away from the pickup. That latter problem in particular had worked against the dual-carb Corvette racers at Sebring in 1956.

On the downside, fuelie Corvette models earned an early reputation for hard starts and finicky

1957

Model availability	convertible, with optional removable hardtop
Wheelbase	102 inches
Length	168 inches
Width	70.46 inches
Height	51.9 inches (top up), 50.98 inches (with hardtop)
Shipping weight	2,796 pounds (with Powerglide), 2,704 pounds (with three-speed manual)
Tread (front/rear, in inches)	57/59
Tires	6.70x15 inches
Brakes	11-inch drums
Wheels	15x5 inches
Fuel tank	16.4 gallons
Front suspension	independent short and long wishbones, coil springs, and stabilizer bar
Rear suspension	solid axle with longitudinal leaf springs; tubular shock absorbers
Steering	recirculating ball, 16:1 ratio
Standard drivetrain	220-horsepower 283-ci V-8 with single four-barrel carburetor, three-speed manual transmission, 3.70:1 axle ratio
Optional engine	245-horsepower 283-ci V-8 with dual four-barrel carburetors, hydraulic lifters
Optional engine	250-horsepower 283-ci V-8 with fuel injection, hydraulic lifters
Optional engine	270-horsepower 283-ci V-8 with dual four-barrel carburetors, solid lifters (optional Powerglide transmission not available)
Optional engine	283-horsepower 283-ci V-8 with fuel injection, solid lifters (optional Powerglide transmission not available)
Optional transmission	Powerglide automatic (with 3.55:1 axle ratio)
Optional transmission	four-speed manual
Optional gear ratios	3.70:1, 4.11:1, 4.56:1, all with Posi-Traction

Air box modifications also involved relocating the tachometer from the dashboard to the steering column, where it could be easily read. Mike Mueller

Widened 15x5.5 wheels (RPO 276) also appeared for 1957. Small dog-dish hubcaps were included with these rims. Mike Mueller

operation. Keeping everything in proper tune was a must, although that was difficult considering that so few local mechanics were qualified to tinker with a fuel-injected Corvette in 1957.

Teething problems aside, the fuel-injected 283 V-8 helped put Corvette performance at the cutting edge in 1957. *Road & Track* testers managed a 0–60 run in a stunning 5.7 seconds. Quarter-mile times, at 14.3 seconds, were equally alarming.

Various fuel-injected 283s were offered that first year, all listed under RPO 579. RPOs 579A and 579C both featured a hydraulic-cam FI V-8 rated at 250 horsepower. The letters "A" and "C" referred to the transmission choice: 579A was the four-speed manual, 579C the Powerglide automatic. RPO 579B was the designation assigned to the fabled 283-horsepower 283 fuelie, a certified screamer with 10.5:1 compression and the solid-lifter Duntov cam. The four-speed manual was the only transmission available behind the 283/283, which Chevrolet promo people have long loved to claim was Detroit's first engine to reach the 1-horse-per-cubic-inch plateau. Sorry, guys, Chrysler beat you to it the previous year with the 300B's optional 355-horse 354 hemi V-8.

A fourth version of the 283 fuelie Corvette was also offered in 1957, this one clearly built with competition in mind. One of the lessons learned during the high-speed runs at Daytona and Sebring in 1956 was the value of allowing the engine to breathe in cooler, denser outside air instead of the hot under-hood atmosphere. Experimentation with cool-air induction setups led to the creation of the "air box" Corvette.

The idea was simple. A plenum box was fabricated and mounted on the fender well panel on the driver's side. At the front, this box mated to an opening in the support bulkhead beside the radiator. Inside the box was an air filter. At the rear was a rubberized duct that ran from that filter to the Ramjet injection unit. Cooler outside air entered through the bulkhead opening into the box, then through the filter into the FI system. The result was the release of a few extra ponies on top end as the air box Corvette's injected 283 sucked in its denser supply of precious oxygen.

Additional air box modifications included moving the tachometer from its less-than-desirable stock spot in the center of the dash to atop the steering

> "When you can whip the stick around from one gear to any other the way you'd stir a can of paint, that's a gearbox that's synchronized," said *Motor Trend*'s Walt Woron.

35

CHAPTER ONE

Fuel injection and a four-speed manual transmission debuted on the Corvette options list in 1957, as did oversized 15.0x5.5 wheels. The latter were adorned with "dog-dish" hubcaps in place of the full wheel covers with their simulated spinners. *Mike Mueller*

column where it could do its job like it should. A round plate then covered the opening at the standard location. Additionally, the radio and heater—items rarely needed at a racetrack—were both deleted. And with a radio not present, ignition shielding wasn't required. This meant plug wires could be run more directly from the distributor to the spark plugs over the valve covers as far away from the hot exhaust manifolds as possible. Plug wires on all other 1957 Corvettes were routed the long way, down along the cylinder heads below the manifolds, because this was the easiest place to mount the static-suppressive shielding.

Other racing-inspired pieces debuted along with the air box equipment in 1957, beginning with a Posi-Traction differential and wide 15x5.5 wheels. The steel wheels, RPO 276, were 1/2-inch wider than stock rims and came crowned with small, plain hubcaps in place of the standard, ornate knock-off wheel covers.

A heavy-duty suspension option, RPO 581, entered the fray early on. Included were beefed springs front and rear; larger, stiffer shocks; a thicker front stabilizer bar; and a quick-steering adapter. Early in the 1957 model run, this option was repackaged under RPO 684, after a heavy-duty brake package was added into the mix along with the race-ready suspension components.

Sounding very much like the equipment found on the four Sebring Corvette racers of 1956, the RPO 684 brakes featured cerametallix linings, finned drums, and vented backing plates with scoops to catch cooling air. Helping deliver this air to the rear wheel scoops was a somewhat odd ductwork arrangement that began at each side of the radiator, ran back through the engine compartment and down around each front wheelwell, and then made its way inside the lower rocker panels. At the trailing end of each rocker was a short, fiberglass deflector duct that directed the airflow inboard toward the scoops on each vented backing plate.

The Posi rear was mandatory with RPO 684, which itself was only available with the 270- and 283-horse engines. Only 51 Corvettes were fitted with this Sebring-inspired performance package in 1957.

1957 SS racer

In Zora Duntov's mind, beating the world's best sports cars at their own game represented the quickest way toward a long and prosperous future for the Corvette. "All commercially successful sports cars were promoted by participation in racing with specialized or modified cars," he explained in his 1953 address to the SAE. Plain and simple, Duntov wanted to take the Corvette to the track, both to prove its merits up against rivals from around the planet and to promote sales at home.

Chevrolet's first big shot at the world's racing elite—John Fitch's assault at Sebring in March 1956—may have supplied the firm with ample advertising fodder. But Fitch's modified production Corvette racers actually did little to convince Jaguar, Ferrari, Porsche, and the rest that America's only sports car had risen to world-class status. Everyone from Ed Cole on down recognized that they either needed to invest full force in an all-out competition Corvette, or stay home.

Enter Harley Earl. A few months after the dust settled at Sebring, Mr. Earl "borrowed" the 12-hour race's third-place finisher, a D-type Jaguar. Earl's Jag went into a GM Styling studio in June 1956 with the plan to fit it with a Chevy V-8 and take

1957 SS

Construction	magnesium roadster body on tubular space frame
Wheelbase	92 inches
Length	168 inches
Height	48.7 inches
Weight	1,850 pounds
Tread	51.5 inches, front and rear
Tires	6.50x15 inches front, 7.60x15 inches rear, Firestone Super Sports
Brakes	12-inch Chrysler center-plane drums, finned for cooling (rears mounted inboard on Halibrand differential housing)
Wheels	15x5 inches, Halibrand cast-magnesium knock-offs
Front suspension	independent short and long wishbones, coil-over shock absorbers, and stabilizer bar
Rear suspension	de Dion axle with coil-over shock absorbers
Steering	Saginaw recirculating ball, 12:1 ratio
Engine	310-horsepower 283-ci V-8 with fuel injection, tubular headers, and sidepipes
Transmission	aluminum-case synchromesh four-speed manual (with 1.87:1 low)
Differential	Halibrand quick-change with aluminum housing (ratios ranged from 2.63:1 to 4.80:1)

it racing as an experimental Corvette. Duntov, of course, would have nothing of such shenanigans. He immediately began work on a proposal for a purpose-built, all-American racing Corvette. Soon afterward, Cole rose from chief engineer to Chevrolet general manager, perhaps in part explaining how Duntov's project, labeled XP-64, was quickly approved in August.

So what about Earl's Jaguar? It was returned to its owner equally as quickly, which may have been part of the far-fetched plan all along. Perhaps the idea was to bluff Duntov and Cole into action. If so, the ploy worked.

Initial paperwork called for "a competition racing car with special frame, suspension, engine, drivetrain, and body." The plan involved building four such machines, one nonrunning mockup for show duty and three functional racers intended for a March 1957 debut at Sebring, then on to Le Mans. But that schedule allowed barely six months to design, build, and test the XP-64 Corvette. Such heavy deadline pressure quickly convinced all involved that four cars were out of the question. Final approval of the project involved only one race car, dubbed the "Corvette SS."

The SS was a Corvette in name alone, save for its familiar toothy grille—wearing two more teeth than its stock counterpart—and contrasting bodyside cove panels. Beneath its slippery, hand-formed magnesium skin was a low-slung tubular space frame, featuring 1-inch chrome-moly tubing. Total weight for the SS frame was only 180 pounds.

Veteran driver John Fitch (at left here taking direction from Zora Duntov) piloted the SS at Sebring in March 1957 along with Piero Taruffi. The magnesium-bodied racer lasted only 23 laps during the legendary 12-hour event, 22 of them with Fitch at the wheel.

In December 1958, the Corvette SS managed an incredible 183 mph during testing at GM's Mesa Desert Proving Ground. Then in February 1959, Duntov himself drove the SS around the new Daytona International Speedway's 2.5-mile high-bank, hitting 155 mph along the way.

CHAPTER ONE

Zora Duntov's beautiful blue SS was hastily constructed during the winter of 1956–1957 to (it was hoped) put the Corvette on the international racing map. It ran once, at Sebring in March 1957, before the Automotive Manufacturers Association's ban on factory racing cut its career short. *Mike Mueller*

Additional pounds were shed thanks to the liberal use of aluminum throughout, including the four-speed gearbox, bell housing, water pump, radiator, and cylinder heads. The engine's special oil pan and the five Halibrand knock-off wheels (a spare was included per international racing rules) were made of magnesium, yet another weight-saving consideration. All told, the Corvette SS tipped the scales at 1,850 pounds, about 100 less than the dominating D-type Jaguars.

Suspension was by coil-over shocks at all four corners. A typical short-arm/long-arm (SLA) design went up front, and a not-so-typical de Dion axle was added in back. Located by four trailing links, the curved de Dion axle wound its way from wheel hub carrier to wheel hub carrier behind a frame-mounted Halibrand quick-change differential housing.

Sending torque to that differential was a modified 283 fuelie V-8. Compression inside this race-ready small block was kept at 9:1 to ensure the powerplant would hang in there for 12 hours at Sebring. The lightweight aluminum heads were treated to some port reworking, and tuned headers were installed to speed spent gases on their way. Output estimates ran from 300 to 310 horsepower.

Although crude compared to its blue running mate, the plastic-bodied mule didn't cook its driver. And it impressed many witnesses with its power and speed during Sebring practices sessions. Here, the mule undergoes testing in wintry Michigan.

Above: Zora Duntov's team had less than six months to create the Corvette SS racer, during which time they built two, the magnesium-bodied star of the show and a fiberglass-bodied test mule. Here, Duntov rolls out the incomplete mule. He logged most of the mule's 2,000+ test miles.

Left: The SS cockpit was both purposeful and pretty, and on race day it gave all new meaning to the term hot seat. *Mike Mueller*

Below: Chevrolet people tried in vain to cure the SS racer's various maladies on race day at Sebring in March 1957. One of these fixes included cutting body panel portions away to help cool things down inside. Holes began appearing up front not long after this photo was taken.

SS brakes were big, burly 12-inch drums (with cerametallix linings) at all four wheels. Rear drums were mounted inboard on the Halibrand quick-change housing to reduce unsprung weight. Completing the package was an innovative vacuum-controlled booster system that kept the rear brakes from locking up during hard deceleration.

The Corvette SS truly was a state-of-the-art racer, at least in American terms. But with only six months between its birth and its Sebring debut in March 1957, there was little chance to iron out bugs. Fortunately, Duntov did manage to also cobble together an SS "mule," a crude fiberglass-bodied running mate to his blue magnesium baby. Despite weighing about 150 pounds more (thanks to that fiberglass shell) than the SS, and armed with a few less horses, the white mule nonetheless impressed witnesses at Sebring with its performance during early test sessions.

CHAPTER ONE

But the star of the show simply was not ready for the race. It literally didn't arrive in Florida until the last minute and by then didn't have a driver. Juan Fangio was the original choice, with Stirling Moss considered as a co-driver. But Moss was already taken and Fangio was released from his Chevrolet commitment once it became clear the SS wouldn't arrive in time for proper testing.

The job was then given to John Fitch, who in turn suggested that Chevrolet call veteran Italian driver Piero Taruffi to serve as co-driver. What both Taruffi and Fitch discovered during their all-too-short practice time was that the SS was an entirely different animal compared to the mule. First off, the mule's fiberglass body helped insulate the driver from heat, while the magnesium SS shell served as a heat conductor. It didn't take long on the track for the SS to cook its rider. Rising cockpit temperatures, coupled with failing brakes and a loose rear suspension, brought the Corvette SS's debut at Sebring to a quick end after only about 20 laps.

Though it looked like a disaster, Duntov and the rest saw this as just a beginning—a hastily concocted one at that. Plans existed for additional testing and reworking, leading up to a resounding return to the track at Le Mans later that summer. But such hopes were dashed after the Automobile Manufacturers Association (AMA) issued its factory racing "ban" in June. We can only wonder about true potential, especially after the SS hit 183 miles per hour at GM's Phoenix test track in December 1958.

The SS never did experience the racing glory Duntov originally envisioned. In 1967, he donated the blue racer to the Indianapolis Motor Speedway Hall of Fame museum, where it was refurbished in 1987. In August 1994, the SS temporarily left its home for Kentucky to help mark the National Corvette Museum's opening. While in Bowling Green, it took its honored place up front in the museum's lobby, representing the first in a long line of Corvette factory racers.

1958

Revised styling made headlines in 1958, at least up front. In keeping with a corporate-wide design trend, the 1958 Corvette sported new quad headlights, and these were joined by enlarged bumpers at each front corner and backed by large aircraft-style air intakes. Purely ornamental in standard form, those openings were put to work sucking in cooling breezes whenever the race-ready heavy-duty brake/suspension equipment (RPO 685) was ordered. Eagle eyes also might have noticed a less-toothy grille with 9 vertical chrome pieces in place of the 13 seen in 1957.

Additional artistic updates for 1958 included 18 simulated louvers on the hood, two parallel chrome bands on the trunk lid, long chrome trim atop each front fender, and nonfunctional vents added to the leading sections of each bodyside cove panel. All these additions overnight transformed

1958

Model availability	convertible with optional removable hardtop
Wheelbase	102 inches
Length	177.2 inches
Width	72.8 inches
Height	51.09 inches (top up); 51 inches (with removable hardtop)
Shipping weight	2,781 pounds
Tread (front/rear, in inches)	57/59
Tires	6.70x15 four-ply
Brakes	11-inch drums
Wheels	15x5K
Fuel tank	16.4 gallons
Front suspension	independent short and long wishbones, coil springs, and stabilizer bar
Rear suspension	solid axle with longitudinal leaf springs; tubular shock absorbers
Steering	Saginaw recirculating ball (16.0:1 ratio; 16.3:1 adapter available option)
Standard drivetrain	230-horsepower 283-ci V-8, three-speed manual transmission, 3.70:1 axle ratio (3.55:1 ratio standard behind optional Powerglide automatic)
Optional engine	245-horsepower 283-ci V-8, dual four-barrel carburetors, hydraulic lifters
Optional engine	250-horsepower 283-ci V-8, fuel injection, hydraulic lifters
Optional engine	270-horsepower 283-ci V-8, dual four-barrel carburetors, solid lifters
Optional engine	290-horsepower 283-ci V-8, fuel injection, solid lifters
Optional transmission	Powerglide automatic (not available behind solid-lifter V-8s)
Optional transmission	four-speed manual
Optional gear ratios	4.11:1, 4.56:1, both with Posi-Traction (Posi differential also available for 3.70:1 axle)

Standard output went up to 230 horsepower in 1958. Painted valve covers were used on the base 283 V-8. *Mike Mueller*

A totally new dashboard appeared for 1958, and that time around the tachometer was located where it belonged—directly in front of the driver. Factory-installed seatbelts were another first; they had been dealer-installed features previously. *Mike Mueller*

the beautiful, understated body of 1956–1957 into a package that *Road & Track* critics called "too fussy." "That supposedly hard-to-sell commodity, elegant simplicity, is gone," added the *R&T* guys concerning the 1958 makeover.

Apparently, Chevrolet stylists too concluded that the 1958 Corvette was a bit overdressed. Although the chrome fender trim and fake cove vents stuck around to the solid-axle era's end, the hood louvers and deck lid chrome were quickly deleted, after appearing for one year only.

Cockpit impressions were new for 1958, as the basic dashboard used since 1953 was finally updated, this after critics had annually complained of the original Corvette's poor instrument layout. All instruments were now located directly in front of the driver in a modern-looking pod layout. An enlarged speedometer (calibrated up from 140 miles per hour to 160) was added, as was a 6,000-rpm tachometer in place of 1957's 7,000-rpm unit. An 8,000-rpm tach replaced the standard piece whenever the optional solid-lifter engine was ordered, and reading either rev-counter was made easier by moving it from down low in the center of the dash to atop the steering column.

Standard output for 1958 rose to 230 horsepower, up 10 ponies from 1957. Top optional

All GM cars got quad headlights in 1958. Corvette updates that year also included simulated hood louvers and two chrome bands on the decklid. The 1958 Corvette's top optional engine was the 290-horse, fuel-injected 283.

CHAPTER ONE

A 1958 body drop from the rear. The chrome bands on the decklid appeared for that year only, as did the dummy louvers on the hood.

Ready to roll—a completed Corvette nears the end of the St. Louis line in 1958.

On June 7, 1958, a specially painted gold 1958 Corvette was honored at the end of the St. Louis assembly line as the 39 millionth automobile produced by Chevrolet since 1911. Standing is Chevrolet general manufacturing manager Edward Kelley, driving is Ann Long, and next to her is her father, Edward, then Missouri's lieutenant governor.

Like those quad headlights, those aircraft-inspired "nacelles" below also were new for the 1958 Corvette. Nine "teeth" now appeared in the grille, down four from 1957.

power increased as the hottest Ramjet-injected 283 V-8 was now rated at 290 horses. Another change beneath the hood involved relocating the generator from the driver's side to the passenger's to help increase the fan belt's grip on the water pump pulley.

1959

Corvette production rose dramatically after 1955: 3,467 in 1956, 6,339 in 1957, and 9,168 in 1958. Though slowing slightly, that upward trend continued the following year as 9,670 cars left the St. Louis line. The original annual production goal mentioned late in 1953 was finally reached in 1960, when Chevrolet rolled out 10,261 'glassbodied two-seaters. Only minor changes were made inside and out for 1959, with the deletion of 1958's hood louvers and chrome deck lid bands representing the most notable change. Ten slots were added to the wheel covers (the same simulated knock-off pieces used from 1956 to 1962) to cool the brakes, and interior updates included reshaped, reupholstered seats, rearranged door panels, and revised instrument lenses (for improved legibility). A convenient storage bin was added to the dash beneath the passenger-side grab bar introduced in 1958, and a new T-handle shifter appeared for the optional four-speed. New as well with 1959's four-speed box was a safety-conscious reverse lockout mechanism.

Mechanical changes involved adding two radius rods in back to attach the live axle more confidently. Shock absorber mounting points also were relocated to improve their damping effect.

> Corvette production rose dramatically after 1955: 3,467 in 1956, 6,339 in 1957, and 9,168 in 1958.

New on the 1959 options list were high-speed 6.70x15 tires and a second heavy-duty brake package that simply added special sintered-metallic linings supplied by GM's Delco Moraine division. The metallic shoes didn't torture the drums' inner surfaces as severely as their eramic-based counterparts, nor did they work as poorly when cold. From the beginning, the cerametallix brakes included in RPO 684 didn't become effective until warmed up.

RPO 684's asking price dropped by nearly half in 1959, as all the brake-cooling ductwork used in 1957 and 1958 was deleted in favor of simpler vented backing plates at all four corners. Meanwhile, spring rates for the suspension half of the 684 package were increased to compensate for

1959

Model availability	convertible with optional removable hardtop
Wheelbase	102 inches
Length	177.2 inches
Width	72.8 inches
Height	51.09 inches (top up); 51 inches (with removable hardtop)
Shipping weight	2,729 pounds (standard drivetrain)
Tread (front/rear, in inches)	57/59
Tires	6.70x15 four-ply
Brakes	11-inch drums
Wheels	15x5K
Fuel tank	16.4 gallons
Front suspension	independent short and long wishbones, coil springs, and stabilizer bar
Rear suspension	solid axle with longitudinal leaf springs; tubular shock absorbers
Steering	Saginaw recirculating ball (16.0:1 ratio; 16.3:1 adapter available option)
Standard drivetrain	230-horsepower 283-ci V-8, three-speed manual transmission, 3.70:1 axle ratio (3.55:1 ratio standard behind optional Powerglide automatic)
Optional engine	245-horsepower 283-ci V-8, dual four-barrel carburetors, hydraulic lifters
Optional engine	250-horsepower 283-ci V-8, fuel injection, hydraulic lifters
Optional engine	270-horsepower 283-ci V-8, dual four-barrel carburetors, solid lifters
Optional engine	290-horsepower 283-ci V-8, fuel injection, solid lifters
Optional transmission	Powerglide automatic (not available behind solid-lifter V-8s)
Optional transmission	four-speed manual
Optional gear ratios	4.11:1, 4.56:1, both with Posi-Traction (Posi differential also available for 3.70:1 axle)

> No one said Chevy people couldn't keep competition-conscious parts coming for supposedly independent racers who took their own Corvettes to the track.

the extra weight the Corvette had gained in 1958 from its quad-headlight nose job.

Another new enhancement, limited-production option (LPO) 1625, was announced in March 1959. Included in this little-known deal was an oversized fuel tank that upped the fuel load from the stock 16.4 gallons to 24. Reportedly, Chevrolet had installed as many as seven enlarged 21-gallon fuel tanks in 1957 Corvette racers, but these cars were meant only for the track. LPO 1625 made it possible for John Q. Public to enhance his Corvette's range in everyday use, although he obviously also could've used this extra travel time to run longer between pit stops on a race

The 1959 Stingray was born after Bill Mitchell asked Larry Shinoda to develop a new racer based on the 1957 SS chassis.

CHAPTER ONE

The Soviet government in Russia celebrated the launch of its Moscow ExpoCenter in 1959 by inviting the United States to show its products, including a beautiful Corvette, outside one of the displays.

Below: One of Bill Mitchell's first concept cars, the 1959 XP-700—which he called the Phantom—set the stage for a bizarre and progressive generation of automotive dreamers. Though the front end of the Phantom never saw production, its backside showed up on 1961 models.

44

C1 1953–1962

course if he so desired. Remember, Chevrolet wasn't supposed to be directly involved in racing, according to the AMA edict of 1957 barring such activities. But no one said Chevy people couldn't keep competition-conscious parts coming for supposedly independent racers who took their own Corvettes to the track.

Filling up more space than usual behind the '59 Corvette's bucket seats, the big-tank option required installation of RPO 419, the removable hardtop, because there was no room left to include the standard folding roof in its typical location. Another mandated modification involved the standard gas cap. The protruding tab supplied as a handle on this cap conflicted with the fuel filler door after the filler neck was extended to compensate for the larger tank. Making things fit was, in assembly-manual words, simply a matter of "removing [the] handle or bending it over to make it flat with [the] surface of [the] cap."

No one knows exactly how many big-tank solid-axle Corvettes were built, with estimates claiming less than 200 hit the streets between 1959 and 1962. For that latter year, the option code changed from LPO 1625 to RPO 488, and it is known that 65 RPO 488 installations were made.

The cap clearance problem was dealt with differently beginning in 1961, when a lengthened neck simply poked through a holed filler door that was sealed to the body. A chrome cap topped that filler neck, making it relatively easy to spot a rare big-tank Corvette built in 1961 or 1962. But very few exposed-cap cars are known, as some owners cut down the neck and reinstalled the working filler door.

1959 Stingray racer

Harley Earl, the Corvette's true father, stepped down as head of GM Styling in December 1958. In his place came William Mitchell, who, like his mentor, loved his toys. He also loved racing, and in 1956 had, again like Earl, commissioned the construction of an SR-2 competition Corvette for his very own. Two years later, endowed with even more executive privilege, Mitchell put his men to work on yet another personalized hot rod, this one based on the Corvette SS mule's chassis, left over after Chevrolet's overt racing projects were shut down by the AMA in June 1957.

During the winter of 1958–1959, Mitchell's right-hand man, Larry Shinoda, fashioned a new fiberglass shell for the SS mule that featured much of the Q-Corvette's shape and thus foretold the upcoming 1963 restyle. Foretelling the future as well was the name chosen for Mitchell's second personalized Corvette: "Stingray." Most mechanicals, at least early on, were SS carryovers, including the aluminum-head 283 fuelie V-8, which in this case was rated at about 280 horsepower. New for the Stingray was a prototype version of the aluminum Harrison radiator that would appear for the hottest Corvette V-8s in 1960.

Bill Mitchell took his promotional obligations to new heights, coordinating his whole suit to match the color schemes of many cars he exhibited. He often drove his Stingray to work, though probably left the suit at home.

The XP-700 concept vehicle appeared in the spring of 1959 with a truly wild nose and a double-bubble plastic top, removed in this photo.

CHAPTER ONE

The 1960 Corvette was the last to feature the curvaceous tail introduced in 1956. *Mike Mueller*

Supposedly financed and campaigned solely by Mitchell himself, the gloriously red Stingray first hit a track in anger at Maryland's Marlboro Raceway on April 18, 1959. The "Flying Dentist," Dr. Dick Thompson, drove the car that day to a fourth-place finish, a nice start in the minds of most witnesses, including the car owner's. Zora Duntov, however, wasn't at all happy about Mitchell's revival of his dream machine and never acknowledged the car in any way. He had envisioned his SS as a world-class champion, not an American road race special. Road racing across the United States nonetheless proved to be Mitchell's plan, as was winning an SCCA championship. But his Stingray was at first too heavy to compete effectively, and those problem-plagued SS brakes also hindered progress. The original cerametallix stoppers were exchanged for sintered-metallic linings during the summer of 1959, and a lighter body (painted metallic silver) was fabricated by November, bringing overall weight down by about 75 pounds. Stiffer springs were added in preparation of the 1960 racing season.

A third-place finish at Marlboro in April 1960 was just the beginning for Dick Thompson and Mitchell's Stingray. By year's end, Thompson had won the SCCA C-Modified division with 48 points—30 more than the runner-up.

The Stingray became a show car the following year, appearing at Chicago's McCormick Place

> Telling the 1960 apart from the 1959 was best done by looking inside, where pleats in the seats switched from horizontal to vertical.

C1 1953–1962

The large aircraft-style openings below the headlights were purely ornamental in most cases. They became functional in 1958, when the heavy-duty brakes and suspension package (RPO 684) was ordered, serving as intakes for the rear-brake cooling duct work. However, all that ducting was deleted early in 1959. A 1960 Corvette nose is shown here. Mike Mueller

in February 1961. Soon, Mitchell was driving it to work, and later improvements included the addition of disc brakes and the substitution of a Weber-carbureted 427-cubic-inch big-block V-8. Red paint temporarily returned before Mitchell opted to bring back the silver metallic paint, in honor of the Stingray's best days as an SCCA racing champion.

1960

"We predict that this will be the year of the big changes for the Corvette," claimed a report in the January 1959 issue of *Road & Track*. "The changes to the car in the last six model years are not so great as we think will come about in 1960." But John Bond and the rest of the *R&T* staff were left wearing a bit of egg on their faces when the 1960 Corvette appeared looking almost identical to its 1959 forerunner. Telling the two apart was best done by looking inside, where pleats in the seats switched from horizontal to vertical.

Prime inspiration for *Road & Track*'s January 1959 forecast had come from the Q-Corvette project, kicked off late in 1957. Apparently, this proposal indeed was considered as a production possibility for 1960, and would have made some radical changes that year. Early mockups featured an innovative driveline incorporating an aluminum V-8 sending torque to a transaxle located at the rear wheels. Such flights of fancy did foretell the future, just not the near future. An all-aluminum V-8 first found its way between fiberglass fenders in 1969, while relocating the transmission to the rear happened nearly 30 years later as part of the C5 redesign.

One Q-code experiment nearly did make its way into production for 1960. Initially, both fuel-

1960

Model availability	convertible, with optional removable hardtop
Wheelbase	102 inches
Length	177.2 inches
Width	72.8 inches
Height	51.09 inches (top up); 51 inches (with removable hardtop)
Shipping weight	2,890 pounds (standard drivetrain)
Tread (front/rear, in inches)	57/59
Tires	6.70x15 four-ply
Brakes	11-inch drums
Wheels	15x5K
Fuel tank	16.4 gallons
Front suspension	independent short and long wishbones, coil springs, and stabilizer bar
Rear suspension	solid axle with longitudinal leaf springs; tubular shock absorbers
Steering	Saginaw recirculating ball (16.0:1 ratio; 16.3:1 adapter option available)
Standard drivetrain	230-horsepower 283-ci V-8, three-speed manual transmission, 3.70:1 axle ratio (3.55:1 ratio standard behind optional Powerglide automatic)
Optional engine	245-horsepower 283-ci V-8, dual four-barrel carburetors, hydraulic lifters
Optional engine	250-horsepower 283-ci V-8, fuel injection, hydraulic lifters
Optional engine	270-horsepower 283-ci V-8, dual four-barrel carburetors, solid lifters
Optional engine	290-horsepower 283-ci V-8, fuel injection, solid lifters
Note	fuel-injected V-8s were initially advertised with aluminum heads, rated at 275 and 315 horsepower; carryover iron heads actually used and 1959's ratings remained
Optional transmission	Powerglide automatic (not available behind solid-lifter V-8s)
Optional transmission	four-speed manual
Optional gear ratios	4.11:1, 4.56:1, both with Posi-Traction (Posi differential also available for 3.70:1 axle)

CHAPTER ONE

Above: The toothy grille used since the beginning last appeared on the 1960 Corvette. Corvette production surpassed 10,000 for the first time that year.

Right: America's Corvettes attracted quite a crowd in France at Le Mans in 1960.

injected 283 V-8s offered that year, RPOs 579 (with hydraulic lifters) and 579D (with solid lifters), featured weight-saving cylinder heads with revised combustion chambers and larger intake valves. Cast from aluminum, these hot heads helped delete 53 unwanted pounds. And, combined with an enlarged injection plenum and more compression (11:1 instead of 10.5:1), they also boosted RPO 579D output from 290 horsepower to 315. Retaining 1959's compression ratio (9.5:1), the aluminum-head hydraulic-cam 283 was uprated from 250 horses to 275.

These bigger numbers, however, appeared only on paper, as casting irregularities shelved the aluminum heads before they made it into production. Duntov's engineers were forced to fall back on the same 250- and 290-horse injected 283s used in 1959. Both fuelie V-8s were limited to four-speed installations in 1960, as the Powerglide automatic was no longer available with RPO 579, as it had been previously.

Despite the cylinder head setback, other aluminum pieces showed up elsewhere on the 1960 Corvette. A new aluminum bell housing cut off 18 pounds in four-speed applications, while a

C1 1953–1962

Zora Duntov began to work for Chevrolet in May 1953 and soon was concentrating on building a future for America's only sports car. He was named Chevy's director of high-performance vehicle design and development in 1956 but didn't get the official title of Corvette chief engineer until 1968. Here he stands next to the CERV I concept vehicle, the machine that laid the groundwork for the 1963 Sting Ray's independent rear suspension.

little more weight melted away from solid-lifter V-8s thanks to the addition of an aluminum Harrison radiator. Hydraulic-cam Corvette V-8s used conventional copper-core radiators in 1960.

Additional mechanical upgrades appeared underneath. Introducing a rear stabilizer bar and thickening its counterpart in front meant stiffened springs were no longer needed. In turn, the heavy-duty brake/suspension package (RPO 684) failed to return for 1960. In its place was RPO 687, which included stiffer shocks, a quick-steering adapter, and finned brake drums with vented backing plates and cooling scoops, called "elephant ears." Gone were the gnarly cerametallix brake linings, replaced in the new 687 deal by the sintered-metallic shoes used in the RPO 686 package. Also included in RPO 687 were 24-blade cooling fans mounted inside each brake drum.

CERV I

With his ill-fated SS racer relegated to the history books and the highly advanced Q-Corvette project derailed, Zora Duntov refocused his innovative efforts as the 1960s dawned on his Chevrolet Experimental Research Vehicle, or "CERV I," a name that only came about after a second CERV testbed appeared in 1963. An open-wheel machine that looked every bit like an Indy-style racer, Duntov's CERV I made its public debut at Riverside, California, in November 1960. Notably not in the business of building race cars, at least not officially, Chevrolet was quick to diplomatically describe CERV I as "a research tool for Chevrolet's continuous investigations into automotive ride and handling phenomena under the most realistic conditions."

Innovative CERV I features included an aluminum V-8 mounted directly behind the driver, a tubular space frame, a specially constructed lightweight fiberglass body, and independent rear suspension (IRS) with inboard aluminum brake drums. The work of Duntov and his senior engineers, Harold Krieger and Walt Zeyte, this relatively simple IRS design featured a three-link arrangement that used the U-jointed half shafts as upper locating members. Typical chrome-moly tubes handled lower linking duties, while a boxed-steel

> An open-wheel machine that looked every bit like an Indy-style racer, Duntov's CERV I made its public debut at Riverside, California, in November 1960.

combination hub carrier/radius arm on each side supplied the all-important horizontal third link, which transferred forward thrust to the frame. IRS advantages include the obvious ability of both wheels to respond separately to changing road

49

CHAPTER ONE

In September 1960, Zora Duntov got his "hillclimber" to Pikes Peak. The summit was closed, though, from early snow, so Duntov only ran in brief spurts, giving him data he couldn't really compare to past performances.

conditions. Depending on the design, an IRS setup also can be adjusted for negative camber, meaning the tires lean in slightly on top. Negative camber translates into better adhesion for the outside tire during hard cornering, since more of the tread remains planted as the car rolls away from the turn. Standard solid-axle suspensions tend to lift the tread's footprint off the road during hard turns, as the outside rear tire is forced to lean out. IRS designs also considerably reduce unsprung weight, since a heavy axle housing is no longer around to tax the springs.

Too much unsprung weight (mass not supported by the springs: tires, wheels, brake components, axle housing, etc.) relative to the amount of sprung weight (body, frame, engine and transmission, passengers, fuel, luggage) translates into severe vertical wheel motion under harsh driving conditions. Simply put, sprung weight must be substantial enough to dampen its unsprung counterpart's natural tendency to react proportionally to bumps, shocks, and body roll. Reducing unsprung weight not only helps handling, it also greatly improves general ride quality.

Duntov hoped to transfer both the CERV I's IRS and its midengine layout into the Corvette equation sometime in the future, but only the former feature carried over, this as part of the Sting Ray redesign for 1963. As for the CERV itself, it did its research thing quietly for a few years, then went through various collectors' hands like so many other unforgettable pieces of Corvette history.

1961

Model	availability convertible, with optional removable hardtop
Wheelbase	102 inches
Length	177.7 inches
Width	70.4 inches
Height	51.6 inches (top up); 51.5 inches (with removable hardtop)
Shipping weight	2,905 pounds (standard drivetrain)
Tread (front/rear, in inches)	57/59
Tires	6.70x15 four-ply
Brakes	11-inch drums
Wheels	15x5K
Fuel tank	16.4 gallons
Front suspension	independent short and long wishbones, coil springs, and stabilizer bar
Rear suspension	solid axle with longitudinal leaf springs; tubular shock absorbers
Steering	Saginaw recirculating ball (16.0:1 ratio; 16.3:1 adapter available option)
Standard drivetrain	230-horsepower 283-ci V-8, three-speed manual transmission, 3.36:1 axle ratio (3.70:1 standard behind optional four-speed manual, 3.55:1 ratio standard behind optional Powerglide automatic)
Optional engine	245-horsepower 283-ci V-8, dual four-barrel carburetors, hydraulic lifters
Optional engine	275-horsepower 283-ci V-8, fuel injection, hydraulic lifters
Optional engine	270-horsepower 283-ci V-8, dual four-barrel carburetors, solid lifters
Optional engine	315-horsepower 283-ci V-8, fuel injection, solid lifters
Optional transmission	Powerglide automatic (available with base V-8 and 245-horsepower 283)
Optional transmission	four-speed manual (available with all V-8s)
Optional gear ratios	4.11:1, 4.56:1

1961

Road & Track wasn't the only source to predict an all-new Corvette for 1960, and many of these incorrect forecasts were inspired by the XP-700, another executive toy fashioned for Bill Mitchell, this one coming early in 1959. Many members of the press figured the XP-700 dream machine represented a styling practice for the new body that was supposedly arriving the next year. Though they were wrong, they weren't 100 percent wrong.

Like Dave McLellan's engineers later on in the 1980s, Bill Mitchell's styling crew let Corvette buyers take a peek into the crystal ball in 1961. Mitchell's men simply grafted the XP-700's rear bodywork onto the existing Corvette shell. From the front, the '61 Corvette carried the same quad headlights, complemented with new painted bezels instead of chromed units. But in back, those familiar curves were traded for a crisper, uplifted tail fitted with two pairs of small, round taillights, a look that would carry over through 1962 and into the Sting Ray era.

That "boat-tail" rear not only appeared more modern, it also backed up its warmed-over form with a well-received functional upgrade—trunk space was increased without even a midgen being tacked on to the existing solid-axle platform's 177.2-inch length. Additional changes in back involved rerouting the dual exhausts, which now dumped out just behind the rear tires. According to *Sports Car Illustrated*, the new exhausts "rumble with a truly musical motorboat tone and beat a tattoo on the sides of the cars you pass." New up front was a toothless grille featuring an anodized rectangular mesh.

> According to *Sports Car Illustrated*, the new exhausts "rumble with a truly musical motorboat tone and beat a tattoo on the sides of the cars you pass."

A new "toothless" grille appeared in 1961. Notice the custom wheelcovers with painted center sections behind their spinners.

CHAPTER ONE

Foretelling things to come in 1963, the 1961 Corvette was fitted with new "boat-tail" styling. Again notice the non-stock wheelcovers with their custom center treatment.

> Like Dave McLellan's engineers later on in the 1980s, Bill Mitchell's styling crew let Corvette buyers take a peek into the crystal ball in 1961.

More optional power appeared in 1961, as the aluminum heads tried the previous year were recast in iron, leading to the belated introduction of the 275- and 315-horsepower fuel-injected V-8s. Revised option codes also were introduced: RPO 353 for the former, RPO 354 for the latter. While the hydraulic-cam, dual-carb 245-horse 283 carried over into 1961 still wearing the RPO 469 tag used the previous year, its solid-lifter 270-horse running mate was retagged RPO 468.

Mounting the optional big tank behind a 1961 Corvette's seat meant the removable hardtop had to be installed because there was no longer room for the standard folding roof. *Mike Mueller*

52

C1 1953–1962

The top engine option in 1961 was the 315-horsepower fuel-injected 283 V-8. *Mike Mueller*

Below: Like XP-700, XP-755 too showed off the boat-tail treatment that became a regular-production feature in 1961. The Shark is pictured in 1995 on display at the National Corvette Museum. *Mike Mueller*

More weight-saving metal surfaced in 1961, as the optional four-speed (RPO 685) was treated to an aluminum case and an aluminum radiator became standard with all engines. Included with the latter equipment was a remote surge tank (in place of the integral header tank used in 1960) that looked like a mini beer keg. Not all 1961 Corvettes had this remote tank, however, as leftover supplies of 1960's header-tank radiators had to be exhausted first.

Mako Shark I

Bill Mitchell just couldn't collect enough toys. After having his Stingray racer built in 1959, he followed it up with yet another personal ride, the Shark, or XP-755, in 1961. Built atop a mildly modified 1961 Corvette frame, the Shark was viewed by some at the time as a precursor to the all-new Corvette then being readied for its 1963 debut. Indeed, like Mitchell's other personal customizations, it did foretell many of the upcoming Sting Ray's lines and curves. But what most witnesses didn't know was that the Sting Ray's form had already been finalized when Larry Shinoda drew up XP-755. In this case, the boss simply wanted to update his collection, and Shinoda complied.

Mitchell's request (make that demand) concerning XP-755's looks were explicit: he wanted it painted to match a mako shark he'd caught off Bimini, which he had mounted on his office wall—thus the "Shark" name. But reproducing that blue-fading-into-white shade proved quite difficult, so much so that the GM paint crew reportedly snuck in one night to respray the fish to match the car, fooling the main man in the process.

Snazzy side exhausts and either wire wheels or Halibrand mags complemented that finish. The curious double-bubble roof used atop the XP-700 Corvette in 1959 was applied, too, with some modifications. Power, meanwhile, came from a supercharged 327 small-block V-8 fed by four side-draft Weber carburetors.

The Shark was completed in only a few months, just in time to take a tour around the Road America track during the summer of 1961. It followed that up with auto show appearances in 1962, while also serving as Mitchell's occasional driver. More than

53

CHAPTER ONE

Created early in 1962, XP-755 originally was called "the Shark." It later was renamed Mako Shark I after the Mako Shark II appeared in 1965.

one upgrade was made over the years, including a switch to big-block power, which the car still carries today.

A name change occurred along the way, after another prototype was created in 1965 sporting a similar paint job. Predicting the third-generation Corvette that debuted for 1968, this Shinoda-drawn machine was at first called the Mako Shark. The official moniker became Mako Shark II after it was decided to retroactively rename XP-755 Mako Shark I.

1962

At $4,038, the 1962 base price was the Corvette's first to bash the four-grand barrier. But in exchange for all that dough, a customer got what ads called the "finest, fiercest yet."

Although exterior refinements were few, they still stood out. A tasteful blacked-out grille led the way up front, and another consolation to simplicity involved the deletion of the chrome trim around the traditional bodyside coves, which in turn meant that two-tone paint was no longer possible. Completing 1962's trim package were revised grilles for the cove panels' simulated vents and new rocker panel moldings. A few updated emblems, restyled door panels, and slightly revised seat upholstery, and there you had it, the 1962 Corvette.

Major changes came beneath the hood, beginning with a bigger small block, a bored and stroked 283 that now displaced 327 cubic inches. Dual carburetors were dropped after 1961, leaving three four-barrel 327s and the top-dog fuelie V-8 as the only power choices.

C1 1953–1962

The '62 Corvette's base 327 used a more potent hydraulic cam, 10.5:1 compression, and small-valve (1.72-inch intakes) cylinder heads to make 20 more horses than its 283 forerunner. Adding a larger Carter four-barrel and big-valve (1.94-inch intakes) heads resulted in the 300-horsepower 327, RPO 583. The most potent 327s both relied on the solid-lifter Duntov cam, big-valve heads, and 11.25:1 compression. The Carter-fed version,

> The Carter-fed version produced 360 horses, all ready, willing, and able, in advertisement's words, to "please the wildest wind-in-the-face sports car type."

Above: The 1962 Corvette was the first model to draw four thousand dollars and more from buyers' bank accounts.

Left: The 1962 Corvette was the last year for the solid-axle models, and the plant cranked out 14,531 of them.

CHAPTER ONE

RPO 396, was rated at 340 horsepower, while its injected counterpart, RPO 582, produced 360 horses, all ready, willing, and able, in advertisement's words, to "please the wildest wind-in-the-face sports car type."

Additional drivetrain upgrades for 1962 included a new weight-conscious aluminum case for the optional Powerglide automatic transmission, RPO 313, which was available only behind the two hydraulic-cam 327s. A three-speed manual was standard issue, and this tame little gearbox again lost ground in the popularity race with its optional counterparts. By the time the tire smoke cleared in 1962, sales of four-speed Corvette models had soared to 11,318, representing 78 percent of the total run. Two different T-10 four-speeds, both listed under RPO 685, were offered in 1962, another first. The existing close-ratio box was available behind the two solid-lifter 327s, while a new wide-ratio four-speed was optional for the two hydraulic-cam V-8s.

Other optional firsts for 1962 included off-road straight-through mufflers (RPO 441), positive crankcase ventilation (RPO 242, mandatory on '62 Corvettes delivered in California), and narrow whitewall tires (RPO 1832) that replaced the antiquated wide whites used up through 1961.

From building only 300 Corvettes in 1953, Chevrolet managed to up the production ante to 14,531 by 1962, the last year for the venerable solid-axle models, and the last year for a Corvette trunk until the C5 convertible debuted in 1998. But nary a tear was shed as the car's first generation came to a close; far better things waited in the wings.

1962

Model availability	convertible, with optional removable hardtop
Wheelbase	102 inches
Length	177.7 inches
Width	70.4 inches
Height	51.6 inches (top up); 51.5 inches (with removable hardtop)
Shipping weight	2,905 pounds (standard drivetrain)
Tread (front/rear, in inches)	57/59
Tires	6.70x15 four-ply
Brakes	11-inch drums
Wheels	5Kx15
Fuel tank	16.4 gallons
Front suspension	independent short and long wishbones, coil springs, and stabilizer bar
Rear suspension	solid axle with longitudinal leaf springs; tubular shock absorbers
Steering	Saginaw recirculating ball (16.0:1 ratio; 16.3:1 adapter available option)
Standard drivetrain	250-horsepower 327-ci V-8, three-speed manual transmission, 3.36:1 axle ratio
Optional engine	300-horsepower 327-ci V-8, four-barrel carburetor, hydraulic lifters
Optional engine	340-horsepower -ci V-8, four-barrel carburetor, solid lifters
Optional engine	360-horsepower 327-ci V-8, fuel injection, solid lifters
Optional transmission	Powerglide automatic (available with base V-8 and 300-horsepower 327)
Optional transmission	four-speed manual (2.54:1 low with base V-8 and 300-horsepower 327; 2.20:1 low with 340-horsepower and 360-horsepower V-8s)
Optional gear ratios	3.08:1, 3.55:1, 3.70:1, 4.11:1, 4.56:1

On February 11, 1962, the FIA added a new season opener to the Daytona Continental 3 Hour Run, its international endurance racing series. Bob Johnson drove the no. 17 Corvette in the front.

C1 1953–1962

Above: Though maybe not as practical a vehicle in the cold northern climate, the 1962 Corvette's larger small-block engine (which now displaced 327 cubic inches) was impossible to ignore, especially at the St. Paul Winter Carnival.

Left: The small-block Chevy V-8 grew again in 1962, this time to 327 cubic inches. The 1962 Corvette's standard engine was this 250-horsepower 327. *Mike Mueller*

CHAPTER TWO

02 Enter the Sting Ray

C2 1963–1967

1963–1967

For years sports car purists pooh-poohed the Corvette, claiming that it no way, no how belonged in the international sporting fraternity. In their humble opinion, the car always has been, among other things, too big, too heavy, too convenient . . . or, plain and simply, too American. But such stones have been thrown less and less often as Chevrolet's fantastic plastic plaything has matured over the years.

- First enclosed coupe body style and hideaway headlights (1963)
- Independent rear suspension (IRS) introduced (1963)
- Wheelbase cut to 98 inches (1963)
- Optional air conditioning introduced (1963)
- Split-window styling appears for one year only (1963)
- New RPO code system introduced (1963)
- Last fuel-injected V-8 until 1982 (1965)
- Four-wheel disc brakes become standard (1965)
- First big-block V-8 (1965)
- Optional teakwood steering wheel and telescopic steering column introduced (1965)
- Big-block V-8 displacement goes from 396 cubic inches to 427 (1966)
- Triple-carburetor induction introduced for 427 V-8 (1967)
- Aluminum heads become an option (RPO L88) for 427 V-8 (1967)

BY THE TIME THE C5 MODEL DEBUTED for 1997, no one doubted the Corvette's proven place among the world's best sports cars, and for the price you simply couldn't do any better. Today's C6 is not only this planet's biggest bang for the buck, it's also able to run circles around performance machines costing many, many thousands more. If only Zora was still here to see it.

Duntov often found himself defending his baby against those slings and arrows during the solid-axle years, and, in many cases, detractors did have a case. While first-generation Corvette models unquestionably were the hottest things running on American streets during the 1950s, they never could quite get over the hump when compared to Europe's best, especially so on the international racing stage. Under-hood affairs matched up well with all but the most exotic foreign sports cars; it was the chassis that needed some serious work.

Enter the Sting Ray—the new model that had Duntov beaming in the fall of 1962. "For the first time, I now have a Corvette I can be proud to drive in Europe," he said, while introducing Chevrolet's second-generation two-seater to the press. Beneath that beautiful body, penned by Larry Shinoda, was a radically revised foundation featuring innovative independent rear suspension (IRS), a layout first tested beneath the CERV I experimental vehicle in 1960.

Similar in layout to the CERV I design, save for springing, the 1963 Sting Ray's IRS relied on each U-jointed half shaft to play the part of an upper locating link running from the differential to the hubs. Typical control rods made up the lateral lower links, from differential to hub carriers, with the latter coming at the end of a pair of boxed-steel trailing arms that supplied the longitudinal third link to the frame on each side. Space constraints ruled out the CERV I's coil-over shock absorbers, forcing Duntov to use what he called an "anachronistic feature"—a transverse multi-leaf spring mounted below and behind the differential. Clearly representing the only feature ever shared by a Model T Ford and a Corvette, this nine-leaf buggy spring may have appeared antiquated at a glance, but it did the job in lieu of the more expensive coil-over shocks.

Combined with improved suspension geometry up front, IRS helped transform the Sting Ray into a more complete sporting package. "This is a *modern* sports car," wrote *Motor Trend*'s Roger Huntington. "In most ways, it's as advanced as the latest dual-purpose sports/luxury cars from Europe. The new Corvette doesn't have to take a back seat to any of them, in looks, performance, handling, or ride."

Above: Grand Sport number 005, owned by Florida collector Bill Tower. The first two Grand Sports built (001 and 002) were converted into roadsters in 1964. *Mike Mueller*

Opposite: A new big-block hood complemented the 1967 Sting Ray's cleaner exterior but was nonfunctional.

59

Adding an extra pair of taillights was a common customizer's trick during the 1960s. But this one-off 1967 Sting Ray was modified that way at the St. Louis plant as a special-order reward for a successful West Coast Chevrolet dealer. The paint stripe and mag wheels also were added in St. Louis. *Mike Mueller*

Sting Ray roots date back to 1957, when Duntov's team began work on the "Q-Corvette," a radical departure initially proposed for regular production in 1960. Q-code features included an innovative rear-transaxle layout with independent suspension and inboard drum brakes. By relocating the transmission to the rear while keeping the engine up front, designers hoped to move closer to a preferred tail-heavy weight distribution, something all but impossible in a car carrying a big V-8 in its nose. Duntov's ideal design, of course, involved a supremely balanced midengine layout, but the Q-code proposal was as close as he could get to such a high-minded goal back in the 1950s.

Duntov planned a unit-body chassis on an extremely short 94-inch wheelbase (down 8 inches from the existing model) for the Q-Corvette. Stylist Bob McLean then fashioned a coupe body standing only 46 inches tall. Slated for production in steel, McLean's startlingly sleek shell debuted full-

"This is a *modern* sports car," wrote *Motor Trend*'s Roger Huntington.

C2 1963–1967

Called the "birdcage," this welded-steel structure meant that the 1963 Sting Ray's body could be molded up using thinner fiberglass panels. It was much stronger and safer than its 1962 predecessor.

Below: To determine airflow, engineers used strands of thread affixed to the body. The ink dots show the air current.

scale in clay in November 1957. Slim and compact with pop-up headlights, a pointed fastback roofline, and stylish bulges atop each wheel opening, the body eventually would carry over into regular-production reality even though the Q-Corvette never made it past the mock-up stage. Chevrolet quickly gave up on the Q concept once retooling costs were determined to be too great.

Though Duntov did have his dreams, he also recognized the budgetary realities that controlled his hand. The Q-Corvette was simply too big a step ahead; improving the car would have to happen in smaller doses, and to that end he wrote the following memo to engineer Harry Barr in December 1957:

"We can attempt to arrive at the general concept of [a new] car on the basis of our experience, and in relationship to the present Corvette. We would like to have better driver and passenger accommodation, better luggage space, better ride, better handling, and higher performance. Superficially, it would seem that the comfort requirements indicate a larger car than the present Corvette. However, this is not so. With a new chassis concept and thoughtful body engineering and styling, the car may be bigger internally and somewhat smaller [externally] than the present Corvette. Consideration of cost spells the use of a large number of passenger car components, which indicates that the chassis cannot become so small that [those components] cannot be used."

61

CHAPTER TWO

Zora Duntov hoped to build as many as 125 Grand Sport racers in 1963, but GM's anti-racing edict, sent down in January that year, squelched his plan. *Mike Mueller*

> "For the first time, I now have a Corvette I can be proud to drive in Europe."—*Zora Duntov*

While Duntov battled with bean counts, Bill Mitchell's stylists were hard at work fashioning the final shape for the next-generation Corvette to come. Public previews showed up in the form of Mitchell's own Stringray racer, the 1958 XP-700, and the 1961 XP-755 Shark, all of which were created in the Q-Corvette's image. The actual working model for the 1963 Corvette form was XP-720, born in the fall of 1959 in a cramped basement area at GM Styling known as "Studio X," home to Mitchell's right-hand man, Larry Shinoda.

A concession to comfort, transforming the Corvette into a coupe also was key to the Sting Ray image. That tapered roof in back served as an extension for Bill Mitchell's stinger concept, a styling queue that began as a blade-shaped bulge on the hood dubbed "Mitchell's phallic symbol" by Shinoda. The bulge was followed by a sharp ridge running up over the roof, and that ridge continued on to the tail, creating the famed split-window theme in the process. Representing a clear case of function falling victim to form, the split rear window didn't work at all for Duntov, who questioned its negative impact on rear visibility. Mitchell, however, was adamant. "If you take that off, you might as well forget the whole thing," Mitchell said. The stinger stayed, but not for long. It was gone, to the delight of most drivers, when the 1964 Corvette debuted.

As for the package as a whole, it wowed witnesses, both in America and overseas, like no Corvette before. And the so-called midyear models stuck around for five years, one more than planned.

Designer Larry Shinoda went to work for GM late in 1956 and quickly became Bill Mitchell's favored son. Shinoda styled Mitchell's Stingray racer and then went to work on XP-720. He also was responsible for CERV I, CERV II, and both Mako Shark show cars. *Mike Mueller*

62

C2 1963–1967

Though the initial design caused extreme front-end lift, Bill Mitchell stuck to "function follows form" with the design of the new Sting Ray.

Mitchell, Duntov, and crew originally had hoped to have a third-generation Corvette up and running for 1967, but technical difficulties delayed that introduction until 1968. Critics who still believe that the '67 Sting Ray represented one of the finest of the breed remain thankful.

1963

Available for the first time in two body styles, coupe or convertible, the 1963 Corvette rolled on a revised ladder-type frame (with five lateral crossmembers) that was 50 percent more rigid than the antiquated X-member unit it replaced. And along with being much stronger, the new frame's widely spaced boxed perimeter rails allowed designers to lower the Sting Ray's floorpan, which in turn meant the roofline could be dropped without cutting headroom.

Wheelbase also was reduced, from 102 inches to 98, and Duntov managed to favorably redistribute the car's weight by relocating the engine and passenger compartment rearward on that shortened hub-to-hub stretch. Better balance and less unsprung weight in back (200 pounds, compared to 300 in 1962), thanks to the new IRS layout, made the Sting Ray the best-handling, nicest-riding Corvette to date.

Steering precision was improved, as were brakes. Although drum diameters stayed at 11

1963

Model availability	coupe and convertible (optional removable hardtop)
Wheelbase	98 inches
Length	175.3 inches
Width	69.6 inches
Height	49.8 inches, in all cases
Tread (front/rear, in inches)	56.25/57
Tires	6.70x15 four-ply
Brakes	four-wheel drums
Fuel tank	20 gallons
Front suspension	parallel A-arms with coil springs
Rear suspension	independent three-link with transverse leaf spring
Standard drivetrain	250-horsepower 327-ci V-8, backed by a Saginaw three-speed manual transmission and 3.36:1 rear axle
Optional engine	300-horsepower 327-ci V-8 (L75)
Optional engine	340-horsepower 327-ci V-8 (L76)
Optional engine	360-horsepower 327-ci fuel-injected V8 (L84)
Optional transmission	M35 Powerglide automatic (limited to base 327 and L75 only)
Optional transmission	M20 wide-ratio four-speed manual with 2.54:1 low (limited to base 327 and L75 only)
Optional transmission	M20 close-ratio four-speed manual with 2.20:1 low (limited to L76 and L84 V-8s only)
Optional gear ratios	ranged from 3.08:1 to 4.56:1, with or without optional Posi-Traction (4.11:1 and 4.56:1 were Posi-Traction only)

CHAPTER TWO

Standard 1963 wheel covers featured simulated knock-off spinners.

Telling the five second-generation Corvette models apart is a simple task if you know your fuel filler doors. This is the 1963 style. *Mike Mueller*

Two body styles were offered for the first time in 1963. Coupe production that year was 10,594, compared to 10,919 convertibles.

Sting Ray doors were cut aircraft-style into the coupe's low roof, an idea that looked cool and made entry and exit a little less hairy.

inches at all four corners, width increased—from 2 inches to 2.75 in front, and from 1.75 inches to 2 in back. Optional power assist (RPO J50) for those brakes debuted for 1963, as did optional power steering (RPO N40). The standard steering ratio was 20.2:1; a faster 17.6:1 ratio was optional. Steering wheel turns, lock to lock, were 3.4 for the former, 2.92 for the latter.

Further enhancing the newfound solid feel was the body's welded-steel birdcage, an internal reinforcing structure that weighed nearly twice as much as the minimal steel skeleton backing up the 1962 Corvette's fiberglass shell. Thoroughly modern styling touches for 1963 included the hideaway headlights that would remain a Corvette trademark up through 2004 and aircraft-style doors that were cut into the roofline to aid entry and exit. The two vents in the hood were nonfunctional, though initial plans had them working. The idea was dropped after engineers determined that the hot under-hood air exiting those vents would flow directly into the passenger compartment's cowl intakes.

Powertrain packages carried over unchanged from 1962, with the 250-horsepower 327 backed by a three-speed manual transmission as standard fare. All the hot performance options of the previous year also returned wearing Chevrolet's new RPO codes: metallic brakes and the Posi-Traction differential were listed under RPOs J65 and G81, respectively. Another new option, the N11 off-road exhaust system, was announced for 1963 but apparently didn't show up until 1964.

Last, but certainly not least, on the 1963 RPO list was Z06, the Special Performance Equipment group package. Chevrolet's original Z06 package (it would return in 2001) was introduced at Riverside, California,

C2 1963–1967

Optional leather seats debuted along with the Sting Ray in 1963. Black plastic doorknobs were used that year. Mike Mueller

> Thoroughly modern styling touches for 1963 included the hideaway headlights that would remain a Corvette trademark up through 2004 and aircraft-style doors that were cut into the roofline to aid entry and exit.

on October 13, 1962, when the first six Z06 Sting Rays built went directly into prominent racers' hands. SCCA production-class competition represented the Z06's reason for being—hence the inclusion of every hot piece on the Corvette parts shelf.

The L84 327, close-ratio four-speed, and Posi-Traction rear were mandatory additions with the Special Performance Equipment package. Initially listed as part of the Z06 deal were a heavy-duty suspension, special heavy-duty power brakes, an oversized 36.5-gallon fiberglass fuel tank, and five cast-aluminum knock-off wheels. Typical road-worthy upgrades made up the track-ready suspension: stiffer shocks, beefier springs, and a thickened front stabilizer bar.

Not so typical was the brake equipment, which included a vacuum-assisted, dual-circuit master cylinder unique to the Z06 application—not the standard J50 setup as is often concluded. Unique too were the Z06's sintered cerametallix brakes shoes. They measured 11.75 inches long, compared to their 11-inch sintered metallic J65 counterparts.

Z06 brakes also featured special cooling equipment. Like the J65 units, the Z06's drums were finned to aid cooling. Additionally, five holes were opened up in the Z06 drums' faces, and special vented backing plates were used. In between, an internal fan kept a steady breeze blowing through those parts. Rubber elephant-ear scoops completed the package, attached to the backing plates' backsides to direct cooling air into the special vents.

Originally priced at a whopping $1,818.45, the Z06 package was trimmed a bit in December 1962 after the knock-off wheels and big tank were dropped from the deal. The tank still could have been added to a '63 Sting Ray after December 1962 by checking off RPO N03. Customers fond of the new knock-offs, however, weren't so lucky. Reportedly the result of Corvette racer and Gulf Oil executive Grady Davis's request for a quick-change road-racing wheel, the rims measured 15x6 inches and included a special hub adapter that allowed attachment by a single three-pronged threaded spinner (early prototypes used a two-eared spinner) in place of the typical five lugs. But nearly all wheels produced for 1963 leaked air, resulting in the option's quick cancellation. Some did make it into public hands before dealers began turning down

continued on page 68

CHAPTER TWO

The interior of the 1963 Corvette was entirely new and featured a dual cowl instrument panel that mounted the clock and radio in a center divider. The tachometer was larger, as well as the speedometer.

Below: The top engine option in 1963 was again the fuel-injected 327 V-8 (RPO L84), rated at 360 horsepower. Standard power came from a 250-horse 327. *Mike Mueller*

66

C2 1963–1967

Output for the L84 fuel-injected 327 V8 was 360 horsepower in 1963. The L84 jumped up to 375 horsepower the following year.

Left: Segmented linings and a cooling fan were incorporated inside the Z06's finned drums. The elephant ear on the ground attached to the inside of the backing plate to funnel cooling air into the brakes. *Mike Mueller*

Right: Along with the L84 327, the Z06 package included special brakes with finned drums for improved cooling. Also notice the holes in the Z06 drum's face—these allowed an escape route for overheated internal air. *Mike Mueller*

67

CHAPTER TWO

continued from page 65
requests for the desired option. Exactly how many remains unknown.

Another change made to the Z06 option in December 1962 involved availability. Like the N03 fuel tank, RPO Z06 was initially only offered for coupes. Plain and simple, the oversized tank could not be located in a topless Sting Ray. That exception disappeared after the big tank was dropped from the Z06 lineup, and thus the Special Performance Equipment package became listed as an option for the convertible, too, according to a distribution bulletin dated December 14. Reportedly, one of the 199 '63 Z 06 Sting Rays released was a drop top.

1963 Grand Sport

Chevrolet wasn't supposed to be involved in racing in 1963, at least according to the infamous ban on factory-backed competition projects announced by the AMA in the summer of 1957. Yet Zora Duntov was again working on a special-edition racing Corvette—a car he hoped would pick up where his ill-fated SS had left off in 1957. Originally known simply as the "lightweight Corvette," Duntov's latest racing machine was officially titled "Grand Sport."

Duntov's engineering team kicked off the Grand Sport project in the summer of 1962, beginning with a tubular-steel frame that weighed 94 pounds less than the stock foundation and incorporated mounting points for a full roll cage. With its transverse leaf spring and three-link layout, the Grand Sport's independent rear suspension looked familiar, but it was beefed up considerably. Unwanted pounds, meanwhile, were trimmed by using modified sheet-steel trailing-arm hub carriers drilled with lightening holes. Other weight-saving components included an aluminum steering box, special sheet-steel A-arms up front, and a set of 15x6 Halibrand magnesium knockoff wheels. Solid 11.75-inch Girling disc brakes with three-piston calipers were mounted at all four corners. Heavy-duty front coil springs, various reinforced steering pieces, and stiff Delco shocks completed the chassis.

On top of that purpose-built platform went a one-piece fiberglass shell with superthin panels to save even more weight. Other measurements here and there—a lower, narrowed roofline and slightly

> Originally known simply as the "lightweight Corvette," Duntov's latest racing machine was officially titled "Grand Sport."

1963 Grand Sport

Model availability	coupe and roadster
Body	specially molded lightweight fiberglass panels with trunk lid for spare tire access
Wheelbase	98 inches
Chassis	tubular steel
Brakes	Girling 11.75-inch discs
Wheels	Halibrand magnesium knock-offs
Fuel tank	36.5 gallons
Front suspension	independent with sheet-steel A-arms and coil springs
Rear suspension	independent with sheet-steel trailing arms, aluminum differential housing, and transverse leaf spring
Engine (proposed)	377-ci aluminum small-block V-8 with four Weber two-barrel carburetors and tube headers
Transmission	four-speed manual

Introduced in October 1962, the Z06 racing package was created just before GM execs ordered Chevrolet to cease all competition activities. Veteran Corvette racer Bob Bondurant was among the first to put a 1963 Z06 Sting Ray into action. The number on his car, 614, was the street address of his sponsor, Washburn Chevrolet.

C2 1963–1967

Texan John Mecom took three Grand Sports to the Bahamas in November 1963 for the Nassau Speed Week races. By then, the cars had been heavily modified with various scoops, vents, and flairs, the latter items added to help house wider wheels and tires.

more windshield slope, for example—differed from stock specs as designers tried to improve on the Sting Ray's surprisingly poor aerodynamics. The car looked slick, but it remained a veritable brick in the wind. Frontal area was much too large for a race car, and body lift at high speeds was akin to an airplane. That unwanted lift plagued all Corvettes built from 1963 to 1967.

Additional differences between the standard Sting Ray shell and its flimsy Grand Sport counterpart included enlarged wheelhouses (to make room for more rubber), a rear deck lid for spare tire access, fixed headlights mounted behind clear plexiglass in place of the heavy hideaway units, and aluminum underbody reinforcement instead of the conventional steel birdcage. All windows, save for the windshield, were made of plexiglass, and Mitchell's stinger was deleted in back.

Initial plans had the Grand Sport powered by an exotic 377-cubic-inch small block made entirely of aluminum and topped by four Weber 58-mm two-barrel carburetors, but the engine wasn't ready in time. The first Grand Sport was fitted with an aluminum version of the Sting Ray's fuel-injected

69

CHAPTER TWO

Initial plans called for this all-aluminum 377-cubic-inch V-8 to power the Grand Sports, but it wasn't ready in time, leaving the cars to race early on with basically stock fuel-injected 327s. Mike Mueller

Zora Duntov hoped to build as many as 125 Grand Sport racers in 1963, but GM's anti-racing edict, sent down in January that year, squelched his plan. Mike Mueller

327 V-8 and sent out for testing at Sebring in December 1962.

Grand Sport number 001 was the first of 125 planned lightweight Corvette racers. That plan, however, barely got off the ground before GM execs issued their own ban on factory racing in January 1963. Those same corporate officers had looked the other way while Chevrolet continued supporting competition projects after the AMA edict had gone into effect, but enough was enough. The Grand Sport was history even before it could make history at the track.

The aluminum 377 V-8 never made it into production and only five Grand Sport chassis were completed before the project was canceled. Or at least that's what the experts thought up until 2003, when evidence surfaced that a sixth might have been built but destroyed in 1964 or 1965 by order of GM officers who weren't happy about even one Grand Sport getting built. Whatever the case, the five Grand Sports known to exist today managed to find their way out of Chevrolet Engineering in 1963 and into supposedly private hands.

Once in the wild, all five Grand Sports began sprouting various louvers, holes, and vents for added cooling. Spoilers, huge hood scoops, and enormous fender flares were also added, and the transparent headlight shields were quickly replaced by solid covers. Various engines were tried over the years, including the 427 Mk. IV big block.

But the most noticeable modification involved the first two coupes, GS numbers 001 and 002. They were later converted into roadsters in the hopes of improving on the lackluster aerodynamics. But the topless Grand Sports didn't get the chance to prove their merits, at least not immediately, after their conversions in early 1964.

GM officials once again laid down the law after reading all about the Grand Sport coupes, which had been making mucho noise at various racing venues in 1963, including a wild foray to the Bahamas for the Nassau Speed Weeks in November. Texan John Mecom (who now claims he bought all six Grand Sports that year) fielded the

> A Grand Sport coupe (number 004) toured a big-time track, Daytona, in anger for the last time in February 1967.

team with much help from Chevrolet. When the dust had cleared, Mecom's Grand Sports had thoroughly embarrassed Carroll Shelby's Cobras. They also had mortified GM execs who were sure they had ordered the Grand Sport Corvette racers scrapped. Another upper office edict came down, this one telling Chevy people to steer clear of the Grand Sports or else, and no other high-profile appearances followed for Duntov's lightweight Corvette racers.

From there, the cars went through various racers' hands and competed with so-so success. The last major on-track appearance for a roadster (number 002) came in June 1966 at Mosport in Ontario. A Grand Sport coupe (number 004) toured a big-time track, Daytona, in anger for the last time in February 1967.

C2 1963–1967

While Grand Sport Corvettes grabbed the big headlines, regular-production Corvettes continued running strong in SCCA production-class competition in 1963.

In John Mecum's Grand Sport, bold fender flairs covered wide experimental tires from Goodyear, and a pair of impressive intake nostrils went on the hood.

1964

The Z06 Corvette didn't return for 1964, even though dealers had begun taking orders for them in the fall of 1963. Instead of the Special Performance Equipment group, Chevrolet simply offered the various components separately, with the heavy-duty suspension listed as RPO F40 and the gnarly metallic brakes found under RPO J56. The knock-off wheels (P48) and 36.5-gallon gas tank (N03) were brought back individually, too, and the off-road exhaust system (N11) mentioned in 1963 was officially introduced for 1964. Another new option, the K66 transistorized ignition, also debuted.

Once again, the 250-horsepower 327 backed by a three-speed manual transmission was standard beneath the hood. A 300-horse L75 V-8 and two improved 327s followed. The peak option was the L84 fuelie, by then rated at 375 horsepower, thanks to revised heads with bigger valves (2.02-inch intakes, 1.60-inch exhausts) and a new mechanical cam with more lift and longer duration. Since the extra lift meant the valves would intrude farther into the combustion chamber, machined reliefs were required for the piston tops, which in turn translated into a slight compression cut to 11:1. Basically an L84 V-8 with a big Holley four-barrel carburetor in place of the Ramjet fuel-injection setup, the L76 small block was advertised at 365 horsepower, the highest output Chevrolet would record for a noninjected 327.

Save for the removal of Mitchell's beloved stinger from the rear window, updates to the 1964 Sting Ray body were minor, and restyled wheel covers

71

CHAPTER TWO

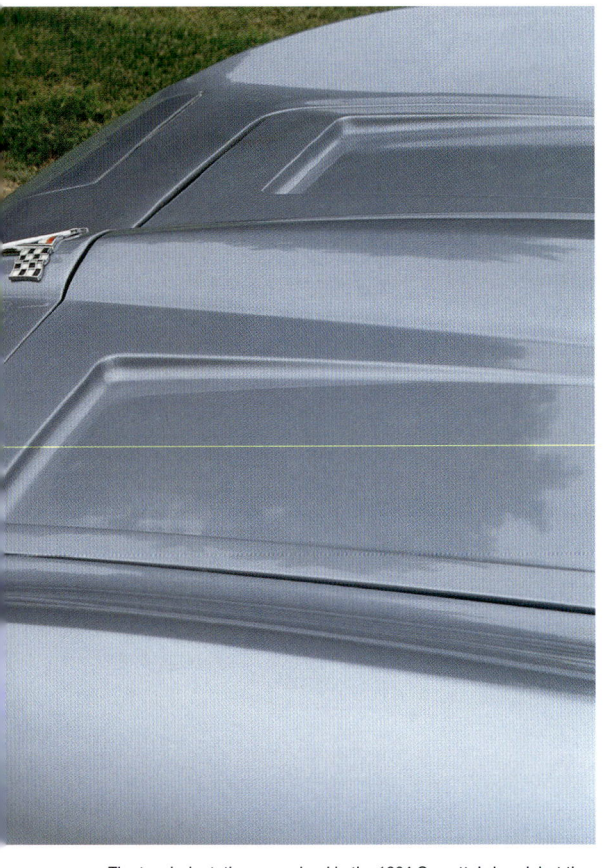

The two indentations remained in the 1964 Corvette's hood, but the fake grilles seen the year before were deleted. *Mike Mueller*

1964

Model availability	coupe and convertible (optional removable hardtop)
Wheelbase	98 inches
Length	175.3 inches
Width	69.6 inches
Height	49.8 inches, coupe and convertible; 49.3 inches, with removable hardtop
Tread (front/rear, in inches)	56.3/57
Tires	6.70x15 four-ply
Brakes	four-wheel drums
Fuel tank	20 gallons
Front suspension	parallel A-arms with coil springs
Rear suspension	independent three-link with transverse leaf spring
Standard drivetrain	250-horsepower 327-ci V-8, backed by a Saginaw three-speed manual transmission and 3.36:1 rear axle
Optional engine	300-horsepower 327-ci V-8 (L75)
Optional engine	365-horsepower 327-ci V-8 (L76)
Optional engine	375-horsepower 327-ci fuel-injected V-8 (L84)
Optional transmission	M35 Powerglide automatic (limited to base 327 and L75 only)
Optional transmission	M20 wide-ratio four-speed manual with 2.56:1 low (limited to base 327 and L75 only)
Optional transmission	M20 close-ratio four-speed manual with 2.20:1 low (limited to L76 and L84 V-8s only)
Optional gear ratios	ranged from 3.08:1 to 4.56:1, with or without optional Posi-Traction (4.11:1 and 4.56:1 were Posi-Traction only)

A restyled, frosted wheel cover appeared for 1964. *Mike Mueller*

Revised seats and chromed doorknobs set the 1964 interior apart from its 1963 predecessor. *Mike Mueller*

C2 1963–1967

By 1964, the coupe had shed its split window and both cars had abandoned the faux-grille on the front hood.

represented a notable change. Harder to spot, the revamped rocker moldings only sported three black stripes instead of eight. The fake grilles were removed from the hood, but their indentations remained. The fake air outlets behind the coupe's doors were restyled too, and the driver-side outlets became functional after an electric motor and ductwork were added to help ventilate the interior.

And let's not forget the gas cap—a restyled lid was introduced each year during the C2 run.

CERV II

Chevrolet wasn't supposed to be racing, but that didn't stop Zora Duntov from building another prototype race car immediately after his Grand Sport plan was shot down by corporate party-poopers in 1963. Originally proposed in the summer of 1962, the CERV II two-seater was built in 1964 with hopes of competing in Sports Prototype classes at Le Mans, Daytona, and Sebring. Like its CERV I forerunner, CERV II featured a midengine layout, but its all-wheel-drive design was truly unique, made possible by installing

An AM/FM radio (RPO U69) was a popular $176.50 option in 1964. U69 installations numbered 20,934 that year. *Mike Mueller*

A new fuel filler door adorned the 1964 Corvette's tail. *Mike Mueller*

CHAPTER TWO

Produced in 1964, the XP-819 concept vehicle featured a rear-mounted V-8 beneath a flip-up tail section. Its curvaceous body was penned by Larry Shinoda, and previewed much of the forthcoming third-generation look of the Corvette.

two torque converters—one for the front wheels, another for the back.

After initials plans for a tubular space frame were cast aside, the CERV II was constructed using boxed aluminum chassis members on a wheelbase measuring 92 inches from hub to hub. Original specifications called for a downsized 240-cubic-inch small-block V-8 to stay within the 4.5-liter displacement limit then in place for international endurance racing. Reportedly, one prototype CERV II was to be followed by six actual competition machines; three would be raced, three would serve as backups.

Only the prototype was built, however, as corporate killjoys again stepped in before Duntov could take the fiberglass-bodied two-seater racing. From then on, the company line read: the lone CERV II was created solely as an engineering testbed. Putting tires to task then briefly became its job before it soon faded into obscurity. Along the way, an all-aluminum 427 big-block V-8 was installed.

Again like its CERV I predecessor, CERV II survived the crusher and has passed through various collectors' hands over the years.

1965

Minor exterior upgrades for 1965 included a revised, blacked-out grille and a reformed hood devoid of those disliked indentations present in 1963 and 1964. Three functional vertical louvers

1965

Model availability	coupe and convertible (optional removable hardtop)
Wheelbase	98 inches
Length	175.3 inches
Width	69.6 inches
Height	49.8 inches, coupe and convertible; 49.3 inches with removable hardtop
Tread (front/rear, in inches)	56.3/57
Tires	7.75x15 four-ply
Brakes	four-wheel discs (drums available as credit option)
Fuel tank	20 gallons
Front suspension	parallel A-arms with coil springs
Rear suspension	independent three-link with transverse leaf spring
Standard drivetrain	250-horsepower 327-ci V-8, backed by a Saginaw three-speed manual transmission and 3.36:1 rear axle
Optional engine	300-horsepower 327-ci V-8 (L75)
Optional engine	350-horsepower 327-ci V-8 (L79)
Optional engine	365-horsepower 327-ci V-8 (L76)
Optional engine	375-horsepower 327-ci fuel-injected V-8 (L84)
Optional engine	425-horsepower 396-ci V-8 (L78)
Optional transmission	M35 Powerglide automatic (limited to base 327 and L75 only)
Optional transmission	M20 wide-ratio four-speed manual with 2.56:1 low (limited to base 327 and L75 only)
Optional transmission	M20 close-ratio four-speed manual with 2.20:1 low (limited to L76, L78, L79, and L84 V-8s only)
Optional gear ratios	ranged from 3.08:1 to 4.56:1, with or without optional Posi-Traction.

C2 1963–1967

Various ideas were tried before the bulging hood for the new-for-1965 big-block Corvette was finalized. This design was fashioned in 1963.

> Minor exterior upgrades for 1965 included a revised, blacked-out grille and a reformed hood devoid of those disliked indentations present in 1963 and 1964.

graced the fenders, new mag-style wheel covers went on at the corners, and restyled trim (with one long black stripe) covered the rocker panels. More-supportive vinyl bucket seats went inside, as did flat-faced instruments (in place of 1964's conical units), seatbelt retractors, and new molded door panels with integral armrests.

Four-wheel disc brakes became standard equipment in 1965. The new system featured 11.75-inch rotors squeezed by four-piston calipers. Buyers who preferred less stopping power could order the old drums (at a $64.50 credit), but only 316 did.

Optional side-mount exhausts debuted for the 1965 Sting Ray. *Mike Mueller*

CHAPTER TWO

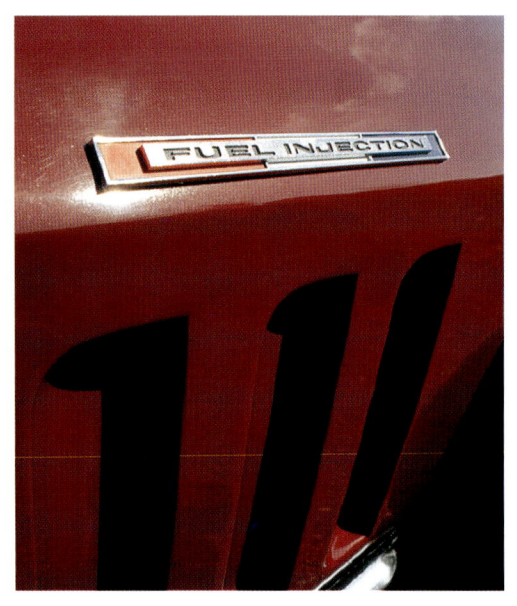

This legendary badge made its final appearance on fiberglass flanks in 1965. *Mike Mueller*

A classy teakwood steering wheel became a Corvette option in 1965. *Mike Mueller*

Cast-aluminum knock-off wheels were first offered as part of the Z06 package in 1963, but quality problems led to their recall that year. They returned, problem-free, in 1964. *Mike Mueller*

Rated at 360 horsepower in 1963, the fuel-injected L84 327 was muscled up to 375 horses the following year. The rating remained the same for the last fuelie V-8, offered for 1965 Corvette models. *Mike Mueller*

76

Making even more headlines was a new optional engine, the 396-cubic-inch Mk. IV V-8. Listed under RPO L78, the big-block baby featured free-breathing cylinder heads with staggered valves laid out in a seemingly haphazard fashion, reminiscent of a porcupine's quills—thus the nickname "porcupine motor." Mk. IV intake valves were located up high near their ports and inclined slightly, making for a straighter flow from intake manifold to combustion chamber. An opposite inclination was applied (to a slightly lesser degree) on the exhaust end to the same effect. Along with exceptional breathing characteristics, the inclined valve setup also offered slightly more room for bigger valves—Mk. IV units were truly big, with 2.19-inch intakes and 1.72-inch exhausts.

A serious solid-lifter cam activated those valves, a big Holley four-barrel fed the beast, and a mandatory K66 transistorized ignition supplied the spark. Inside the bores, impact-extruded aluminum pistons squeezed the mixture to an octane-intensive 11:1 ratio. The sum of the parts equaled a potent 425-horsepower big block that helped many among the fiberglass faithful forget all about the 327 fuelie V-8, which was then in its last year.

Another new engine appeared in 1965, the L79 small block. More or less an L76 327 with a slightly milder hydraulic cam in place of the solid-lifter stick, the L79 was rated at 350 horsepower. The L79 was identical to the L76 on the outside with its chromed air cleaner (atop the same Holley four-barrel) and finned cast-aluminum valve covers. Compression, at 11:1, was also the same. Priced a little lower than the L76, the L79 was easier to live with in everyday operation thanks to those hydraulic lifters.

Mako Shark II

Bill Mitchell first got serious about the next-generation Corvette image midway through 1964. He instructed Larry Shinoda to have a suitable preview ready in time for the New York International Auto Show, scheduled for April 1965. Shinoda's team rolled out a full-size mockup for press release photography in March—rolled out because, even though the hood said "Mark IV 396," there was no engine residing beneath. Originally labeled "Mako Shark," the moniker was soon changed to "Mako Shark II" after it was decided to retroactively rename the XP-755 prototype "Mako Shark I." A fully functional Mako Shark II followed, fitted with a 427-cubic-inch Mk. IV big-block V-8.

With or without an engine, the car looked fast. A sharply pointed prow led the way, followed by a domed hood that signified power. The bulging wheelhouses at all four corners looked ready to explode, their flared openings barely able to contain the fat Firestones within. Adding to the mockup's image were cast-aluminum side exhausts that exited the empty engine room halfway up the front fenders. The finned sidepipes were

1965 Mako Shark II

Body style	two-door fastback coupe with lift-up roof and tilt front end
Wheelbase	98 inches
Length	184.5 inches
Width	69.1 inches
Height	46.5 inches
Tires	8.80/9.15
Wheels	7.50x15 cast-aluminum knock-offs
Engine	427-cid V-8
Transmission	Turbo Hydra-matic automatic

Above: Designer Larry Shinoda's Mako Shark II concept car debuted at the New York International Auto Show in 1965, setting the stage for the C3 Corvette three years later.

Left: Electric switches abounded inside the Mako Shark II. Along with power locks and windows, the car also featured hideaway wipers and rear window louvers that were both controlled electrically.

Like the Mako Shark I, the Mako Shark II, unveiled in 1965, was styled by Larry Shinoda.

painted in crackle black with the edges of the fins remaining polished bright. The paint was removed after the New York showing, and the pipes were fully polished. Side exhausts were then deleted altogether on the functional Mako Shark II once its 427 big block went in place.

Measuring 3 inches lower than the existing Sting Ray, the Mako Shark II featured a sleek roofline that rolled back into a tapered exclamation point. A pronounced ducktail brought up the rear, and the whole works were painted dark down to light (Firefrost Midnight Blue, to lighter blue, to light gray).

The seats were fixed inside, as the foot pedals adjusted electrically to match the driver's reach. Equally unconventional, the aircraft-style steering wheel looked like a rectangle squeezed in slightly on the bottom.

Like those impractical side exhausts, the futuristic steering wheel was left off of the Mako Shark's functional alter ego, which was already in the works while the mockup was wowing New Yorkers in April 1965. When the 427-powered Mako was shown to the press on October 5, it had a typical round steering wheel. In place of the sidepipes was a conventional full-length exhaust system exiting through two highly stylized rectangular tips at the car's tail.

Gizmos and gimmicks, most of them electrical, abounded throughout the car. In all, there were 17 remote power units added to control all the toys, which included two spoilers in back that could've

> Measuring 3 inches lower than the existing Sting Ray, the Mako Shark II featured a sleek roofline that rolled back into a tapered exclamation point.

been raised some 4 inches at the flick of a switch. Also predicting things to come were digital readouts for the clock, fuel gauge, and speedometer.

After its press introduction, the 427-powered Mako Shark II was flown to France for an appearance at the Paris Automobile Salon on October 7, 1965. From there, it went to London, Turin, Brussels, and Geneva, before returning to America for the New York Auto Show in April 1966. Though some critics at home didn't think much of the car's eccentricities, overall opinions around the world were positive, and the sensationally sexy Mako Shark II image evolved into the third-generation Corvette, now known as the "Shark" model.

1966

Powertrain updates again made the most noise in 1966, as the 250-horsepower 327 was dropped in favor of a new standard engine, the 300-horse L75. Behind the V-8 was a new, stronger Saginaw three-speed manual transmission.

The optional 365-horse L76 too was cancelled, but the more civilized L79 327 remained. The two small blocks were then joined by a new big block, as the 396 was bored out to 427 cubic

Below: Big-block Corvettes wore the same hood in 1966, but beneath the bulge that year was the 427-cubic-inch V-8, created by boring out the 396. *Mike Mueller*

1966

Model availability	coupe and convertible (optional removable hardtop)
Wheelbase	98 inches
Length	175.3 inches
Width	69.6 inches
Height	49.8 inches, coupe and convertible; 49.3 inches with removable hardtop
Tread (front/rear, in inches)	56.8/57.6
Tires	7.75x15 four-ply
Brakes	four-wheel discs
Fuel tank	20 gallons
Front suspension	parallel A-arms with coil springs
Rear suspension	independent three-link with transverse leaf spring
Standard drivetrain	300-horsepower 327-ci V-8, backed by a heavy-duty Saginaw three-speed manual transmission and 3.36:1 rear axle
Optional engine	350-horsepower 327-ci V-8 (L79)
Optional engine	400-horsepower 427-ci V-8 (L36)
Optional engine	425-horsepower 427-ci V-8 (L72)
Optional transmission	M35 Powerglide automatic (limited to base 327 only)
Optional transmission	M20 wide-ratio four-speed manual with 2.52:1 low (available behind all engines but the L72)
Optional transmission	M21 close-ratio four-speed manual with 2.20:1 low (available behind all engines but the base 327)
Optional transmission	heavy-duty M22 Rock Crusher four-speed manual
Optional gear ratios	ranged from 3.08:1 to 4.56:1, with or without optional Posi-Traction

CHAPTER TWO

Extra pleats were added to the Sting Ray's bucket seats in 1966. *Mike Mueller*

> "Chevrolet insists that there are only 425 horses in there, and we'll just have to take their word for it," explained a *Car and Driver* report.

inches. Two 427 options were offered in 1966: the comparatively mild, hydraulic-cam L36, rated at 390 horsepower, and the truly awesome L72. Like its L78 forerunner, the L72 427 was stuffed full of 11:1 pistons, compared to 10.25:1 slugs for the L36. The L72 mechanical cam, however, was even lumpier than that of the L78, meaning even more horsepower was waiting beneath a '66 big-block hood. After being initially rated at an unheard-of 450 horses, the L72 was given a token 425-horsepower label. But anyone who believed that number needed only to drop the hammer one time on an L72 Corvette.

"Chevrolet insists that there are only 425 horses in there, and we'll just have to take their word for it," explained a *Car and Driver* report, which went on to mention the only numbers that really counted, anyway: 0–60 in 5.4 seconds and the quarter-mile in 12.8 seconds at 112 miles per hour. "Unreal" was *Car and Driver*'s understated conclusion.

Among other new options for 1966 were the J56 brake package and F41 sport suspension. Included in J56 were metallic linings and a dual-circuit master cylinder with power assist. F41 parts consisted of typically beefed-up springs and a thickened front stabilizer bar.

Basic exterior changes were limited to different rocker trim, restyled wheel covers, and a revised grille, this one featuring an attractive egg-crate design. Less-noticeable upgrades included adding "Corvette Sting Ray" script to the hood, and designers finally deleted the roof vents behind the doors, dropping the electric interior ventilation system too. Reupholstered seats, a vinyl-covered foam headliner in place of the previously used fiberboard, chrome door-pull handles, and standard backup lights were new as well for 1966.

If you identified this fuel filler door as the 1966 edition, you'd be right on. *Mike Mueller*

C2 1963–1967

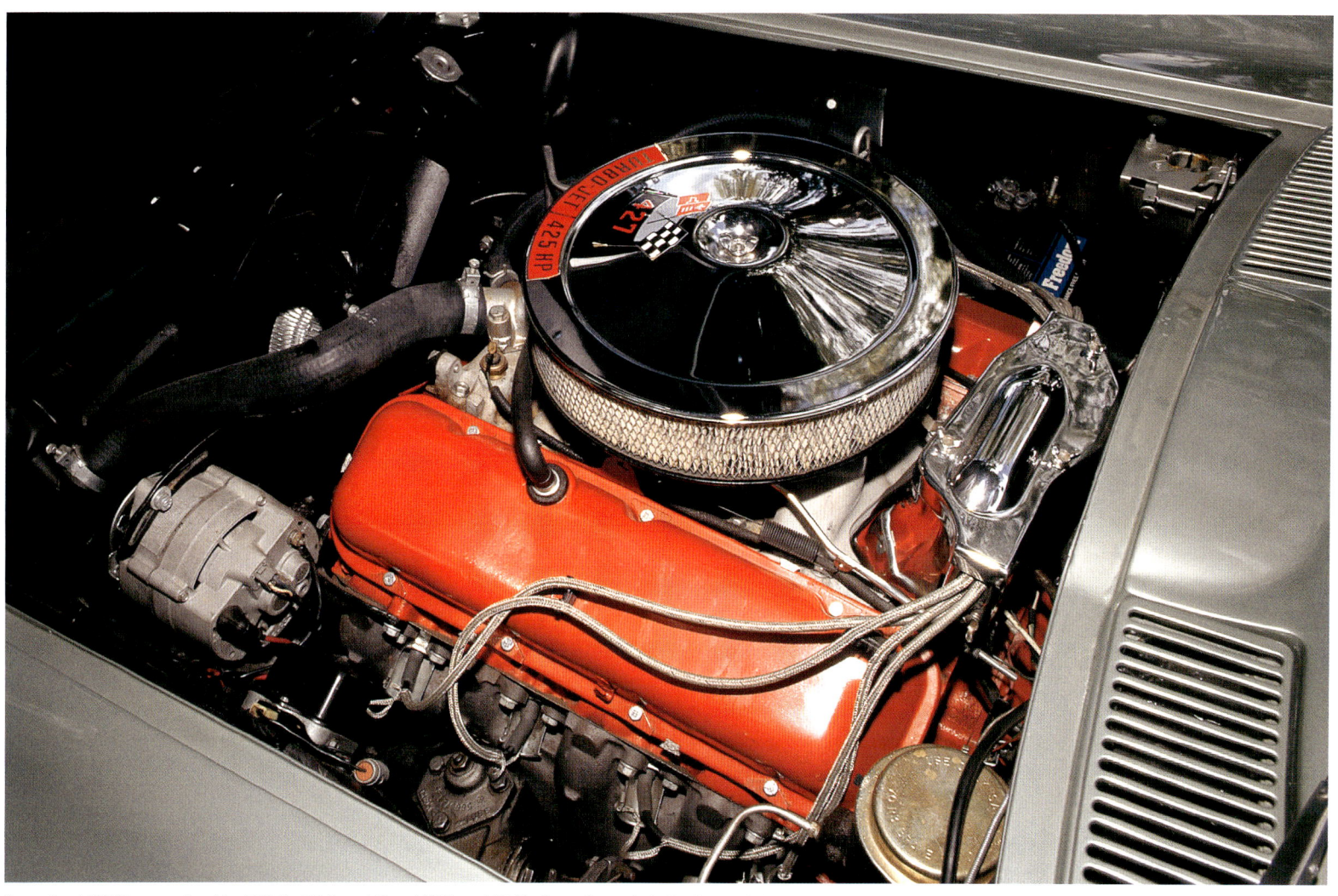

Two optional 427 V-8s were offered for 1966: the 390-horse L36 and 425-horse L72. The former featured a hydraulic cam, the latter solid lifters. *Mike Mueller*

For the 1966 model year, Chevrolet offered two engine displacements: the 327 sat under the standard hood, while the 427 required a power bulge to accommodate its air cleaner.

CHAPTER TWO

1967

Originally scheduled for 1966, the C2's closing act had most critics raving in 1967. This unintended encore resulted in what many still feel was the best Corvette of the short midyear run—as well as one of the best of all time. Even *Road &Track*'s ever-critical testers approved: "The Sting Ray is in its fifth and probably last year with that name and body style, and it finally looks the way we thought it should have in the first place."

Exterior modifications included stripping all identification off the fenders, which looked cleaner thanks to small, modernized louvers with a slight forward rake. Along the body's lower edge were new rocker moldings that were nearly completely blacked out. The traditional crossed-flag emblem up front was smaller, and the Corvette script seen on 1966 hoods was deleted. The whole works were complemented by a new standard wheel, the popular Rally rim with its brake cooling slots, trim ring, and center cap.

But easily the most prominent upgrade was the bold, new big-block hood, a marked improvement over the somewhat crude bulges used in 1965 and 1966. Complementing its sleek stinger shape was a set of paint stripes that varied in color depending on the exterior and interior shades chosen. For example, a black '67 Sting Ray with a black or red interior received red striping, while a black car with saddle or green appointments inside was adorned with a white scoop. In addition to the red and white stripes, hood scoops were also painted black, dark teal blue, or medium bright blue. And thanks to problems perfecting the painting process, some early '67 427 Corvettes were delivered with no stripes at all. Another problem—this one involving a mishap with the molds for the familiar, flat small-block hood—resulted in some stinger scoops appearing on 327 Corvettes in 1967. Apparently some dealers made this swap too.

Standard power beneath a '67 Corvette hood was once more the 300-horsepower 327, and the L79 small block and L36 big block (then rated at

> "The Sting Ray is in its fifth and probably last year with that name and body style, and it finally looks the way we thought it should have in the first place."—*Road & Track*

1967

Model availability	coupe and convertible (optional removable hardtop)
Wheelbase	98 inches
Length	175.1 inches
Width	69.6 inches
Height	49.8 inches, coupe and convertible; 49.3 inches with removable hardtop
Tread (front/rear, in inches)	57.56/58.36
Tires	7.75x15 four-ply
Brakes	four-wheel discs
Fuel tank	20 gallons
Front suspension	parallel A-arms with coil springs
Rear suspension	independent three-link with transverse leaf spring
Standard drivetrain	300-horsepower 327-ci V-8, backed by a heavy-duty Saginaw three-speed manual transmission and 3.36:1 rear axle
Optional engine	350-horsepower 327-ci V-8 (L79)
Optional engine	390-horsepower 427-ci V-8 (L36)
Optional engine	400-horsepower 427-ci 3x2 V-8 (L68)
Optional engine	435-horsepower 427-ci 3x2 V-8 (L71)
Optional engine	430-horsepower 427-ci V-8 (L88)
Note	aluminum heads (L89) were available for the L71; output did not change with this installation
Optional transmission	M35 Powerglide automatic (available behind base 327, L36 and L68 427s)
Optional transmission	M20 wide-ratio four-speed manual with 2.52:1 low (available behind all engines except the L71 and L88)
Optional transmission	M21 close-ratio four-speed manual with 2.20:1 low (available behind all engines except the base 327)
Optional transmission	heavy-duty M22 Rocker Crusher four-speed manual (for L88 427)
Optional gear ratios	ranged from 3.08:1 to 4.56:1, with or without optional Posi-Traction

The "Bat 'Vette" was a modified 1967 coupe designed by Chuck Jordan, shown here.

C2 1963–1967

New for 1967, the 435-horsepower L71 427's triple Holley carburetors used vacuum operation for a truly smooth power boost.

390 horsepower) were again optional. But an even meaner 427 was available in 1967, topped by three Holley two-barrel carburetors and crowned by a triangular air cleaner. This 3x2 L71 V-8 was rated at 435 horsepower, the highest rating ever stuck on a Corvette big block. Everything that made the L72 so tough—four-bolt main bearing caps, a lumpy solid-lifter cam, 11:1 compression, big valve heads, K66 transistorized ignition—was in the L71 package, save, of course, for that big 780-cfm Holley four-barrel.

Chevrolet's 3x2 induction setup began with an aluminum intake manifold. The center carb atop that manifold served as the primary fuel-feeder during normal operation, as throttle plates in the two secondary Holleys' venturi remained closed until revs reached roughly 2,000 rpm. Then a mass-air vacuum signal sent directly from the primary carb's venturi (as opposed to a typical central vacuum source) brought the front and rear two-barrels into the fray in a progression relative to the weight of the driver's foot. By the time engine speeds hit 4,000 rpm, all three throats were wailing away. The best published quarter-mile time for the L71 Corvette was a sizzling 12.90 seconds.

Chevrolet offered two triple-carb 427s in 1967, the other being the milder 400-horse L68. Despite a major price difference ($437.10 for 435 horses, $305.50 for 400), the L71 outsold the L68 by nearly a two-to-one margin. Production totals for the L68 and its L71 big brother amounted to 2,101 and 3,754, respectively.

Much rarer was another new big-block option, L88. Only 20 of these were built in 1967, and all were meant to go right to the track despite GM's claim that it wasn't involved in racing. Zora Duntov simply wouldn't be denied, and he had kicked off yet another competition-conscious project in 1965 based on a collection of regular-production options. At the heart of it all was the fabled L88 427, conservatively rated at 430 horsepower. Introduced in the spring of 1967, the first L88 Corvette didn't exactly set the racing world afire, thanks primarily to early durability problems involving overheating and an inherently weak lower end. Nonetheless, it was no slouch when it came to scorching a racetrack. At Le Mans in June 1967, an L88 topped 170 miles per hour on the Mulsanne Straight before a

Convertibles still dominated in 1967; production was 14,436. Coupe production was 8,504. *Mike Mueller*

connecting rod disconnected during the night to end the Corvette's pursuit of the legendary 24-hour laurels.

Plain and simple, this uncivilized beast was never intended for use on the street, although some brave souls did try. What these Walter Mitty types probably noticed first were the blank covers in place of both the radio and heater/defroster controls normally found in the center of a '67 Corvette's dash. RPO C48, the heater/defroster delete credit introduced in 1963, was included as part of the L88 deal. Options like a radio and power windows were off-limits, and a convenient automatic choke was not installed, although a retrofit hand-choke kit was available from Chevrolet.

Another feature common to street cars, a fan shroud to aid cooling, was also missing on the L88, as was any semblance of emissions controls. A typical PCV valve wasn't even present; instead, the L88 427 used an obsolete road-draft tube that vented crankcase vapors directly into the atmosphere via the driver-side valve cover. Anything that wasn't needed on a racetrack and anything that added unwanted extra pounds wasn't included in the L88 package, an off-road option if there ever was one.

C2 1963–1967

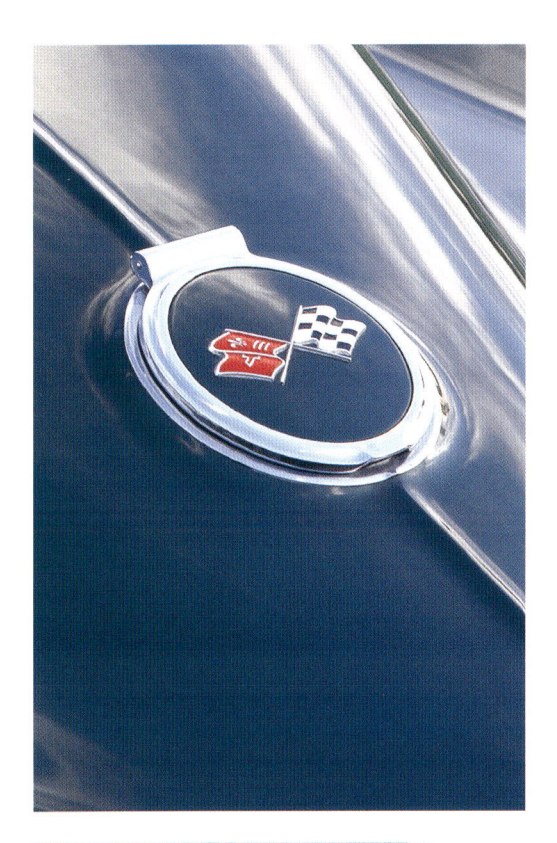

Far left: Notice that the 1967's fuel filler door was color-keyed to match the exterior paint choice. *Mike Mueller*

Left: Conventional bolt-on cast-aluminum wheels replaced the knock-off option offered in 1966. An ornamental center hub cap hid the bolt-on apparatus. *Mike Mueller*

> While L88 production would carry over into 1968, the midyear Corvette retired at the end of 1967 to, in many minds, a standing ovation.

The 1967 big-block hood was functional in the L88's case. The air cleaner fit into ductwork on the hood's underside that allowed cooler, denser atmosphere an easy route from the base of the car's windshield to the carburetor below. *Mike Mueller*

85

CHAPTER TWO

A huge 850-cfm Holley four-barrel carburetor fed the L88 427 through a special aluminum open-plenum manifold. *Mike Mueller*

Both the radio and the heater were deleted as part of the L88 deal. *Mike Mueller*

What was included was an impressive list of purpose-built, heavy-duty hardware. For starters, the L88 427 big block featured weight-saving aluminum cylinder heads atop a typical Mk. IV four-bolt iron block. The L88's crankshaft was specially forged out of 5140 alloy steel, then cross-drilled for sure lubrication and Tuftrided for hardness. Attached to that crank by shot-peened, Magnafluxed connecting rods were eight forged-aluminum pop-up pistons that squeezed the air/fuel mixture at a molecule-mashing 12.5:1 ratio. As it turned out, the L88's connecting rods were its weakest links. High-rpm failures convinced engineers to reinforce the L88's lower end in 1968 with beefier rods featuring heavier 7/16-inch bolts.

Those lightweight heads were fitted with big valves: 2.19 inches on the intake side, 1.84 inches on the exhaust. The same aluminum heads were also available in 1967 as RPO L89, which was offered exclusively atop the L71 427. Only 16 pairs of L89 heads were sold in 1967.

Supplying air/fuel to the L88 427 was a huge 850-cfm Holley four-barrel. The special high-rise aluminum intake below that four-holer had its internal partition machined down to create an open plenum for maximum high-rpm performance. Designed by Denny Davis, the L88's solid-lifter cam was radical to say the least, with 337 degrees intake duration, 340 degrees exhaust. Valve lift was 0.5365-inch intake, 0.5560 exhaust. Pushrods were thick 7/16-inch pieces with hardened ends. On top, special long-slot stamped-steel rockers rocked on heat-treated, hardened ball studs. Heavy-duty valve springs were held in place by beefed-up retainers and locks.

On the L88's breathing end, a unique "air cleaner" (using the term very loosely) fit into the hood's underside. Looking little like an air cleaner at all, the L88 unit featured an open pan mounted atop that big Holley. In the center of the pan was a small screen; around its edge was a foam gasket. With the hood closed, the gasket sealed the pan to a duct than ran back to the rear edge of the Corvette's normally nonfunctional stinger scoop. By making that scoop functional, engineers allowed the L88 to breathe in from the denser air mass that typically gathers at the base of a car's windshield at speed—a clever idea that had proven itself under Chevrolet hoods during the Daytona 500 in 1963.

Bolted up between the L88 427 and M22 trans was an appropriate heavy-duty clutch mated to a small 12.75-inch-diameter flywheel, the latter added in the best interests of increasing rpm potential by reducing reciprocating mass. A Harrison heavy-duty cross-flow radiator handled cooling chores.

The L88's sky-high compression left Chevrolet officials no choice but to warn owners about their use of fuel. "This unit operates on Sunoco 260 or equivalent gas of very high octane," read delivery paperwork. "Under no circumstances should regular gasoline be used." Another label inside the car repeated this caveat. "Warning: Vehicle must operate on a fuel having a minimum of 103 research octane and 95 motor octane or engine damage may result." Remaining standard features contributed further to a nasty nature. All L88s built in 1967 were fitted with such mandatory options as the stiff F41 suspension, G81 Posi-Traction differential, J56 power-assisted metallic brakes, and M22 Rock Crusher four-speed manual transmission. RPO K66, the transistorized ignition, was also a specified option. Living with the J56 brakes alone in everyday operation was not for the faint of heart. The fade-resistant clampers worked famously once warmed up. But when cold, they were about as effective as Fred Flintstone's feet.

While L88 production would carry over into 1968, the midyear Corvette retired at the end of 1967 to, in many minds, a standing ovation.

Dana Chevrolet entered this 1967 L88-equipped coupe in the 24 Hours of Le Mans on June 10-11, 1967.

87

CHAPTER THREE

03 Third Time's A Charm

C3 1968–1982

1968–1982

Like your best girl in school, the Corvette has never been able to make it downstairs on time for the big date. Consider the C5. Initial plans called for a 1993 debut, but various gremlins helped push that intro back to 1997. And who can forget 1983? Chevrolet decision-makers apparently did—they delayed the C4's unveiling by more than six months, resulting in a model-year jump from 1982 directly to 1984.

YET ANOTHER PERCEIVED DELAY had occurred some 25 years earlier when the rumor mill had the public anxiously awaiting a radically redesigned second-generation Corvette for 1960. But Zora Duntov and Bill Mitchell simply would sell no Sting Ray before its time, and that moment didn't arrive until 1963.

As exciting as the first Sting Ray was, it still was doomed to a short life from the outset. The first-generation solid-axle Corvette had carried on for 10 years—far longer, in many critics' minds, than it should have. No way Duntov, Mitchell, and the gang were going to let things go that stale again. They began working on the next best 'Vette yet even as the paint was drying on the first of the second-generation midyear models. Four years and out was the plan, as was topping a classic. Literally. Developing the original Sting Ray had already cost a boatload; busting the bank again only a few years later was out of the question. Developing the C3 was then left primarily to Mitchell's stylists, as placing an exciting new body atop the existing platform became the only choice. And the coming-out party for this new look initially was scheduled for 1967.

Wouldn't you know it?—late again. The revamped C3 didn't debut until 1968, leaving the midyear Sting Ray to make an encore appearance for 1967. Not that that was bad— many critics still consider the '67 Sting Ray to be a classic among classics. But Corvette watchers then were anxiously awaiting the next step, especially after many had caught a glimpse of Larry Shinoda's Mako Shark II, which had debuted at New York's International Auto Show in April 1965. *Car Life*'s editors asked if this was the "next Corvette"? "Let's hope Chevrolet's Special Engineering boys land this one by 1967," added *Hot Rod*'s Eric Dahlquist.

While the Mako Shark II was busy turning heads on the show circuit, Shinoda and the rest of Chevrolet's styling team were themselves hustling to finalize the form that would indeed emerge as the C3 Corvette, or "Shark," the nickname devotees prefer. Early prototypes mirrored the Mako Shark II image right down to the tapered roofline, which carried on in the best tradition of the original Sting Ray coupe. A one-piece, lift-off Targa top—an idea unashamedly borrowed from the European sports car scene—crowned things.

Making the Mako dream a reality, however, hit a hitch once Duntov's engineers began their tasks of mating function to form. The prototype C3 looked great, but it floundered in real-world tests as driver visibility, engine ventilation, and aerodynamics fell well below acceptable standards. Shinoda's shape

- C3 generation was the Corvette's longest (15 model runs)
- Zora Arkus-Duntov officially made Corvette chief engineer (1968)
- Stingray name tag (now as one word) returns after one-year hiatus (1969)
- Last L88 built (1969)
- All-aluminum ZL-1 V-8 built for one year only (1969)
- Annual production surpasses 30,000 for first time (1969)
- Base price surpasses $5,000 for first time (1970)
- Original LT-1 V-8 introduced (1970)
- Original ZR-1™ options package introduced (1970)
- Advertised horsepower drops for first time (1971)
- Original LS6 V-8 introduced (1971)
- Crash-proof plastic front bumper debuts (1973)
- Crash-proof plastic rear bumper follows (1974)
- Catalytic converters appear (1974)
- Last big-block Corvette built (1974)
- Dave McLellan becomes second Corvette chief engineer (1975)
- Last Corvette convertible built until 1986 (1975)
- Stingray badge last used (1976) until 2014
- Annual production surpasses 40,000 for first time (1976)
- St. Louis plant builds Chevrolet's 500,000th Corvette (1977)
- All 1978 Corvettes were 25th Anniversary models
- Two-tone Silver Anniversary paint option offered (1978)
- Corvette paces the Indianapolis 500 for first time (1978)
- Base sticker price surpasses $10,000 for first time (1979)
- All-time production record (53,807) established (1979)
- St. Louis assembly line closes, Bowling Green plant opens (1981)
- Two-tone paint option returns (1981)
- Collector Edition offered (1982)

Above: Coupes were again the most popular in 1970, with 10,668 built. Convertible production was 6,648.

Opposite: New fender louvers and flared body panels appeared in 1970. Mike Mueller

CHAPTER THREE

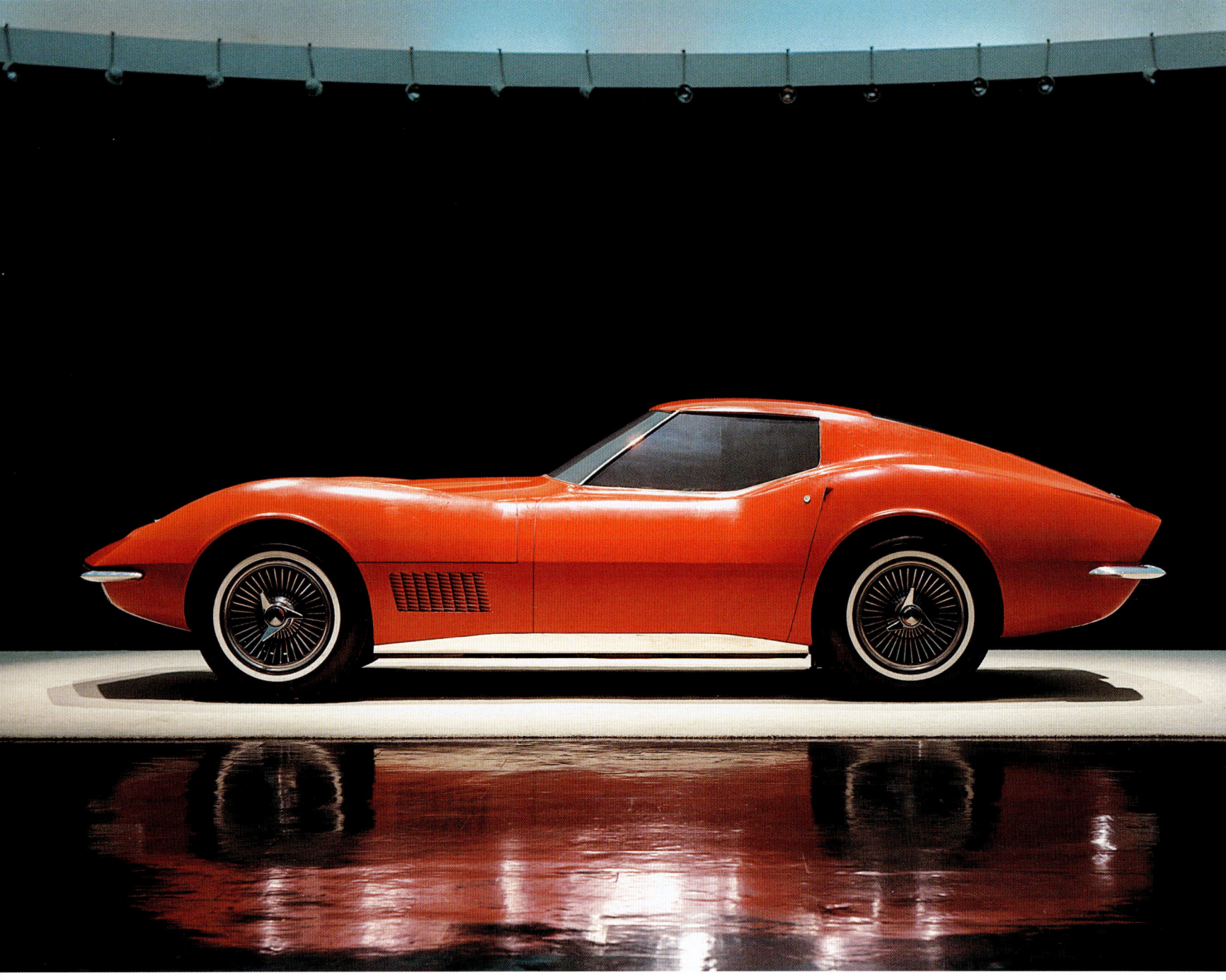

This model entered the Design Center auditorium for review in September 1965 and emerged, after a few tweaks, as the 1968 production Corvette.

was too curvaceous, too bulging. Those tall fender tops, that low tapered roof, and the big duck tail in back worked in concert to make it almost impossible to see anything other than low-flying airplanes from behind the wheel. The slinky shell was also too close to the ground in front, where its sharp-edged beak limited the amount of cooling air able to reach the engine.

Smoothing things out beneath that oversexed skin required more time than was allotted, thus Duntov requested an extra year to properly prepare the C3 for production. One of the first changes made was to trade that boat-tail roof for a more practical, more pleasing design also borrowed from Europe—in this case, from the Porsche 904. Vertical rear glass was sandwiched between two parallel flying-buttress C-pillars, a style known as a tunneled window.

A removable roof remained in place atop prototypes until the last minute. But with no fixed roof structure in place to help stiffen the platform, the Targa-top Corvette flexed too much. This pesky twisting not only made the top creak, it also compromised the car's weather-sealing capabilities. Designers had no choice but to add a central reinforcing strut to join the windshield header to the rear roof arch. This in turn meant the one-piece roof had to be separated into two sections—presto, a "T-top."

Additional fixes included cutting down the front fenders to allow the driver a safer look ahead. The rear quarters and roofline were also modified to improve rearward visibility. Downsizing the rear spoiler into a molded-in lip further enhanced the view. Engine cooling was aided by a reshaped nose that allowed the radiator a more prominent location up front. An air-dam lip that ran beneath the car and up around the front wheel openings was added to better direct cool air toward that radiator. Finally, large fender vents were incorporated to help hot air escape more easily from beneath the hood. Those last two modifications also helped decrease high-speed lift—a problem Duntov had been battling for years.

Though it looked sleek, the original Sting Ray was an aerodynamic disaster when it took off in 1963. In Zora's own words, the second-generation body possessed "just enough lift to be a bad airplane." The Mako Shark II was even worse, leaving the C3 prototype little chance of playing down the Corvette's high-flying reputation.

A third-generation test vehicle was touring GM's Milford Proving Grounds as early as the fall of 1965. A new '65 Corvette also hit the track to serve as a measuring stick. Duntov's "bad airplane" tended to lift at both ends at high speeds. At 120 miles per hour, the '65 Sting Ray's nose rose 2.25 inches, the rear a half-inch. In comparison, Engineering's C3 prototype hunkered down in back at speed, thanks to that large rear spoiler. At 120 miles per hour, its tail dropped a quarter-inch. This depression, in turn, helped raise the nose, a task the car could already handle well enough on its own. Front-end lift at 120 miles per hour measured 3.75 inches.

> Now the new 1968 Corvette was ready to take on the world.

Above: The third-generation Shark appeared in 1968 with a nicely styled body mounted atop the same basic platform introduced in 1963—even powertrain choices carried over. *Mike Mueller*

Below: A T-top roof was standard for the C3 "Shark." Notice the blank front fenders on this 1968 coupe— a "Stingray" badge would appear above those vents in 1969.

C3 1968–1982

91

CHAPTER THREE

Above: Total big-block Corvette production for 1968 was 12,627. Four different 427 V8 options were offered that year, including the aluminum-head L88.

Right: The beat went on at St. Louis in 1968 as the Corvette's third generation began. Notice the Turbo Hydra-matic automatic transmission, which replaced the old Powerglide on the RPO list that year.

Reducing this lift was first achieved by venting the front fenders. These vents (which, as mentioned, also aided cooling) allowed trapped airflow up front a quicker exit, which helped bring the prototype's aerodynamics down close to stock '65 Sting Ray levels. Adding a chin spoiler up front decreased top-end lift even further, to a measly five-eighths of an inch.

Case closed? Not exactly.

Excessive under-hood heat continued to plague the Shark right up to its press introduction in August 1967. All Duntov needed was one tour in the blue big-block coupe then being readied for journalists' scrutiny to recognize that the car would never keep its cool under the magnifying glass. Big-block Corvettes had always run hot, and this particular prototype was no exception. In truth, it was even more so due to the fact that all that cast iron was stuffed into stuffier confines. Cooling air still couldn't find its way into the radiator.

Duntov's quick fix saved the day. He opened up two oblong vents beneath the car's low-slung nose just ahead of the chin spoiler, then enlarged that spoiler to help increase the pressure, forcing the airflow up into those openings. From there, the rush of air could only flow through the radiator as all gaps to either side were closed up. The big-block prototype ran all day long in 85-degree heat at the Milford Proving Grounds during the press introduction, and the temperature gauge remained calm. Now the new 1968 Corvette was ready to take on the world.

1968

Corvette customers were no strangers to fiberglass imperfections, as body-finish quality had fluctuated considerably during the midyear run. But it had never been as bad as it was in 1968. Yet even with its obvious quality-control problems, the first Shark by no means flopped. Corvette sales for 1968 hit 28,566, a new record. And even though more than one press source was willing to point out the body's blemishes, other critics looked past those pimples to the C3's inner beauty.

Car Life called the 1968 Corvette "the excitement generator," and rightly so. Along with quickening pulses every bit as easily as the Mako Shark II had done three years before, Chevy's newest two-seater also did a decent job of cheating the wind, although most critics pointed out that much of the car's aerodynamic performance

92

1968

Model availability	sport coupe and convertible (optional removable hardtop)
Wheelbase	98 inches
Length	182.5 inches
Width	68.9 inches
Height	47.76 inches (sport coupe), 47.88 inches (convertible, top up), 47.78 inches (convertible, with hardtop)
Shipping weight	3,055 pounds (coupe), 3,070 pounds (convertible)
Tread (front/rear, in inches)	58.23/59
Tires	F70-15
Brakes	11.75-inch discs
Wheels	15x7 Rally rims with trim rings and center caps
Fuel tank	20 gallons
Front suspension	parallel A-arms with coil springs
Rear suspension	independent three-link with transverse leaf spring
Steering	recirculating ball, 20.2:1 ratio (17.6:1 with fast steering adjustment)
Standard drivetrain	300-horsepower 327-ci V-8 with single four-barrel carburetor, three-speed manual transmission, 3.36:1 axle ratio
Optional engine	350-horsepower 327-ci V-8 (L79) with four-barrel carburetor, hydraulic lifters (four-speed manual transmission only)
Optional engine	390-horsepower 427-ci V-8 (L36) with four-barrel carburetor, hydraulic lifters
Optional engine	400-horsepower 427-ci V-8 (L68) with three two-barrel carburetors, hydraulic lifters
Optional engine	435-horsepower 427-ci V-8 (L71) with three two-barrel carburetors, solid lifters (close-ratio four-speed transmission only)
Optional engine	430-horsepower 427-ci V-8 (L88) with four-barrel carburetor, aluminum heads, solid lifters (heavy-duty M22 four-speed transmission only)
Optional transmission	Turbo Hydra-matic automatic (with 3.08:1 axle ratio)
Optional transmission	wide-ratio four-speed manual
Optional transmission	close-ratio four-speed manual
Optional transmission	heavy-duty four-speed manual (M22), L88 only
Optional gear ratios	2.73:1, 3.55:1, 3.70:1, 4.11:1, 4.56:1

Widened 15x7 Rally wheels became standard at the corners in 1968. Red-stripe F70x15 tires were optional. *Mike Mueller*

remained a mirage; less so, however, than in 1967. Test figures for drag and high-speed lift were, according to *Sports Car Graphic*'s Paul Van Valkenburgh in 1970, "very respectable considering that the shape was dictated by GM Styling, and Chevrolet engineers had to sweat acid trying to keep the nose on the ground at speeds over 150."

Although the C3 gained some high-speed abilities, it lost a little ground on the scales, where it weighed in at about 3,440 pounds, roughly 200 more than the '67 Sting Ray. It was also 7 inches longer overall at 182.5 inches. At 69 inches, width was two-tenths less than in 1967, while height dropped from 49.6 inches to 47.8. Those last two measurements, working in concert with the radically increased tumble-home of the rounded Shark body sides, translated into a considerable reduction in interior space.

A sleek shape and claustrophobic cockpit weren't the only things to carry over from the Mako Shark II. Among features shared by the show car and the production model were a fiber-optic warning-light system and hidden windshield wipers. The latter rested below a vacuum-operated panel that popped open on demand. Like the Mako Shark II, the '68 Corvette also arrived without vent windows. In their place was Chevrolet's new Astro ventilation system, which routed fresh breezes in through the cowl, around interior airspace, and out through grilles located in the rear deck right behind the back window. Like those T-tops, that rear window also could be removed for maximum ventilation.

Most engineering features were carryovers as well. Beneath the '68 Corvette was basically the same chassis introduced in 1963, with its 98-inch wheelbase and independent rear suspension. Though familiar to any '63–'67 Sting Ray owner, the '68 suspension did receive a couple of nice tweaks, these performed to address another problem inherent to midyear models. Prior to 1968, mashing the go-pedal on a Corvette instantly pitched the nose up, which then threw front wheel geometry out of whack. The end result was a tendency for the car to wander at a time when precise control was preferred more than ever. To correct this, spring rates were stiffened, and the rear roll center was dropped from 7.56 inches above the ground to 4.71 inches by lowering the inner pivot points of the lateral suspension arms.

To compensate for the increased understeer dialed in by these changes, widened 7-inch Rally wheels were added to allow the use of fatter F70x15 tires. More rubber on the ground meant more resistance to lateral g forces. Maximum lateral acceleration measured 0.84 g for the Shark, compared to 0.74 for the midyear Corvette.

Did these modifications do the trick? Even *Road & Track*'s ever-critical critics were impressed: "No question about it, the Corvette is one of the best-handling front-engine production cars in the world."

The '68 Corvette's standard drivetrain was a complete midyear carryover: the same tried-and-true, 300-horse 327 small block backed by a three-speed manual gearbox. And 1968's optional engines list was identical, too. At the bottom was the L79 327, rated at 350 horses. Next came the 427 big blocks: the 390-horsepower L36 and its triple-carb running mates, the 400-horse L68 and

CHAPTER THREE

The 390-horsepower L36 427 big-block V-8 was a $200.15 option in 1968. L36 production that year was 7,717. *Mike Mueller*

435-horse L71. Topping things off were the L88 and the L89 aluminum-head package for the L71.

All 427s, save for the L88, were fitted with new low-rise intake manifolds designed to allow the carburetor (or carburetors) ample clearance beneath the third-generation Corvette's low, low hood. Efficient intake flow was preserved by sinking the manifold's underside into the big block's lifter valley.

Changes to the outrageous L88 427 in 1968 included the addition of smog controls, including a PCV valve and Chevrolet's air injection reaction (AIR) system. Beefier connecting rods and various new cams also were installed, as were Chevy's famed open-chamber heads midway into the year. L88 production for 1968 jumped to 80, all again featuring M22 Rocker Crusher four-speeds.

Optional transmissions rolled over from 1967 with one exception. The Powerglide automatic was finally replaced by Chevrolet's three-speed Turbo Hydra-matic, introduced in 1965. Bringing up the rear in 1968 was the Posi-Traction (RPO G81) differential in most cases. The G81 axle was a mandatory extra-cost choice behind all engine/trans combos, save for the manual-shifted 327s.

XP-882

From his earliest days working with Chevrolet's sporty two-seater, Zora Duntov envisioned an ideal design based on a chassis that located its engine amidships. Advantages to the midengine idea vary, not the least of which involves getting all that weight off the nose. Reducing pressure on the front wheels lightens up steering effort. This means faster manual steering ratios can be used without overtaxing the driver's arms. And moving the engine to the middle not only better balances the load, it also can translate into a preferred lower center of gravity, because all that motive mass can be mounted closer to the road with no steering or suspension components barring the way. The cockpit, too, can be lowered, primarily because the pilot doesn't have to look over a big V-8.

Duntov reportedly first proposed relocating the Corvette's engine rearward in 1954. His idea wasn't really considered until 1959, but no one on the street paid all that much attention when the midengine CERV I appeared the following year. Much the same could've been said when the all-wheel-drive CERV II followed in 1964.

Duntov proposed another midengine racer, the GS 3, in April 1964, and at the same time began drawing up a production-based counterpart. Meanwhile, R&D engineer Frank Winchell was hard at work on his rear-engine XP-819 to prove that Corvair technology could work wearing Corvette carb. Track tests proved otherwise.

Winchell returned to the drawing board in 1967, resulting in the sleek XP-880. This midengine machine was running by February 1968, and was

The XP 882 concept was the first of three midengine design and engineering prototypes to explore the idea of a mid-engine Corvette.

C3 1968–1982

Beneath the '68 Corvette was basically the same chassis introduced in 1963, with its 98-inch wheelbase and independent rear suspension.

The thoroughly modernized 1968 interior featured Astro ventilation, which drew cooling breezes in through the cowl and directed them back out through grilles located in the rear deck behind the rear window. *Mike Mueller*

officially named "Astro II" in preparation for a public introduction at the New York Auto Show that April.

Picking up where Winchell's XP-880 left off was XP-882, which Duntov began putting together early in 1968. Unlike previous prototypes, XP-882 featured a transverse-mounted engine, in this case a 400-cubic-inch small-block V-8. Power was transferred by a chain to an automatic transmission, where it was redirected 90 degrees to a differential. Wheelbase was a scant 95.5 inches, and total vehicle weight was a tidy 2,595 pounds.

This unique vehicle was nearing testing stages early in 1969 when new Chevrolet General Manager John DeLorean cancelled the project. He directed designers instead to try a less-expensive course using a more conventional platform based on the Camaro® chassis then being readied for 1970.

Fortunately, this "Camaro" idea was itself quickly cancelled after it was learned that both Ford and American Motors would be showing up at the 52nd annual New York Auto Show in April 1970 with their own midengine proposals, the Pantera and AMX/3, respectively. In response, Chevrolet revived the XP-882 project and put together its own 1970 exhibit labeled simply "Corvette prototype."

"We'll stake our reputation on this being the Corvette of the future," announced a July 1970 *Road & Track* report, "but don't expect it until 1972 at the earliest." Six months later, *R&T*'s Ron Wakefield explored the midengine proposal further. "We have now established beyond a doubt that the car was indeed a prototype for future production—1973, to be exact—and can report full details on the 1973 Corvette," he wrote in the January 1971 issue.

XP-882, however, was just the first in a short line of midengine experiments that never made it past the dream stage.

> Far and away the meanest, nastiest engine to ever make an RPO list, the exotic ZL-1 was purely and plainly a racing mill let loose on the street.

95

CHAPTER THREE

1969

Like the midyear models it replaced, the third-generation Corvette's initial body style rolled on in essentially identical form through five model runs, and two more five-year plans followed the first before the C3 run finally came to a close. Among other things, the basic profile, those cool T-tops, and the coupe's removable rear window all carried over unchanged from 1968 to 1972. Much the same could be said for the original Shark's standard wheels. Though widened from 7 inches to 8 in 1969, the Corvette's 15-inch Rally rims retained the same style each year up through 1972, as did the optional full wheel covers.

Most casual witnesses still find it difficult to readily differentiate between those first five Sharks. Perhaps the easiest clues in 1969 involved badging. Although the Sting Ray nameplate used from 1963 to 1967 didn't reappear in 1968, it did show up on 1969 fenders, this time spelled as one word. Furthermore, push-button door releases, individual square backup lights, and a dash-mounted ignition switch were all unique to 1968 models. In 1969, the ignition was moved to the steering column, the doors' pushbuttons were deleted, and the backup lights were moved up into the center of the inner pair of taillights. Additional 1969 giveaways included two noticeable options offered for that year only: side-mount exhausts (RPO N14) and bright trim (RPO TJ2) for the car's fender louvers.

Much less noticeable in 1969 was a new standard engine, a 350-cubic-inch small block originally created in 1967 by stretching the 327's 3.25-inch stroke to 3.48 inches while retaining the 4.00-inch bore. All other numbers carried over from the Corvette's 327 to the new base 350 as compression remained at 10.25:1 and output stayed at 300 horsepower.

A new code on the options list, RPO L46, emerged to mark the popular 350-horse small block's evolution from 327 cubic inches to 350—out with L79, in with L46. A second code first seen in 1969, ZL-1, identified an entirely new breed of big block, an all-aluminum 427. All the other optional 427s—they with their cast-iron blocks—rolled over from 1968, as did the L89 aluminum-head option.

Far and away the meanest, nastiest engine to ever make an RPO list, the exotic ZL-1 was purely and plainly a racing mill let loose on the street. Once again, Chevrolet wasn't supposed to be in racing, but there was performance products chief Vince Piggins lobbying for an all-aluminum big-block engine in 1968, with intentions being to support, among others, Bruce McLaren's successful Canadian-American (Can-Am) Challenge Cup team. Small-block Chevy-powered McLaren racers had begun their domination of the Can-Am series in 1967. Then Jim Hall built an aluminum big block for his Chaparral race team,

1969

Model availability	sport coupe and convertible (optional removable hardtop)
Wheelbase	98 inches
Length	182.5 inches
Width	69 inches
Height	47.8 inches (sport coupe), 47.9 inches (convertible, top up), 47.8 inches (convertible, with hardtop)
Shipping weight	3,091 pounds (coupe), 3,096 pounds (convertible)
Tread (front/rear, in inches)	58.7/59.4
Tires	F70-15
Brakes	11.75-inch discs
Wheels	15x8 Rally rims with trim rings and center caps
Fuel tank	20 gallons
Front suspension	parallel A-arms with coil springs
Rear suspension	independent three-link with transverse leaf spring
Steering	recirculating ball, 20.2:1 ratio (17.6:1 with fast steering adjustment)
Standard drivetrain	300-horsepower 350-ci V-8 with single four-barrel carburetor, three-speed manual transmission, 3.36:1 axle ratio
Optional engine	350-horsepower 350-ci V-8 (L46) with four-barrel carburetor, hydraulic lifters (close- or wide-ratio four-speed manual transmissions only)
Optional engine	390-horsepower 427-ci V-8 (L36) with four-barrel carburetor, hydraulic lifters
Optional engine	400-horsepower 427-ci V-8 (L68) with three two-barrel carburetors, hydraulic lifters
Optional engine	435-horsepower 427-ci V-8 (L71) with three two-barrel carburetors, solid lifters (optional aluminum-head L89 available)
Optional engine	430-horsepower 427-ci V-8 (L88) with aluminum heads, four-barrel carburetor, solid lifters
Optional engine	430-horsepower 427-ci V-8 (ZL-1) with aluminum heads and cylinder block, four-barrel carburetor, solid lifters
Note	370-horsepower 350-cid LT-1 V-8 initially listed but wasn't offered until 1970
Optional transmission	Turbo Hydra-matic automatic (with 3.08:1 axle ratio) available with all engines except L46 350
Optional transmission	wide-ratio four-speed manual
Optional transmission	close-ratio four-speed manual
Optional transmission	heavy-duty four-speed manual (M22)
Optional gear ratios	2.73:1, 3.55:1, 3.70:1, 4.11:1, 4.56:1

Coupe production for 1968 was 9,936. Base price was $4,463. *Mike Mueller*

The familiar Stingray (now one word) badge returned to Corvette fenders in 1969 after a one-year hiatus.

inspiring McLaren to threaten to look to Ford for a comparable lightweight big block to power his 1968 Can-Am cars. Piggins stepped in, promised Bruce his aluminum big blocks, and the rest is racing history. Armed with the ZL-1, McLaren destroyed all Can-Am comers from 1968 to 1971, winning 32 of 37 events.

Thoroughly baptized by fire on the 1968 Can-Am circuit, the ZL-1 427 then made its way in 1969 onto the Corvette's RPO list, as well as into GM's central office production order (COPO) pipeline. Although some references identified the ZL-1 V-8 as a "special L88," creating the king of the 427s was by no means a simple matter of trading cast iron for aluminum to make the cylinder block match those lightweight heads. While the ZL-1 and L88 wore similar heads in 1969, the block was a truly unique piece of engineering.

Duntov turned to Fred Frincke for the expertise needed to fashion a high-performance engine completely out of aluminum. Casting was Frincke's forte; he knew his way around a foundry. Winters Foundry was responsible for casting the ZL-1's block, heads, and intake. Machining work and assembly were then completed at Tonawanda under a "100-percent parts inspection" policy in production areas that Duntov described as being "surgically clean."

Frincke chose heat-treated 356 T-6 alloy for the block, which was cast with thickened walls

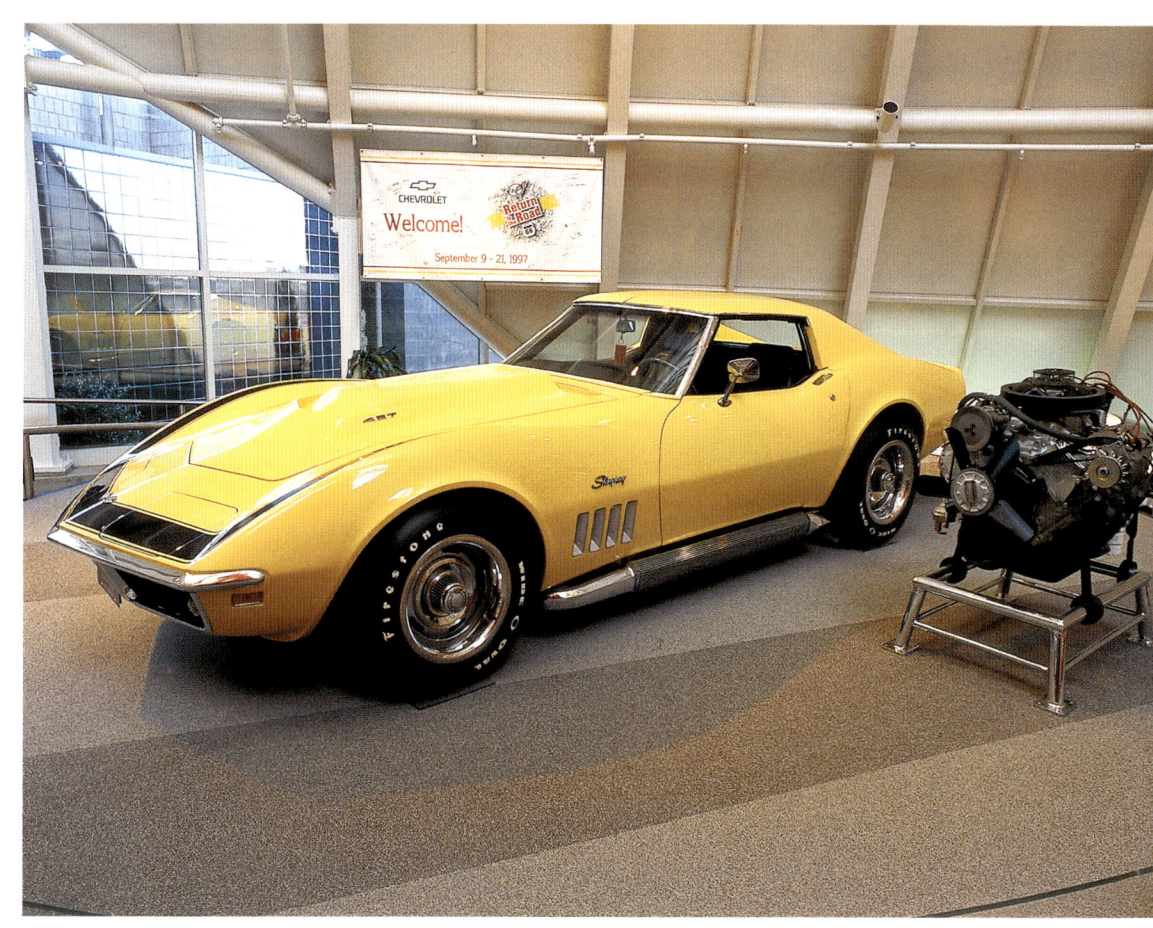

Only two remaining 1969 ZL-1 Corvettes are known—this one belongs to Florida collector Roger Judski. At right is the awesome, all-aluminum ZL-1 427 V-8, token-rated at a laughable 430 horsepower 30 years back. *Mike Mueller*

CHAPTER THREE

Corvette enthusiasts were thrilled to see this familiar name returned in 1969.

and beefed-up main webs to compensate for the aluminum's weaker nature. The heads trapped the eight cast-iron cylinder sleeves in the aluminum block's bores. At the bottom end was a fully nitrided, forged-steel crank held in place by four-bolt main bearing caps. The same Magnafluxed connecting rods, introduced midyear in 1968 for the L88, were used, with their beefed 7/16-inch bolts, full-floating wrist pins, and Spiralock washers.

ZL-1's heads, also cast from 356 T-6 aluminum, were based on the open-chamber unit introduced midway through 1968 for the L88. Combustion chambers in those heads were opened up (thus the name) around the spark plug. Results of this change included a drop in compression (from 12.5:1 to 12:1) for the "second-design" L88 because chamber volume increased. Breathing, on the other hand, went up by a reported 30 percent, thanks to the revised open chamber's closer relation to the exhaust port.

Optional side-mount exhausts were offered for the third-generation Corvette in 1969. *Mike Mueller*

Only two documented 1969 ZL-1 Corvettes are known. *Mike Mueller*

Revised ports also contributed greatly to the open-chamber head's superior breathing. Although the large rectangular intake passages remained the same size as the first-generation L88's, they were recontoured internally to help speed the air/fuel mixture into those open chambers. Exhaust ports were radically reshaped from rectangles into round passages to match up to the tube headers that racers would quickly bolt up in place of the mismatched, rectangular-passage iron manifolds delivered from the factory.

Like the open-chamber L88 head, the ZL-1 unit featured 2.19-inch intake valves and enlarged 1.88-inch exhausts. The ZL-1, however, was fitted with new TRW forged-aluminum pistons with extra-thick tops and strengthened pin bosses. These beefier slugs not only proved more durable than the L88 units, they also reinstated the aluminum-head 427's original 12.5:1 compression by way of increased dome area.

The ZL-1 427's solid-lifter cam was even more radical than the L88's, at least as far as lift was concerned. Intake valve lift was 0.560 inch, exhaust was 0.600. Tests, however, demonstrated that decreased durations cooperated better with those free-flowing, sewer-sized ports. The resulting shorter-duration cam, working in concert with the various reinforced internals, helped the ZL-1 wind out like no big block on this planet. Seven grand on the tach was no problem, and Chevrolet engineers claimed short bursts to 7,600 were within reason. Keeping the juices flowing during those high-rpm trips was a huge 850-cfm Holley four-barrel.

As was the case with the L88, the ZL-1 was given a bogus output rating of 430 horsepower. Reportedly, 525 horsepower was more like it, and engineers claimed 600 horses were possible with only a little tweaking.

Not until the new Z06 was unleashed in 2006 has Chevrolet dared deliver so much raw power to the people. Yes, John Q. Public could have walked into his nearest Chevy dealership in 1969 and rolled out in a street-legal, emissions-controlled, aluminum-engine Corvette able to break not simply into the 13-second bracket, nor the 12—breaking the sound barrier was more like it.

"The ZL-1 doesn't just accelerate, because the word 'accelerate' is inadequate for this car," wrote *Motor Trend*'s Eric Dahlquist after a wild ride in the most outrageous Corvette ever built. "It tears its way through the air and across the black pavement like all the modern big-inch racing machines you have ever seen."

The great white whale of a Corvette that Dahlquist raved about screamed through the quarter-mile in 12.1 seconds at 116 miles per hour. Shorter rear gears easily would have translated into 11 seconds in the quarter-mile, if not less. A second ZL-1 test mule present on the day Dahlquist took his ride managed a 10.89-second

CHAPTER THREE

After building only 20 L88 Corvettes in 1967, Chevrolet came back with 80 in 1968 and 116 in 1969. This is the 1969 iteration. *Mike Mueller*

pass at 130 miles per hour. Not only was that ridiculously fast, it was ridiculously easy. "The fact that almost anybody who knows how to drive could jump in and duplicate this run after run may be the most shattering aspect of all," concluded a *Motor Trend* report entitled "The 10-second Trip."

Why anyone in his right mind would want to travel from point A to B that quickly also begged the question of what type of maniac in 1969 would fall in line to buy a ZL-1 Corvette. "First, he will have a lot of money," answered Duntov. The all-aluminum 427 big block alone wore a $4,718.35 price tag. A complete ZL-1 Corvette cost more than $9,000, and it was this intimidating total that explains why only two documented examples are known today.

Explaining how others appeared over the years is easy enough. Reportedly, as many as a dozen mules or executive toys might've been built. The race-ready ZL-1 427 also was offered on its own, in a crate, and any number of these could have found their way into street-going Corvettes. According to Fred Frincke, Chevrolet's Mk. IV engine facility in Tonawanda, New York, built 154 ZL-1 427s.

Tonawanda plant man and diehard big-block researcher Fran Preve claims a different score. His search through Chevrolet's official "Summary of Engines Shipped" papers uncovered 94 ZL-1s manufactured for Y-body Corvette applications: 80 for four-speeds, 14 for M40 automatics. Add to that another 90 all-aluminum 427s built for F-body Camaros. Sixty-nine ZL-1 Camaros rolled out of Chevrolet's back door in 1969 thanks to Vince Piggins' clever use of the COPO loophole. Along with the two recorded factory installations in Corvettes, one other ZL-1 big block went into the Mako Shark II when it was restyled into the Manta Ray show car in 1969.

Like 1969's engine lineup, that year's transmission list looked familiar, with one notable change as the tough Turbo Hydra-matic (RPO M40) became available behind the two solid-lifter 427s, and not a moment too soon in some minds. "The Turbo Hydro is the best thing that's happened to big-engined Corvettes since high-octane gas," wrote *Hot Rod*'s Steve Kelly. "Those who can overcome the four-speed mystique are in for a surprise," claimed *Car Life*'s Corvette fans. "The Turbo Hydra-matic fitted to the high-performance 427s is magnificent. It slips from gear to gear in traffic without so much as a nudge. Power tightens the shifts into a series of iron hands, strong enough to light the tires at every change."

When ordered behind the base 350 that year, the M40 trans was priced at $221.80. It cost $290.40 when mated to the L71 or L88. "In the mild engine, the M40 was set to shift up quickly," continued *Car Life*'s July 1969 L88/automatic review. "In the wild engines, the transmission stays in the lower gear until the driver lifts his foot, right up to redline." Of the 116 L88 Corvettes built for 1969, 17 featured the Turbo Hydra-matic.

Manta Ray show car

Once off the auto show circuit, the Mako Shark II paced a few races and also, like other Corvette show cars, served time as Bill Mitchell's personal driver. Then, in 1969, it took on a new identity when it was restyled into the Manta Ray.

Already measuring some 9 inches longer than a standard Stingray, the Mako Shark length grew even more after the Manta Ray conversion. Extra inches came by way of a restyled, stretched tail that took on a tapered look from a profile perspective. The point of that tail was protected by a body-colored Endura bumper. Among other updates were a chin spoiler up front and a repaint that played down the shark shading. Sidepipes were later added, as were small mirrors mounted up high on the windshield pillars. The biggest news, though, was the Manta Ray's new power source. In place of the iron-block 427 used by the Mako Shark II was Chevrolet's all-aluminum ZL-1 427, an exotic mill more befitting of such an exotic, one-of-a-kind show car.

Estimates put the price tag for the original 427-powered Mako Shark II at as much as $2.5 million. Reportedly, that bill may have even hit $3 million by the time the ZL-1 Manta Ray transformation was complete.

> *Motor Trend*'s Chuck Koch almost couldn't believe his eyes. "The Corvette was just as fast, if not faster, through the corners as the Porsche."

1970

Model availability	sport coupe and convertible (optional removable hardtop)
Wheelbase	98 inches
Length	182.5 inches
Width	69 inches
Height	47.8 inches (sport coupe), 47.9 inches (convertible, top up), 47.8 inches (convertible, with hardtop)
Shipping weight	3,153 pounds (coupe), 3,167 pounds (convertible)
Tread (front/rear, in inches)	58.7/59.4
Tires	F70-15
Brakes	11.75-inch discs
Wheels	15x8 Rally rims with trim rings and center caps
Fuel tank	20 gallons (California emissions tank, 17 gallons)
Front suspension	parallel A-arms with coil springs
Rear suspension	independent three-link with transverse leaf spring
Steering	recirculating ball, 20.2:1 ratio (17.6:1 with fast steering adjustment)
Standard drivetrain	300-horsepower 350-ci V-8 with single four-barrel carburetor, four-speed manual transmission, 3.36:1 axle ratio
Optional engine	350-horsepower 350-ci V-8 (L46) with four-barrel carburetor, hydraulic lifters (close- or wide-ratio four-speed manual transmissions only)
Optional engine	370-horsepower 350-ci V-8 (LT-1) with four-barrel carburetor, solid lifters (close- or wide-ratio four-speed manual transmission only)
Optional engine	390-horsepower 454-ci V-8 (LS5) with four-barrel carburetor, hydraulic lifters.
Note	460-horsepower (or 465-horsepower) 454-cid LS7 V-8 initially listed but cancelled before reaching production
Optional transmission	Turbo Hydra-matic automatic (with 3.08:1 axle ratio), not available with L46 or LT-1 350-ci V-8
Optional transmission	wide-ratio four-speed manual
Optional transmission	close-ratio four-speed manual
Optional transmission	heavy-duty four-speed manual (M22), ZR1 option only
Optional gear ratios	2.73:1, 3.55:1, 3.70:1, 4.11:1

1970

Corvette production first surpassed the 20,000 level in 1963 and stayed there up through 1969, the year the annual figure topped 30,000 for the first time on the way to a new all-time high of 38,762 cars. It needs to be said, however, that this record came about partially because Chevrolet General Manager John DeLorean allowed the 1969 model year to run over into December after a strike had delayed production for two months earlier in the year. This extension in turn then took a bite out of the following model run, resulting in only 17,316 1970 Corvettes being released by the St. Louis plant, representing its lowest effort since 1962.

The 1970 Corvette included flares added to the wheel openings to reduce bodywork damage created by debris tossed up by the fat tires in 1968 and 1969. Revised fender vents wearing crosshatch grilles appeared in 1970 and stuck around through 1972, as did rectangular exhaust tips, which replaced the round units used previously. The round turn signal lamps and small side marker lights seen in 1968 and 1969 were replaced by larger side markers and rectangular turn signals, and these items carried on unchanged through 1972.

Posi-Traction and the M20 wide-ratio four-speed transmission were added into the standard package. G81 installations had been on the rise throughout the 1960s and by 1968 were found on 94.5 percent of that year's cars. Meanwhile, the reverse was happening concerning deliveries of

CHAPTER THREE

Coupes were again most popular in 1970, with 10,668 built. Convertible production was 6,648.

In 1970, big-block hood bulges were styled like the LT-1, with "454" badging highlighted on the side.

the standard three-speed manual trans, which had made up 67.5 percent of the mix in 1957, the first year for the four-speed option. That figure fell to a microscopic 4.3 percent in 1963, followed by only 1.1 percent five years later. No more three-speeds were seen after 1969.

Both the close-ratio M21 manual box and the Turbo Hydra-matic automatic could've been ordered in place of the M20 at no extra cost in 1970. Yet another former option, tinted glass all around, also was tossed in as part of the standard deal that year.

Of course, adding extra standard equipment meant that Chevrolet had to ask more for base Corvette coupes and convertibles in 1970. After increasing only 2.5 percent from 1968 to 1969, the Corvette coupe's bottom line jumped 8.6 percent in 1970, surpassing $5,000 for the first time in the process. The convertible's standard sticker first breached the five-grand plateau in 1971.

Standard power again came from the 300-horse 350, and the 350-horse L46 returned on the options list. But all other optional engines were dropped in favor of two new V-8s—one small block, one big block, respectively known as "mouse" and "rat" motors among the Chevy faithful. The new rat came about after engineers stroked the Mk. IV V-8 (from 3.76 to 4.00 inches) to raise displacement to 454 cubic inches. At 4.25 inches, the LS5's bore remained constant, as did compression (10.25:1) and base output (390 horsepower). Torque, on the other hand, was a whopping 500 ft-lbs, 40 more than the L36 427 turned out in 1969.

While it could melt a Wide Oval with the best of 'em and gulp a gallon of ethyl quicker than you could pump 'er in, the high-compression LS5 still could've been considered user friendly, thanks in part to its hydraulic cam. "It is by far the most tractable big-engine Corvette unit we've tried," claimed *Road & Track*.

As for the new optional small block, it too impressed witnesses, and to much greater degrees. Labeled LT-1, this 350-cubic-inch mouse motor relied on a solid-lifter cam to help produce a whopping 370 horsepower, meaning it easily could slug it out with engines displacing many more cubes. Built from 1970 to 1972, the LT-1 Corvette combined rat-like muscle with the nimbleness no big-block driver ever knew. "As you would expect, the personalities of the LS5 and the LT-1 are worlds apart," explained *Car and Driver*. "In performance, however, they are neck and neck."

RPO LT-1 was initially listed in Corvette paperwork in 1969 but didn't actually appear until the following year, priced at $447.60, compared to $289.65 for the LS5 big block. Among LT-1 features were big-valve heads (2.02-inch intakes, 1.60 exhausts), an aggressive cam, 11.1:1 forged-aluminum TRW pistons, and a forged crank held in place by four-bolt main bearing caps. On top was an aluminum high-rise dual-plane intake mounting an 800-cfm Holley four-barrel.

All these hot parts translated into a 0–60 run of 5.7 seconds according to *Car Life*. Quarter-mile time was 14.17 seconds. But pure speed wasn't necessarily the goal. Supreme overall performance was, and to that end the LT-1 also came standard with a stiffened suspension, inspiring *Car Life* to call it "the best of all possible Corvettes."

"Corvette handling is superior with any engine," continued *Car Life*, "and the LT-1 is the best of the bunch. The weight balance is a perfect 50/50 with the small-block engine." After pairing the LT-1 up against a Porsche 911E, *Motor Trend*'s Chuck Koch almost couldn't believe his eyes. "The

CHAPTER THREE

New fender louvers and flared body panels appeared in 1970, the latter modifications made to reduce damage from road debris thrown up by the tires in 1968 and 1969.

Corvette was just as fast, if not faster, through the corners as the Porsche."

Neither air conditioning nor the Turbo Hydra-matic could be installed when the 370-horse 350 was ordered. LT-1 buyers in 1970 could have chosen between two four-speeds: the wide-ratio M20 or its close-ratio M21 cousin.

The M22 Rock Crusher was also available but only by way of Chevrolet's original ZR-1 package, which cost an intimidating $965.95 in 1970. Along with the LT-1 V-8 and M22 gearbox, this expensive deal included heavy-duty power brakes, an even stiffer suspension, and an aluminum radiator with metal fan shroud.

Three items were exclusive to the ZR-1 Corvette, and all three were carryovers from the L88s of 1967–1969: the M22 trans, F41 special suspension, and J56 heavy-duty brakes. The F41 suspension consisted of shorter, stiffer coils up front, a seven-leaf spring in back, and heavy-duty stabilizer bars at both ends, at least on paper.

The 1970 facelift extended to the flared fenders and new front fender vents.

C3 1968–1982

Parking lights were now square instead of round in 1970.

The production count for Chevrolet's 1970 Corvette convertible was 6,648. Base price was $4,849

105

CHAPTER THREE

In 1969, the 350-horsepower L46 V-8 picked up where the L79 left off after the Corvette small block grew from 327 cubic inches to 350. The L46 350 cost $131.65 in 1970. Production that year was 12,846. *Mike Mueller*

A 1970 L88-powered Corvette competed at the 1970 24 Hours of Le Mans race. This Corvette was driven by Henri Greder and Jean-Paul Rouget, who finished sixth overall.

Officially listed early on, that rear stabilizer bar apparently didn't make it beneath any ZR-1 cars in 1970.

The special J56 brakes included the J50 power booster, heavy-duty Delco Moraine four-piston calipers, and fade-resistant metallic linings. Front pads were fixed more firmly in place by two mounting pins each, compared to the rear pads that typically only used one pin. Cast-iron caliper mount braces were added up front to restrict vibration during hard stops. This brake system was all but identical to the L88's, save for the fact that its dual-circuit master cylinder didn't incorporate a proportioning valve.

Although it did finally appear in Corvette brochures the following year, the purposeful ZR-1 option wasn't promoted at all in 1970. Offered along with the LT-1 up through 1972, it attracted only 25 buyers in 1970, followed by 8 in 1971 and 20 in 1972.

1971

From 1965 to 1969, the Corvette reigned supreme as Chevrolet's most powerful production car. And it would have remained so again in 1970 had the awesome LS7 454 V-8 appeared as planned. More or less an enlarged L88, this aluminum-head big block was actually listed in brochures and assembly manuals early in the year, rated at 460 or 465 horsepower, depending on your source. One LS7 Corvette was even road tested by *Motor Trend* and

1971

Model availability	sport coupe and convertible (optional removable hardtop)
Wheelbase	98 inches
Length	182.5 inches
Width	69 inches
Height	47.8 inches (sport coupe), 47.9 inches (convertible, top up), 47.8 inches (convertible, with hardtop)
Shipping weight	3,153 pounds (coupe), 3,167 pounds (convertible)
Tread (front/rear, in inches)	58.7/59.5
Tires	F70-15
Brakes	11.75-inch discs
Wheels	15x8 Rally rims with trim rings and center caps
Fuel tank	18 gallons
Front suspension	parallel A-arms with coil springs
Rear suspension	independent three-link with transverse leaf spring
Steering	recirculating ball, 20.2:1 ratio (17.6:1 with fast steering adjustment)
Standard drivetrain	270-horsepower 350-ci V-8 with single four-barrel carburetor, four-speed manual transmission, 3.36:1 axle ratio
Optional engine	330-horsepower 350-ci V-8 (LT-1) with four-barrel carburetor, solid lifters (close- or wide-ratio four-speed manual transmission only)
Optional engine	360-horsepower 454-ci V-8 (LS5) with four-barrel carburetor, hydraulic lifters
Optional engine	425-horsepower 454-ci V-8 (LS6) with four-barrel carburetor, aluminum heads, solid lifters
Optional transmission	Turbo Hydra-matic automatic (with 3.08 1 axle ratio), not available with LT-1 350-ci V-8
Optional transmission	wide-ratio four-speed manual
Optional transmission	close-ratio four-speed manual
Optional transmission	heavy-duty four-speed manual (M22) included with ZR1 and ZR2 options, available with LS6 V-8
Optional gear ratios	3.36:1, 3.55:1, 3.70:1, 4.11:1

Sports Car Graphic, the latter reporting quarter-mile performance of 13.8 seconds at 108 miles per hour, "with full fuel, passenger, and luggage for two." Not to mention full exhausts, street tires, a smog pump, radio, heater, etc.

"We close with the old Texas proverb," concluded *SCG*'s Paul Van Valkenburgh after his thrilling LS7 ride. "If you care who's quickest, don't get caught shoveling manure behind someone else's 465 horses."

But it was Duntov's engineers who ended up doing the shoveling, as they were forced to send the LS7 to an early grave before it ever got the chance to live it up on American streets. What happened? Coupled with GM efforts to tone down performance in 1970 were Chevy officials' decisions to cut back on costly options that complicated assembly lines. "De-pro" was the company jargon

> "If you care who's quickest, don't get caught shoveling manure behind someone else's 465 horses."
> —Paul Van Valkenburgh, *Sports Car Graphic*

The 1971 Corvette looked essentially identical to its 1970 forerunner. Standard small-block power dropped from 300 horsepower to 270. Engine options included the 330-horse LT-1 350, 365-horse LS5 454 (shown), and 425-horse LS6 454. *Mike Mueller*

CHAPTER THREE

The custom interior trim was a $158 option in 1971. Chevrolet's C60 air conditioning added another $459 to the bottom line.

Below: Differences between the 1970 LS6 and its 1971 Corvette counterpart included aluminum heads (instead of cast-iron) for the latter. Output dropped from 450 horsepower in 1970 to 425 horses for the 1971 Corvette. *Mike Mueller*

for this de-proliferation program, and both the LS7 and L46 small block ended up being de-pro victims in 1970. While Duntov's greatest disappointment came in the mid-1970s when GM finally squelched his midengine proposals, he later expressed every bit as much dismay concerning the LS7's cancellation in 1970.

With the LS7 gone, its 450-horsepower LS6 454 brother became Chevy's hottest V-8 in 1970. But it was only installed in SS 454 Chevelles that year, resulting in what many still believe was the muscle car era's supreme machine. Like the LS7, a second-edition LS6 Chevelle was road tested and then cancelled early on in 1971. Meanwhile, the hottest Corvette for 1970 was the 390-horse LS5.

And if that wasn't enough, new for 1971 were the first power cutbacks in Corvette history. Standard output dropped from 300 horses to 270 as compression was slashed (from 10.25:1 to a tidy 8.5:1) as part of GM's response to Washington's crackdown on engine emissions. The LT-1, too, lost ground, to 330 horsepower, as its compression also was cut to 9:1. Same for the LS5 big block: compression went from 10.25:1 to 9:1, dropping advertised output to 365 horsepower.

Fortunately, not all was lost. While the LS6 was dropped from the A-body Chevelle line, it was

108

Chevrolet offered the big, bad LS6 V-8 only as a Chevelle SS 454 option in 1970. LS6 then was dropped from the Chevelle's RPO list in 1971 but reappeared as a Corvette option. This red LS6 coupe is one of 188 built for 1971. *Mike Mueller*

added to the Y-body Corvette lineup for 1971. It didn't escape the compression axe (the squeeze went from 11.25:1 to 9:1), but it remained strong at 425 horsepower, putting the Corvette back on top where it belonged.

Another difference between the 1970 LS6 and the 1971 edition involved the cylinder heads—they were made of iron for A-body applications, aluminum for Y-body installations. Valve sizes were 2.19 inches intake, 1.88 exhaust; pistons were TRW forged-aluminum pieces, and rods were forged-steel units with big 7/16-inch bolts. A potent solid-lifter cam went inside, and a 780-cfm Holley four-barrel mounted on a low-rise, dual-plane aluminum intake went on top.

Like its Chevelle predecessor, the LS6 Corvette wasn't cheap. The aluminum-head 454's asking price was a hefty $1,220, leaving little wonder why only 188 customers chose the LS6 in 1971. "It's Duntov's favorite engine and he's tortured because few customers can afford it," claimed a *Car and Driver* report. But this time Zora's pain may have been self-inflicted. "Maybe for street engine I make mistake," he admitted to *Car and Driver*. "Aluminum heads are expensive and that weight doesn't matter on the street."

Even more expensive was RPO ZR2, the special-purpose turbo-jet 454 option, which mirrored the ZR-1 racing package offered for the 1970–1972 LT-1 small block. The bottom line for RPO ZR2 was $1,747. A mere 12 were sold in 1971.

Even without the track-ready ZR2 package, the LS6 was a big winner on the street, running 0–60

Left: This deluxe wheel cover was a $63 option (RPO PO2) in 1971. *Mike Mueller*

CHAPTER THREE

in 5.3 seconds, according to a *Car and Driver* test. Quarter-mile performance was listed at 13.8 seconds at 104.65 miles per hour; quite notable considering the car was equipped with a less-than-desirable 3.36:1 economy axle.

Car Craft magazine went a step further, adding headers, stump-pulling 4.56:1 gears, and racing slicks, resulting in a 12.64-second quarter-mile pass at 114.21 miles per hour. Not long after this sizzling pass, one of *Car Craft*'s less experienced lead-footers put a couple of the LS6's rods through its oil pan—a somewhat fitting exclamation point for a story entitled "Goodbye Forever, LS6." *Car Craft*'s crew knew even as they were flogging one of the strongest Corvettes ever built that they would probably never see such speed and power again. Chevrolet officials had made it clear that the option wouldn't return for 1972.

XP-895

Further toying with the existing midengine XP-882 chassis in 1971 resulted in the more pleasing XP-895. But as much as this steel-bodied prototype looked like the next step into the future, the XP-895 still weighed every bit as much as a regular-production Corvette. Cutting the car's weight was

Rolling pieces of Corvette history come and go as the National Corvette Museum refreshes its main display. In 1994, the aluminum-bodied model XP-895 could be seen, one of many midengine proposals that Zora Duntov hoped would one day make it to regular production. *Mike Mueller*

John Greenwood put together some of the hottest racing Corvettes during the 1970s. This car was built from a 1969 recycled Corvette convertible, and was powered by an aluminum ZL1 engine. Randy Whitten from GM Design Staff was responsible for the American flag paint scheme.

110

The last LT-1 Corvette appeared in 1972. *Mike Mueller*

significant to the midengine prototype ideal, so to breach this hurdle, John DeLorean's people turned to the Reynolds Metals Company in 1972. Reynolds created an aluminum copy of the XP-895 body, painted silver like its heavier sibling, and delivered it to Chevrolet Engineering. A year or so later, Chevy engineers used that lightweight shell to create an XP-895 variant that weighed about 500 pounds less than the original.

According to Reynolds' officials, the running XP-895 machine represented "an important milestone in the application of aluminum in autobody construction." And beneath that lightweight skin was much of the same innovative mechanical makeup demonstrated earlier by XP-882. But the costs of producing the "Reynolds Corvette" in suitable numbers proved prohibitive, and this experiment also ended up on the shelf.

1972

While ever-tightening federal emissions standards would cut horsepower even more soon enough, advertised outputs for all GM engines plummeted further in 1972 when gross ratings were traded for SAE net figures per an industry-wide trend. The '72 Corvette's standard 350 was listed at a paltry 200 horsepower. New net ratings for the two optional

1972

Model availability	sport coupe and convertible (optional removable hardtop)
Wheelbase	98 inches
Length	182.5 inches
Width	69 inches
Height	47.8 inches (sport coupe), 47.9 inches (convertible, top up), 47.8 inches (convertible, with hardtop)
Shipping weight	3,215 pounds
Tread (front/rear, in inches)	58.7/59.5
Tires	F70-15
Brakes	11.75-inch discs
Wheels	15x8 Rally rims with trim rings and center caps
Fuel tank	18 gallons
Front suspension	parallel A-arms with coil springs
Rear suspension	independent three-link with transverse leaf spring
Steering	recirculating ball, 20.2:1 ratio
Standard drivetrain	200-horsepower 350-ci V-8 with single four-barrel carburetor, four-speed manual transmission, 3.36:1 axle ratio
Optional engine	255-horsepower 350-ci V-8 (LT-1) with four-barrel carburetor, solid lifters (close- or wide-ratio four-speed manual transmission only)
Optional engine	270-horsepower 454-ci V-8 (LS5) with four-barrel carburetor, hydraulic lifters
Optional transmission	Turbo Hydra-matic automatic (with 3.08:1 axle ratio), not available with LT-1 350-ci V-8
Optional transmission	wide-ratio four-speed manual
Optional transmission	close-ratio four-speed manual
Optional transmission	heavy-duty four-speed manual (M22), included with ZR1 option
Optional gear ratios	3.36:1, 3.55:1, 3.70:1, 4.11:1

CHAPTER THREE

Coupe production in 1972 reached 1,741. Mike Mueller

This seat upholstery pattern was unique to the 1972 Corvette. Mike Mueller

> In June 1971, Ed Cole gave the go-ahead to the XP-897GT project. Wearing a Pinanfarina body atop that same transverse-engine chassis, this car, the so-called Two-Rotor Corvette, debuted in September 1973.

de-pro survivors, the LT-1 small block and LS5 big block, were 255 and 270 horsepower, respectively.

That latter figure apparently embarrassed engineers so much they failed to stick an output label on the LS5 air cleaner lid, making it the first time that a big block went out in public incognito. California customers were none the wiser because they never even got a look at a 1972 LS5. Chevrolet officials that year didn't bother to put the LS5 through that state's stringent emissions testing, meaning it failed to meet certification for sale there. This wouldn't be the last time that a Corvette engine would be banned on the West Coast.

Among various far-less-obvious changes made for 1972 was the deletion of transistorized ignition and the optic-fiber warning light system. In exchange for the latter, buyers received a standard horn-honking burglar alarm system, which was previously an option (RPO UA6) for 1968–1971 Corvettes.

LT-1 buyers were treated to one last surprise as their favorite ride rolled on in its final year. With its lowered compression, the LT-1 350 was not nearly the same hothead it had been in 1970, so engineers were able to add optional air conditioning into the equation. The cool C60 option was installed on as many as 240 LT-1 Corvettes during the last four months of 1972 production. The LT-1 production tally for the three-year run read 1,287 in 1970, 1,949 in 1971, and 1,741 in 1972. Some press reports mentioned a fourth LT-1 Corvette for 1973, but most witnesses, critical or otherwise, had already noticed the handwriting on the wall. Two years later, it was painfully clear that the days of clattering lifters and big, thirsty Holley carbs were done. The mouse that roared would roar no more.

Rotary Corvettes

Duntov's midengine ideal was struck down more than once, but his XP-882 platform just wouldn't die. Yet another resurrection followed after GM bought out the patent rights to the Wankel rotary engine in November 1970. In June 1971, Ed Cole gave the go-ahead to the XP-897GT project. Wearing a Pinanfarina body atop that same transverse-engine chassis, this car, the so-called Two-Rotor Corvette, debuted in September 1973. Duntov never did like the rotary Corvette idea, but he had no choice in the matter.

"Ed Cole was enamored with the Wankel engine," said Zora in a 1980 *AutoWeek* interview. "And he kept twisting my arm. 'What about a rotary Corvette?' Then DeLorean comes to Styling and looks at the midengine Corvette. He knows already that the decision has been made to produce this Corvette, but with the Wankel engine. I told him it was not powerful enough, and he lost his composure. 'You're some genius!' he shouted. 'Invent something!'"

Duntov turned to Gib Hufstader, who then did the Two-Rotor job two better. Hufstader's much

C3 1968–1982

The 1972 Corvette's standard drivetrain featured this 200-horsepower 350 small-block backed by a four-speed manual transmission. *Mike Mueller*

Below: LT-1 V-8 output dropped to 255 net-rated horses in 1972. Notice the optional air conditioning—1972 was the only year it could have been combined with RPO LT-1. Also note the autographed air cleaner lid. *Mike Mueller*

CHAPTER THREE

Dave Heinz and Bob Johnson drove this recycled '69 convertible to a fourth overall finish at Sebring in 1972. Here, the car is in action at the 6 Hours of Watkins Glen race in July 1972.

Below: Dick Henderson, GM show car fabrication manager (left), stands next to Dave Holls, GM designer and creator of the Wankel-powered "4-rotor" prototype. Zora Arkus-Duntov never warmed to the rotary idea.

more powerful "Four-Rotor" Corvette debuted one month after the XP-897GT. Using two Wankel engines coupled together, this truly fast gullwinged beauty was, according to *Car and Driver*, "the betting-man's choice to replace the Stingray." All bets were off, however, after Cole announced in September 1974 that GM was postponing the use of the Wankel rotary engine after running into problems getting it emissions certified.

Two years later, the Four-Rotor Corvette's Wankel was replaced by a conventional smallblock V-8, as the name was changed to "Aerovette." The body remained the same, as did those tired, old rumors. According to a February 1977 *Road & Track* prediction, the Aerovette would become the 1980 Corvette. Too bad Chevrolet was still selling conventional Corvettes like there was no tomorrow.

Duntov's dreams for the perfectly balanced, lightweight Corvette had been dashed about three years before that last *Road & Track* prophecy hit the stands. GM execs' opinion of his plan was plain and simple: why fix something that wasn't broken? Chevrolet held a captive audience for its fiberglass two-seater (with its front-mounted engine) during

Corvette sport coupe production in 1972 was 20,496, compared to 6,508 convertibles.

the 1970s, and everyone from DeLorean on up knew it. Even so, Duntov was allowed to dream on almost right up to his retirement early in 1975.

When GM once and for all squelched the midengine Corvette proposal in 1974, it resulted in easily the greatest defeat Duntov encountered during his stay at GM. "Until 1970, 90 percent of what I intended to do, I accomplished," added the Corvette's first chief engineer in that *AutoWeek* interview. "In 1972, a midship car was touch and go. It was all designed." But it still wasn't to be.

1973

This year's Corvette became the first of the breed to break the 30,000-unit sales barrier—in a standard 12-month model run, that is. Remember, John DeLorean had cheated in 1969. Two months after the former Pontiac chief took over as Chevrolet general manager on February 1, the St. Louis assembly line was shut down by a strike. Once the Corvette line restarted, DeLorean decided to make up for lost ground. Normally, 1970 production would have begun in September

1973

Model availability	sport coupe and convertible (optional removable hardtop)
Wheelbase	98 inches
Length	184.7 inches
Width	69 inches
Height	47.7 inches (sport coupe), 47.9 inches (convertible, top up), 47.7 inches (convertible, with hardtop)
Shipping weight	3,407 pounds
Tread (front/rear, in inches)	58.7/59.5
Tires	GR70-15
Brakes	11.75-inch discs
Wheels	15x8 Rally rims with trim rings and center caps
Fuel tank	18 gallons
Front suspension	parallel A-arms with coil springs
Rear suspension	independent three-link with transverse leaf spring
Steering	recirculating ball, 20.2:1 ratio
Standard drivetrain	190-horsepower 350-ci V-8 (L48) with single four-barrel carburetor, four-speed manual transmission, 3.36:1 axle ratio
Optional engine	250-horsepower 350-ci V-8 (L82) with four-barrel carburetor
Optional engine	275-horsepower 454-ci V-8 (LS4) with four-barrel carburetor
Optional transmission	Turbo Hydra-matic automatic (with 3.08:1 axle ratio)
Optional transmission	close-ratio four-speed manual
Optional transmission	wide-ratio four-speed manual
Optional gear ratios	3.36:1, 3.55:1, 3.70:1

1969, but he put 1970's startup on hold and let the 1969 run continue until December. After 1970's limited run, annual production didn't drop again until 1978.

That year commenced the third of the Shark's three five-year design trends. Recognizing this trio is easy enough: the first, spanning 1968 to 1972, had traditional chrome bumpers at both ends, while the last, running from 1978 to 1982, showed off a large, fastback glass in back. In the middle, the 1973 to 1977 Corvettes introduced body-color, crash-proof bumper systems, first at the nose in 1973, then at the tail beginning in 1974.

Trading that classic chrome front bumper for a urethane-covered, energy-absorbent nose was the result of new federal automotive safety standards that specified that all 1973 cars be able to bounce back from 5-mile-per-hour impacts. Further safety enhancement was found inside the '73 Corvette's doors, where steel guard beams were added to protect occupants from side impacts.

Amazingly, all this extra reinforcement, steel or otherwise, didn't add on nearly as many pounds as most detractors feared early on. Reportedly, that

Below: 1973 marked the first year that Corvette owners couldn't remove the rear window; Zora Duntov felt it caused buffeting, so Chevrolet made it permanent.

CHAPTER THREE

A urethane-covered nose appeared in 1973. The body-colored bumper system could survive 5-mile-per-hour impacts, per federal automotive safety specifications. The nose helped stretch overall length by 3 inches and added 35 pounds to total weight.

plastic-covered bumper system up front—which stretched total length by 3 inches—only increased overall weight by 35 pounds in 1973. Typical 1973–1977 curb weights went up about 250 pounds compared to those of 1968–1972.

The 1973 nose job also included a new hood that incorporated duct work to draw in cooler, denser air from the high-pressure area at the base of the windshield. Once a heavy foot depressed the throttle linkage beyond a certain point, a switch activated a solenoid, which in turn opened a flap hidden beneath a grille at the hood's trailing edge. Opening this flap allowed denser air to whistle directly into the air cleaner, which was sealed to the hood's underside by a rubber doughnut.

This new standard hood also did away with a Shark feature that had had many witnesses shaking their heads from its introduction in 1968. The 1973 hood ran all the way back to the windshield uninterrupted, hiding the wiper arms below its rear lip. Gone was the clunky pop-up panel that didn't always work and added unwanted extra weight.

Gone, too, was the removable rear window included on all 1968–1972 Corvettes. Duntov claimed this deletion was done not to cut costs

> This year's Corvette became the first of the breed to break the 30,000-unit sales barrier—in a standard 12-month model run.

but to eliminate an unwanted back draft that would occur at high speeds with windows up and roof panels removed.

Among other upgrades were larger mufflers (to tone down the exhaust note) and extra sound deadener beneath the hood and in the interior cabin's floor and side panels. New rubber-cushioned body mounts were added to reduce the vibrations transferred from the road to the driver through the frame, and ride harshness was reduced by making steel-belted radial tires standard.

Customers in 1973 again had three V-8s to choose from. Joining the base 190-horse 350 small block and the optional LS4 454 big block was a new RPO code—L82. Though it filled in for the LT-1, the L82 350 was actually a descendant of the hydraulic-lifter L46. Featuring 9:1 compression, big-valve heads, and a relatively aggressive hydraulic cam, the L82 produced 250 horses. L82 compression never slipped during the 1973–1977 run and in fact topped all Corvette engines built in those years. Even with its low 8.25:1 compression, the LS4 remained the top power choice at 275 horsepower.

New on the 1973 options list was the off-road suspension and brake package, RPO Z07. Offered up through 1975, the Z07 deal included considerably stiffer springs and shocks at both ends, a thicker front sway bar (a rear sway bar was also listed), and heavy-duty power brakes with dual-pin-mounted pads up front and fade-resistant metallic linings all around. The M21 close-ratio four-speed was mandatory, and neither air conditioning nor the base 350 could've been included along with RPO Z07.

1974

Model availability	sport coupe and convertible (optional removable hardtop)
Wheelbase	98 inches
Length	185.5 inches
Width	69 inches
Height	47.7 inches (sport coupe), 47.9 inches (convertible, top up), 47.7 inches (convertible, with hardtop)
Shipping weight	3,532 pounds
Tread (front/rear, in inches)	58.7/59.5
Tires	GR70-15
Brakes	11.75-inch discs
Wheels	15x8 Rally rims with trim rings and center caps
Fuel tank	18 gallons
Front suspension	parallel A-arms with coil springs
Rear suspension	independent three-link with transverse leaf spring
Steering	recirculating ball, 20.2:1 ratio (17.6:1 with optional power steering)
Standard drivetrain	195-horsepower 350-ci V-8 (L48) with single four-barrel carburetor, four-speed manual transmission, 3.36:1 axle ratio
Optional engine	250-horsepower 350-ci V-8 (L82) with four-barrel carburetor
Optional engine	270-horsepower 454-ci V-8 (LS4) with four-barrel carburetor
Optional transmission	Turbo Hydra-matic automatic (with 3.08:1 axle ratio)
Optional transmission	close-ratio four-speed manual
Optional transmission	wide-ratio four-speed manual
Optional gear ratios	3.36:1, 3.55:1, 3.70:1

1974

More crash protection was added in 1974 when 1973's energy-absorbing nose design was repeated in back. In place of that familiar duck-tail rear and twin chrome bumpers was another resilient plastic cap, this one molded in two pieces. Further evidence of the Corvette's ever-softening nature appeared in the exhaust system, where two small resonators were added to additionally tone down the car's growl.

Standard output rose slightly to 195 horses, while the optional L82 remained at 250 horsepower and the LS4 dipped to 270. This was the big block's swan song, as the painful realities of 50-cent gallons of low-lead and insurance premiums as heavy as car payments forced the 454 into retirement at 1974's end. New options for 1974 initially included a lightweight aluminum wheel deal (RPO YJ8) to complement the radial tires and help offset some of that federally mandated, safety-supplying excess fat. Each YJ8 rim would have cut off 8 pounds of unsprung weight, but manufacturing gremlins forced a recall after a couple hundred poor-quality sets were cast. Chevy paperwork showed that as many as four 1974 Corvettes apparently were delivered with YJ8 wheels before the option was withdrawn.

A 190-horsepower 350 small-block V-8 was standard for the 1973 Corvette. Functional cowl induction was standard, too, atop all three available engines that year. *Mike Mueller*

CHAPTER THREE

An energy-absorbing body-colored bumper was introduced for 1974. It also was the last year for an optional Corvette big block. Shown here is one of 3,494 454 Corvettes built in 1974. *Mike Mueller*

Above: The last Corvette big-block V-8, the 1974 LS4 454, was rated at 270 horsepower. Its price remained at $250. *Mike Mueller*

Left: The custom interior package in 1974 typically included leather seat trim, wood-grain accents, and carpeted lower door panels. The option cost $154 that year. *Mike Mueller*

Below: Along with a custom interior, this 1974 Corvette features optional power windows—thus no hand crank. *Mike Mueller*

CHAPTER THREE

1975

Model availability	sport coupe and convertible (optional removable hardtop)
Wheelbase	98 inches
Length	185.2 inches
Width	69 inches
Height	48.1 inches (sport coupe)
Shipping weight	3,532 pounds
Tread (front/rear, in inches)	58.7/59.5
Tires	GR70-15
Brakes	11.75-inch discs
Wheels	15x8 Rally rims with trim rings and center caps
Fuel tank	18 gallons
Front suspension	parallel A-arms with coil springs
Rear suspension	independent three-link with transverse leaf spring
Steering	recirculating ball, 20.2:1 ratio (17.6:1 with optional power steering)
Standard drivetrain	165-horsepower 350-ci V-8 (L48) with single four-barrel carburetor, four-speed manual transmission, 3.36:1 axle ratio
Optional engine	205-horsepower 350-ci V-8 (L82) with four-barrel carburetor
Optional transmission	Turbo Hydra-matic 400 automatic (with 3.08:1 axle ratio)
Optional transmission	close-ratio four-speed manual
Optional transmission	wide-ratio four-speed manual
Optional gear ratios	2.73:1, 3.36:1, 3.55:1, 3.70:1

More plentiful was another new option, the gymkhana suspension package (RPO FE7), which simply added a thickened front sway bar and higher-rate springs. No ordering restrictions or mandatory equipment were specified when FE7 was chosen.

1975

An end of an era came in 1975, as 65-year-old Zora Arkus-Duntov retired on January 1 after 21 years and seven months with General Motors. Two decades before, he had boldly written Ed Cole about a job after seeing the Corvette prototype on GM's Motorama stage; after "fiddling on the side" (his words) at Chevrolet in 1953 and 1954, Duntov was named the division's director of high-performance vehicle design and development in 1956. He wasn't officially tabbed the Corvette's chief engineer until 1968, this after some of the loop late in the C3 development process. By then, nonetheless, he already was well known as the "father of the Corvette."

For those who've always wondered, Zora got his surname as a result of having two dads. Born Zora Arkus, his name was lengthened after his Russian mother divorced and remarried Josef

Convertible production for 1975 was 4,629. Catalytic converters appeared on all models that year.

C3 1968–1982

> An end of an era came in 1975, as 65-year-old Zora Arkus-Duntov retired on January 1 after 21 years and seven months with General Motors.

Duntov. Though his hyphenated moniker was official, he was most often called Duntov, or, more affectionately, simply Zora. A bold race driver, a dashing, unforgettably handsome man who knew how to show off with style, an engineering genius with few equals—if there was anything bad to say about him no one was talking following his death in April 1996.

"I was impressed with his continental poise, sophistication, and his honesty and dedication to performance and to his work as an engineer," began fellow Chevy engineer Gib Hufstader's respectful homage. "He appreciated people who were very dedicated to doing a good job, to getting the job done. For some of us, it was a dream come true to work with him."

How did a mere mortal pick up where a living legend left off in 1975? The man handed the task of keeping the dream alive was David Ramsay McLellan, who basically had been groomed for the job after joining GM in 1959. He spent most of 1973 and 1974 at the Massachusetts Institute of Technology's Sloan School of Management on GM's dime, then, with his master's degree in hand, he returned to Chevrolet as one of Duntov's staff engineers. Six months later, Zora stepped down and McLellan rose to the post that he already knew was his for the taking. His influence, however, wouldn't be truly noted until the C4 debuted eight years later.

In the meantime, the Shark sailed on in typical fashion. Changes were few for 1975, with a new plastic honeycomb framework beneath that soft nose (to supply additional low-speed cushioning) leading the way. In back, a new body-colored end cap was redone in one piece without that telltale seam down the middle. Beneath the solid cap, a new aluminum bumper bar was attached to the frame with twin hydraulic cylinders. Chevrolet's breaker-less high-energy ignition—with its hotter, more-reliable spark—also became standard.

With the 454 in the archives, the engine lineup was left with only the standard 350 and the optional L82 in 1975—the first time in 20 years that only two power sources were offered. Output

Dwindling demand helped convince Chevrolet to give up on the convertible Corvette after 1975. This Bright Yellow droptop was one of 4,629 topless models built that year, compared to 33,836 coupes. Mike Mueller

Inspection general foreman Bob Shell drives the last topless 1975 Corvette out of the St. Louis plant on July 31 of that year.

121

CHAPTER THREE

The rear bumper of the 1975 models was a single piece, reinforced with inner shock absorbers for impact. The bumper pads were molded in.

for the former was now a paltry 165 horsepower, 205 for the latter. These power cuts were due to the inclusion of contaminant-burning catalytic converters, introduced to help meet more-restrictive emissions standards.

The 1975 Corvette's revised exhaust system featured a single large-capacity converter after a design featuring two smaller units had failed durability tests. Two Y-pipes were used to at least preserve the appearance of a sporty dual-exhaust system. The first one funneled exhaust flow from both sides of the engine together into one tube to enter the converter. From there, a reversed Y-pipe did the opposite in back to deliver the cleansed spent gases to typical twin mufflers at the tail. True duals wouldn't make a comeback until the LT5-powered ZR-1 debuted for 1990. Mainstream Corvettes wouldn't be refitted with a real dual-exhaust system until the second-generation LT1 appeared for 1992.

Horsepower wasn't the only thing to fade away in 1975. One year after the last big-block Corvette rolled into the sunset, the same thing happened to the convertible model. GM's explanation this time involved both safety concerns and nose-diving demand. The convertible's hunk of the Corvette pie sank from 38.4 percent in 1970 to 16.2 in 1973, and the last 4,629 ragtops built in 1975 represented a mere 12 percent of that year's total production run. Another end of an era? Fortunately not: topless Corvettes returned to center stage in 1986.

1976

Modifications in 1976 did away with the C3 hood's cowl flap, which apparently whistled too loudly for most drivers. In place of the solenoid-activated induction setup was a simpler system that rammed in airflow through a duct that ran forward over the radiator support to pick up some of the radiator's cooling breezes. Even though this hood no longer used the cowl-induction equipment, it still kept the intake grille. This opening wasn't deleted until 1977.

A partial underpan made of steel was added in 1976 to increase rigidity and improve heat insulation, and a lighter (13 pounds so), maintenance-free Delco Freedom battery also joined the standard equipment list. The Astro ventilation system used since 1968 was dropped in 1976, meaning the vents previously found behind the rear window were deleted too. A new sport steering wheel, shared with, of all things, Chevrolet's compact Vega, was added inside.

A little tinkering helped standard output rise to 180 horsepower, and the optional L82 also improved to 210 horses. The L82, however, wasn't offered in California in 1976 and 1977 because it didn't meet that state's tougher emissions standards. Four-speed transmissions were also banned out west for those years.

C3 1968–1982

1976

Model availability	sport coupe
Wheelbase	98 inches
Length	185.2 inches
Width	69 inches
Height	48.1 inches
Shipping weight	3,445 pounds
Tread (front/rear, in inches)	58.7/59.5
Tires	GR70-15
Brakes	11.75-inch discs
Wheels	15x8 Rally rims with trim rings and center caps
Fuel tank	18 gallons
Front suspension	parallel A-arms with coil springs
Rear suspension	independent three-link with transverse leaf spring
Steering	recirculating ball, 20.2:1 ratio (17.6:1 with optional power steering)
Standard drivetrain	180-horsepower 350-ci V-8 (L48) with single four-barrel carburetor, four-speed manual transmission, 3.36:1 axle ratio
Optional engine	210-horsepower 350-ci V-8 (L82) with four-barrel carburetor
Optional transmission	Turbo Hydra-matic 350 automatic (with 3.08:1 axle ratio), L48 only
Optional transmission	Turbo Hydra-matic 400 automatic 350 (3.36:1 axle ratio), L82 only
Optional transmission	close-ratio four-speed manual
Optional transmission	wide-ratio four-speed manual
Optional gear ratios	3.08:1, 3.55:1, 3.70:1

The automatic transmission limited to Californians in 1976 and 1977 wasn't the Turbo Hydra-matic 400 previously used behind all Corvette engines. Mandated behind the base 350 during those years was the mundane TH 350 automatic, as product planners opted not to waste the more expensive, heavy-duty Turbo Hydra-matic on engines that didn't put out enough punishment to merit its use. The TH 400 remained the weapon of choice whenever an L82 buyer forked over the extra cash for the M40 automatic transmission in 1976 and 1977, while the lighter TH 350 was a no-cost option behind the 210-horse 350.

The lightweight YJ8 rims were finally perfected in 1976, and this package included four aluminum wheels (supplied by Kelsey-Hayes) with a standard steel spare to keep the cost down. Listed as options early on, power steering, power brakes, and the custom interior trim group (with its leather seats) became part of the base package midyear,

Below: The 1976 model year was Corvette's first without a convertible model since its introduction in 1953. That fact didn't affect sales, however—Chevrolet manufactured 46,558 coupes, a record for the time. Chevrolet also added a new hood that eliminated the cowl-induction ductwork, as the cooler, denser outside air was funneled into the carburetor through a scoop at the front of the engine compartment.

CHAPTER THREE

Right: As early as 1976, Chevrolet engineering experimented with turbocharging the 350-cubic-inch Corvette V-8.

Below left: The new standard steering wheel for 1976 was shared with the low-buck Chevy Vega—a fact that didn't sit well with Corvette fans. *Mike Mueller*

Below middle Chevrolet engineers were no longer proud to display how many horsepower a Corvette engine made in 1976. The standard L48 350 V-8 was rated at 180 horsepower that year. *Mike Mueller*

Below right: Corvette speedometers first started showing metric conversions (kilometers per hour) in 1975. Shown here is the 1976 speedometer. *Mike Mueller*

mandating a bottom-line increase. All Corvettes built for 1976 featured power brakes, while only 173 hit the streets without power steering.

The 1976 Corvette also was the last to wear the familiar Stingray badge.

1977

It took Chevrolet 15 years to build its first quarter-million Corvettes. Hitting the half-million mark required only eight more years. The 250,000th, a Riverside Gold convertible, was built on November 7, 1969. The 500,000th, a white coupe, was driven off the line by Robert Lund on March 15, 1977.

It seemed nothing could deter the fiberglass faithful as popularity continued growing stronger even as the car's price soared. In 1974, the base sticker had edged beyond $6,000 for the first time, and this was only the beginning. It jumped to $6,810.10 in 1975 and then hit $7,604.85 the following year. In 1977, it was a whopping $8,647.65, due to the inclusion of so many standard comfort and convenience features.

Production, meanwhile, kept rolling on, with 33,836 Corvettes built for 1975, followed by a whopping 46,558 in 1976 to establish a new record, which was then broken by 1977's 49,213 tally.

Most Corvettes built during the 1970s wore even higher price tags as popularity of various options also went on the rise. When introduced in 1963, air conditioning attracted only 278 customers, equal to 1.3 percent of the first Sting Ray run. By 1968, C60 installations made up 19.8 percent of the production run, and that figure was 82.8 percent by

124

1977

Model availability	sport coupe
Wheelbase	98 inches
Length	185.2 inches
Width	69 inches
Height	48.1 inches
Shipping weight	3,448 pounds
Tread (front/rear, in inches)	58.7/59.5
Tires	GR70-15
Brakes	11.75-inch discs
Wheels	15x8 Rally rims with trim rings and center caps
Fuel tank	18 gallons
Front suspension	parallel A-arms with coil springs
Rear suspension	independent three-link with transverse leaf spring
Steering	recirculating ball, 20.2:1 ratio (17.6:1 with optional power steering)
Standard drivetrain	180-horsepower 350-ci V-8 (L48) with single four-barrel carburetor, four-speed manual transmission, 3.36:1 axle ratio
Optional engine	210-horsepower 350-ci V-8 (L82) with four-barrel carburetor
Optional transmission	Turbo Hydra-matic 350 automatic (3.08:1 axle ratio), L48 only
Optional transmission	Turbo Hydra-matic 400 automatic (3.55:1 axle ratio), L82 only
Optional transmission	close-ratio four-speed manual (3.70:1 axle ratio)
Optional transmission	wide-ratio four-speed manual (3.70:1 axle ratio)
Optional gear ratios	3.08:1, 3.55:1, 3.70:1 (with L82 four-speeds only)

Chevrolet General Manager Robert Lund drove the 500,000th Corvette, a white coupe, off the St. Louis assembly line on March 15, 1977.

1975. In 1977, 45,249 of the 49,213 cars built were air conditioned, adding an extra $553 to the sticker.

Assisted side glass installations, too, went on the rise. Introduced in 1956, power windows still were found on only 17.6 percent of the Corvettes built for 1967. But by 1973, the cut was 46 percent, followed by 90.1 in 1977, when the A31 option cost $116. Also popular that year were the tilt-telescopic steering wheel (46,487 sold at $165) and no-cost M40 automatic transmission (41,231 specified).

Only 2,060 four-speed transmissions were sold in 1977, and production of the optional L82 V-8 (still rated at 210 horsepower) was a mere 6,148. While buyers seemed willing to spend tons on frills, they apparently couldn't justify shelling out an extra $495 for 30 more horses.

1978

Demand dropped slightly in 1978, as the St. Louis plant rolled out 46,774 Corvettes, all fitted with new swooping rear glass that *Car and Driver* called "a universally appreciated Good Move." "The large rear window freshened up the Corvette's profile, and it also added space and light to help relieve the claustrophobia inside this, the most tightly coupled car known to man."

The new window improved rearward visibility as well as glass area, expanded from 392 square inches up to a whopping 1,425. But while extra cargo space represented a benefit, accessing that space still required reaching behind the seats. Making that huge glass sheet a hinged, lift-up roof would have been cool. Dave McLellan, however, opted not to complicate matters, at least not until he rolled out his Collector Edition hatchback model for 1982.

New for 1978 were special 25th Anniversary badges on all cars. And along with those badges, a customer could have further honored a quarter-century of Corvette history by checking off RPO B2Z, the Silver Anniversary paint option, which added an exclusive two-tone paint scheme done in light silver over dark silver. Dual sport mirrors and aluminum wheels were required extra-cost features on all 1978 Silver Anniversary Corvettes.

A second special model, the Limited Edition Corvette, helped commemorate both the car's 25 years and its appearance in 1978 as the prestigious pacer for the Indianapolis 500. Along with being a first for Chevrolet's two-seater, this year's Indy 500 performance also represented the first time that an unmodified, truly stock machine paced the annual Memorial Day race. The Corvette that led the way

> "The large rear window freshened up the Corvette's profile, and it also added space and light to help relieve the claustrophobia inside this, the most tightly coupled car known to man."
> —*Car and Driver*

for Tom Sneva, Danny Ongais, Rick Mears, and the 30 other drivers on May 28, 1978, was all but identical to the car John Q. Public drove that year, save for the two openings made in back for the requisite flag poles.

When the checkered flag dropped that day, Al Unser Sr. had collected his third Indy 500 win. At the same time, Chevrolet officials were

CHAPTER THREE

1978

Model availability	sport coupe
Wheelbase	98 inches
Length	185.2 inches
Width	69 inches
Height	48 inches
Shipping weight	3,401 pounds
Tread (front/rear, in inches)	58.7/59.5
Tires	P225/70R-15; P255/60R-15 included with Indy Pace Car replica
Brakes	11.75-inch discs
Wheels	15x8 Rally rims with trim rings and center caps; 15x8 cast-aluminum wheels included with Silver Anniversary and Indy Pace Car models
Fuel tank	23.7 gallons
Front suspension	parallel A-arms with coil springs
Rear suspension	independent three-link with transverse leaf spring
Steering	recirculating ball, 20.2:1 ratio (17.6:1 with optional power steering)
Standard drivetrain	185-horsepower 350-ci V-8 (L48) with single four-barrel carburetor, four-speed manual transmission, 3.36:1 axle ratio.
Note	175-horsepower V-8 with Turbo Hydra-matic 350 automatic transmission and 3.55 1 axle ratio only available in California and high altitudes
Optional engine	220-horsepower 350-ci V-8 (L82) with four-barrel carburetor, not available in California and high altitudes
Optional transmission	Turbo Hydra-matic 350 automatic (with 3.08:1 axle ratio with L48, 3.55:1 ratio with L82 and California/high-altitude V-8s)
Optional gear ratios	3.36:1, 3.70:1, with L82 four-speeds only

All Corvettes built for 1978 featured commemorative exterior badges . . . and all were fitted with anniversary horn buttons. *Mike Mueller*

busy capitalizing on the event by, among other things, calling May "Pacesetter Month." Dealers, meanwhile, put on pacesetter promotions, with the star of their shows being the limited-edition pace car replica.

As the name implied, Chevrolet's original plan was to make this package a limited-edition collectors' piece. Building only 300 was at first the goal. The *Wall Street Journal* even went so far as to print a cover story in March 1978 touting the

The base price for the 1977 Corvette was $8647.65; it jumped almost $1,000 from the previous year due to so many standard comfort and convenience features.

C3 1968–1982

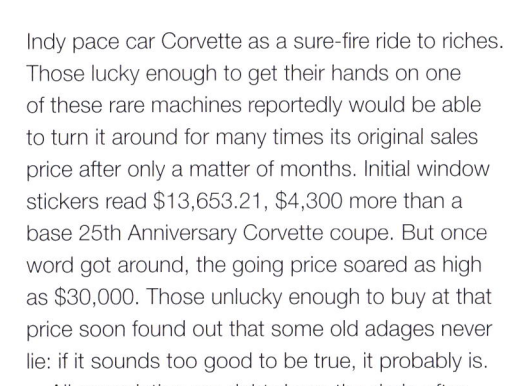

Above: Dave McLellan's engineers created this show car in 1969 to experiment with a new Corvette power source: turbocharging. About 290 horsepower resulted.

Right: In 1978, 6,502 Limited Edition models were sold to mark the Corvette's first appearance as the prestigious pace car for the Indianapolis 500. *Mike Mueller*

Indy pace car Corvette as a sure-fire ride to riches. Those lucky enough to get their hands on one of these rare machines reportedly would be able to turn it around for many times its original sales price after only a matter of months. Initial window stickers read $13,653.21, $4,300 more than a base 25th Anniversary Corvette coupe. But once word got around, the going price soared as high as $30,000. Those unlucky enough to buy at that price soon found out that some old adages never lie: if it sounds too good to be true, it probably is.

All speculation ran right down the drain after Chevy's litigation-shy decision-makers chose to build at least one pace car replica for every dealer in America. Apparently, the idea was to avoid any lawsuits from potential buyers or dealers (translated: opportunistic exploiters) left in the lurch with nothing save for complaints of unfair, monopolistic sales practices. Whatever the case, the final "limited-edition" count for the '78 Indy pace car reached 6,502. Most (if not all) of those who originally jumped on the Limited Edition Corvette bandwagon 20-plus years ago are still waiting to make hay today.

Included in the Limited Edition's original astronomical price was a long list of options combined with a heavy dose of special treatments. On the outside was an exclusive black-over-silver two-tone paint scheme accented by red pinstriping. Official Indy 500 decals were expected. A front air dam, rear spoiler, aluminum wheels (also with red pinstripes), and glass T-tops were part of the deal as well. Inside was an exclusive leather interior done in a silver-gray shade called "smoke." Lightweight buckets were also exclusive to the Limited Edition's interior. Additional extras included power windows, door locks, and antenna; a rear window defogger; air conditioning; sport mirrors; a tilt-telescopic steering column; white-letter P225/60R15 tires; a heavy-duty battery; AM/FM eight-track stereo with dual rear speakers; and the Convenience Group. Included in that last collection, RPO ZX2, were a dome light delay, headlight warning buzzer, under-hood light, low-fuel warning light, interior courtesy light, floor mats, intermittent wipers, and passenger-side vanity mirror on the visor.

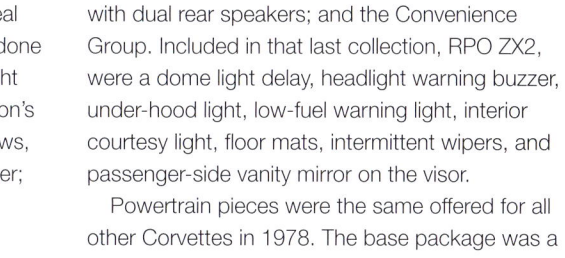

Powertrain pieces were the same offered for all other Corvettes in 1978. The base package was a 185-horsepower 350 backed by a wide-ratio four-

An exclusive leather interior, done in a silver-gray shade called "smoke," was standard inside the Limited Edition Corvette, as were lightweight bucket seats, power windows, door locks, an antenna, air-conditioning, a tilt-telescopic steering column, and am AM/FM eight-track stereo. *Mike Mueller*

127

CHAPTER THREE

Chevrolet commemorated 25 years of Corvette history with a special Silver Anniversary model in 1978. Exclusive paint, sport mirrors, and aluminum wheels were included in the package.

speed manual. The top-shelf 220-horse L82 was optional, as was the close-ratio M21 four-speed and MX1 TH 400 automatic. Adding the FE7 sport suspension allowed a Limited Edition buyer to outfit his replica in identical fashion to the actual Indy pace car.

Increasing L82 output from 210 to 220 horsepower was done by adding a less restrictive exhaust system and a dual-snorkel air induction setup that improved breathing on the top end. And even the standard L48 350 got five more horses, but L48s sold in California and high altitudes that year were rated at 175 horsepower instead of 185.

Either way, the quarter-century-old two-seater was a real hot rod once more, or so claimed *Car and Driver*. "After a number of recent Corvette editions that prompted us to mourn the steady decline of both performance and quality in this once-proud marque, we can happily report the twenty-fifth example of the Corvette is much improved across the board. Not only will it run faster now—the L82 version with four-speed is certainly the fastest American production car, while the base L48 automatic is no slouch—but the general drivability and road manners are of a high order as well." L48 performance was quoted at 7.8 seconds for the 0–60 run, 123 miles per hour on the top end. The L82 reportedly could reach 133 miles per hour.

A 185-horsepower 350 small-block was standard for the 1978 Corvette. *Mike Mueller*

128

1979

Model availability	sport coupe
Wheelbase	98 inches
Length	185.2 inches
Width	69 inches
Height	48 inches
Shipping weight	3,372 pounds
Tread (front/rear, in inches)	58.7/59.5
Tires	P225/70R-15
Brakes	11.75-inch discs
Wheels	15x8 Rally rims with trim rings and center caps
Fuel tank	23.7 gallons
Front suspension	parallel A-arms with coil springs
Rear suspension	independent three-link with transverse leaf spring
Steering	recirculating ball, 20.2:1 ratio (17.6:1 with optional power steering)
Standard drivetrain	195-horsepower 350-ci V-8 (L48) with single four-barrel carburetor, four-speed manual transmission, 3.36:1 axle ratio.
Note	California and high-altitude V-8 also rated at 195 horsepower; both only available with Turbo Hydra-matic 350 automatic transmission
Optional engine	225-horsepower 350-ci V-8 (L82) with four-barrel carburetor, not available in California and high altitudes
Optional transmission	Turbo Hydra-matic 350 automatic (with 3.55:1 axle ratio)
Optional gear ratios	3.36:1, 3.70:1, with L82 four-speeds only

1979

Corvettes were rolling off the line at dizzying paces during the late 1970s. "The St. Louis plant is operating two nine-hour shifts daily and working overtime two Saturdays a month just to meet sales demand," said Chevrolet General Manager Robert Lund in March 1977. "Current demand is running more than 29 percent ahead of last year." Predicting yet another Corvette sales record that year was easier than spotting an SUV today.

But once again that record didn't stick around long. After dipping a bit in 1978, Corvette production soared the following year to 53,807, a feat that undoubtedly will never be repeated, or even approached.

That year's Corvette also established a new base price record by surpassing $10,000 for the first time. As before, adding more features into the standard deal contributed to this all-time high. The Limited Edition Corvette's high-back bucket seats and the previously optional AM/FM radio became standard equipment in 1979, but that was just a start, as product planners once again took note of customer preferences and responded accordingly.

Corvette production peaked at 53,807 in 1979, a record that may never be broken.

CHAPTER THREE

Besides the record production numbers, the 1979 model was also the first year the base coupe cost more than $10,000, clocking in at $10,220.

Corvette introduced a high-backed seat in 1979 that carried through into 1980 and provided better lateral support.

> After dipping a bit in 1978, Corvette production soared the following year to 53,807, a feat that undoubtedly will never be repeated, or even approached.

1980

Base output was 190 horsepower in 1980, except on the West Coast. California's tough smog standards had limited Corvette customers in that state to one engine only, the L48, since 1976. Then that state's extra-strict emissions standards tightened even further for 1980. Chevrolet didn't even bother to certify the 350 V-8 for use there that year, meaning a stand-in was needed.

In place of the L48, and in exchange for a $50 credit, California Corvettes in 1980 were fitted with the 305-cubic-inch LG4 V-8. Although it came right off the mundane passenger car parts shelf, the LG4 still produced only 10 fewer horses than the L48, thanks to the use of stainless-steel tubular headers and a "computer command control" system that automatically adjusted carburetor mixture and ignition timing on demand.

The L82 was boosted to 230 horsepower for 1980, but this was as good as it got for the Corvette's last optional "high-perf" V-8. Engineers were never able to certify this so-called high-compression (9:1) small block for California use, and demand dropped off considerably in remaining states after 1979. Thus, the L82 was cancelled at 1980's end.

In other news, the 1980 body was treated to front and rear body caps featuring integral spoilers, as opposed to the optional add-on spoilers offered

Of the 46,774 Corvettes sold in 1978, 37,858 had tilt-telescopic steering columns, 37,638 were fitted with air conditioning, and 36,931 featured power windows. All told, these three options cost $910 in 1978 and $966 the following year. Then, effective May 7, 1979, this trio was made part of the standard package, raising the base price by $706. Various other additions eventually helped hike the 1979 bottom line to $12,313.23.

Two other pace car features, front and rear spoilers, became optional (listed together under RPO D80) for all 1979 Corvettes and went a long way toward reducing unwanted drag. The optional L82 V-8 was tweaked up to 225 horsepower, and its dual-snorkel induction and open-flow mufflers were passed down to the L48, boosting it to 195 horses, a rating shared by the L48s sold in high altitudes and on the West Coast in 1979.

> California's tough smog standards had limited Corvette customers in that state to one engine only, the L48, since 1976.

C3 1968–1982

Various weight-saving measures made the 1980 Corvette lighter on its feet; this after the car had steadily put on pounds during the 1970s.

the year before. Reportedly, these modifications improved the drag coefficient from 0.503 (for the spoiled 1979 model) to 0.443.

Another performance upgrade involved weight savings, something McLellan paid special attention to. A lighter differential housing and corresponding frame crossmember, both made of aluminum, were introduced in 1980. And adding the L82's aluminum intake manifold to the base L48 cut a few more pounds.

1980

Model availability	sport coupe
Wheelbase	98 inches
Length	185.2 inches
Width	69 inches
Height	48 inches
Shipping weight	3,206 pounds
Tread (front/rear, in inches)	58.7/59.5
Tires	P22570R-15
Brakes	11.75-inch discs
Wheels	15x8 Rally rims with trim rings and center caps
Fuel tank	23.7 gallons
Front suspension	parallel A-arms with coil springs
Rear suspension	independent three-link with transverse leaf spring
Steering	recirculating ball, 20.2:1 ratio (17.6:1 with optional power steering)
Standard drivetrain	190-horsepower 350-ci V-8 (L48) with single four-barrel carburetor, four-speed manual transmission, 3.07:1 axle ratio
Standard California drivetrain	180-horsepower 305-ci V-8 (LG4) with single four-barrel carburetor, Turbo Hydra-matic automatic transmission, 3.07:2 axle ratio
Optional engine	230-horsepower 350-ci V-8 (L82) with four-barrel carburetor, not available in California
Optional transmission	Turbo Hydra-matic 350 automatic (with 3.07:1 axle ratio)

131

CHAPTER THREE

Corvette production fell to 40,614 in 1980—still an impressive result. Front and rear integral spoilers were new that year, and power windows, air conditioning, and a tilt-telescopic steering wheel became standard for the 1980 Corvette.

1981

They passed like those two proverbial ships in the night: the Corvette's old plant and its new Kentucky home. Roughly 27 years and eight months after it started working, the Corvette assembly line in Chevrolet's Fisher Mill Building, located on Natural Bridge Avenue in St. Louis, rolled out its last 'glass-bodied two-seater. Earlier, on June 1, a new line began moving Corvettes out the door in Bowling Green, and for two months the plants worked in tandem to meet some of the heaviest demand ever experienced during the car's long career. Of the 40,606 Corvettes built for 1981, 8,955 began life in the Bluegrass State.

Rumors of a move to a larger, more modern plant began circulating along the Mississippi River as early as 1973. Reportedly, this talk was the result of an increasing flow of bad blood between labor

1981

Model availability	sport coupe
Wheelbase	98 inches
Length	185.2 inches
Width	69 inches
Height	48 inches
Shipping weight	3,206 pounds
Tread (front/rear, in inches)	58.7/59.5
Tires	P225/70R-15
Brakes	11.75-inch discs
Wheels	15x8 Rally rims with trim rings and center caps
Fuel tank	23.7 gallons
Front suspension	parallel A-arms with coil springs
Rear suspension	independent three-link with transverse leaf spring
Steering	recirculating ball, 20.2:1 ratio (17.6:1 with optional power steering)
Standard drivetrain	190-horsepower 350-ci V-8 (L48) with single four-barrel carburetor, four-speed manual transmission, 2.72:1 axle ratio
Optional transmission	Turbo Hydra-matic 350 automatic (with 2.87:1 axle ratio)

and management. GM officials denied such rumors and also shot down claims that they were looking for a way out. "We can say categorically that we have no plans for closing the assembly plant in St. Louis, and any rumors to that effect are without foundation," wrote GM Vice President Robert Magill in a letter to a Missouri congressman in the early 1970s. Whether or not Magill's office truly didn't have a relocation plan in mind then, it was fairly obvious to many that the St. Louis line couldn't be long for this world.

Age was clearly a factor. Completed in 1920, Chevrolet's St. Louis plant was not only archaic, it was also lacking in size and scope. Expansion on the site was out of the question. Things, of course, were different in late 1953, when GM officials opted to relocate Corvette production from its temporary assembly line in Flint, Michigan, to a more promising home in Missouri.

"We selected St. Louis as the exclusive source of Corvette manufacture because the city has a central location and excellent shipping facilities, and we have always found here an ample supply of competent labor," said Edward Kelly, Chevrolet Motor Division's general manufacturing manager. The Flint line ceased operation on December 24, 1953. After Christmas, workers on the new line down in Missouri began assembling their first Corvette on December 28.

Below: Slightly restyled emblems set the 1981 Corvette apart from its 1980 predecessor. Rally wheels were still standard in 1981. *Mike Mueller*

CHAPTER THREE

Above: 1981 was the first and only year Corvettes were built in two locations—here, the first production Corvette (two-tone!) for the new facility in Bowling Green rolls off the line on June 1, 1981.

Right: Corvette production relocated from Missouri to Bowling Green, Kentucky, in 1981. The last fiberglass two-seater rolled out of the old, cramped St. Louis assembly facility on June 31, 1981.

    About a quarter-century later, the rumor mill began churning early in 1979, this after Environmental Protection Agency (EPA) clean-air cops started hounding St. Louis officials about their paint facilities, which were in violation of federal air-quality standards. GM people then did admit that they were considering relocating Corvette production to a plant able to house a roomier, modernized painting section. Official announcement of an impending move finally came down from Michigan on March 26, 1979.

    What GM execs didn't announce was that two St. Louis plant managers had been quietly sent to Bowling Green in 1978 to shop for the Corvette's new home. There they toured a 550,000-square-foot complex formerly used by the Fedders Corporation and the Air Temp Division

134

> "We selected St. Louis as the exclusive source of Corvette manufacture because the city has a central location and excellent shipping facilities, and we have always found here an ample supply of competent labor," said Edward Kelly, Chevrolet Motor Division's general manufacturing manager.

of Chrysler Corporation. After a deal was cut, the plant was expanded to 1 million square feet, and improved state-of-the-art assembly equipment was installed. Furthermore, the facility would be dedicated solely to one vehicle, unlike the St. Louis plant, which was home to various Chevrolet models during its lifetime.

"The Bowling Green facility, which will build Corvettes exclusively, is an investment in Corvette's future," claimed a 1981 brochure. "It represents the experience and knowledge learned over all those years [in St. Louis]."

Kentucky Governor John Brown drove the first Corvette off the Bowling Green line in 1981. A few months later, the new plant was producing Corvettes at a rate of 15 per hour. The best the St. Louis line had ever managed was 10 cars an hour.

Quantity, however, was not the main priority. Higher standards of quality were established in Bowling Green, most importantly concerning finishing the product. New base-coat/clear-coat enamel paint was put into use at the Kentucky plant instead of the lacquer paint used at St. Louis right up until the end. Advantages of the clear-coated enamel finish included superior clarity and mirror-like gleaming depth. The days of streaky, uneven metallic finishes were over.

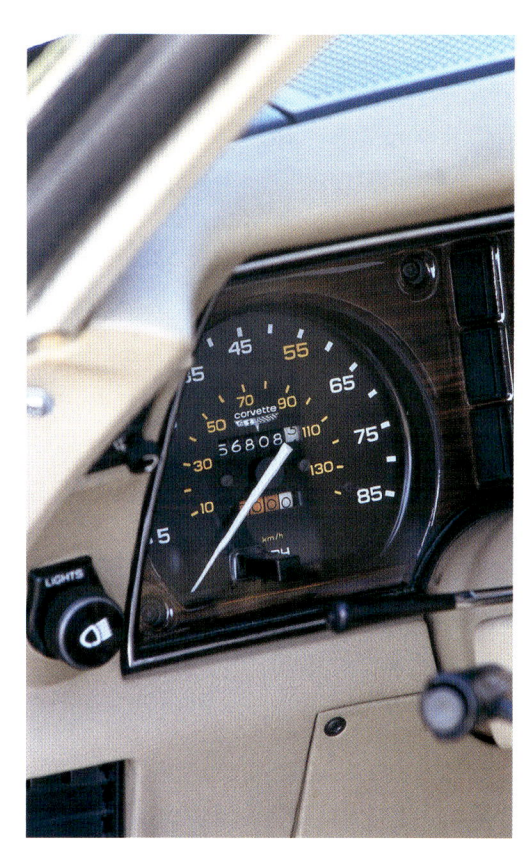

How times had changed. Chevrolet began phasing in 85-mile-per-hour speedometers in 1979—by 1981, the wimpy unit was an accepted reality. *Mike Mueller*

The 190-horsepower L81 350 V-8 was the only available power source in 1981. *Mike Mueller*

CHAPTER THREE

"The Bowling Green facility, which will build Corvettes exclusively, is an investment in Corvette's future," claimed a 1981 brochure. "It represents the experience and knowledge learned over all those years [in St. Louis]."

1982

Model availability	sport coupe (Collector Edition featured exclusive hatchback rear glass)
Wheelbase	98 inches
Length	185.2 inches
Width	69 inches
Height	48 inches
Shipping weight	3,213 pounds (3,222 pounds, Collector Edition)
Tread (front/rear, in inches)	58.7/59.5
Tires	P225/70R-15; P225/60R-15 included with Collector Edition
Brakes	11.75-inch discs
Wheels	15x8 Rally rims with trim rings and center caps; 15x8 cast-alloy turbine rims included with Collector Edition
Fuel tank	23.7 gallons
Front suspension	parallel A-arms with coil springs
Rear suspension	independent three-link with transverse leaf spring
Steering	recirculating ball, 20.2:1 ratio (17.6:1 with optional power steering)
Standard drivetrain	200-horsepower 350-ci V-8 (L83) with Cross-Fire Injection, four-speed Turbo Hydra-matic 700-R4 automatic transmission, 2.72:1 axle ratio (2.87:1 axle ratio with Collector Edition)

Chevrolet officials were proud of their new paint processes at Bowling Green, so much so that they chose to show them off with a special presentation. A distinctive two-tone paint option, RPO D84, was offered for the 1981 Corvette. The majority of the cars built that first year in Kentucky featured two-tone paint.

Four combinations were offered—Beige/Dark Red, Silver/Dark Blue, Silver/Charcoal, and Autumn Red/Dark Claret—with sales of those combinations totaling 5,352. Another 4,871 D84 cars followed in 1982 (in White/Silver, Silver/Dark Claret, Silver Blue/Dark Blue, and Silver/Charcoal) before the option was cancelled, due both to lukewarm popularity and less-than-complimentary critical reviews.

Only one engine was offered in 1981, the 190-horsepower L81 350, a cleaner-running small block that finally made the emissions grade in all 50 states. No California-specific V-8, no optional engines—1981 was the first time since the Corvette's earliest days that all models built relied on the same power source. And the L81 also could've been mated to a four-speed in California, meaning manual-trans Corvettes could be sold in that state for the first time since 1975.

Making the L81 legal in California was simply a matter of incorporating the lightweight tube headers (with oxygen-sensor smog controls) and computer controls used by the LG4 V-8 in 1980. Weight-saving magnesium valve covers were also added, and additional fat was trimmed by introducing a mono-leaf fiberglass spring in back. Added to automatic-transmission Corvettes with standard suspensions, this plastic spring weighed only 8 pounds, compared to the 44-pound steel leaf spring used in 1980.

Choosing between the automatic or manual transmissions required no extra dollars in 1981, after which time the four-speed was temporarily dropped. Not since 1954 had a complete model-year run been limited only to automatic installations, as was the case in 1982.

1982

The fiberglass mono-leaf spring seen in 1981 was one of various upcoming C4 features that Dave McLellan let leak out beforehand. Easily the most prominent of these came in 1982, when an intriguing new drivetrain appeared for the final C3.

"It's a harbinger of things to come," said McLellan about the car that preceded his sport coupe de grace. "For the 1982 model is more than just the last of a generation; it's stage one of a two-stage production. We're doing the power team this year. Next year, we add complete new styling and other innovations."

While Corvette fans would have to wait more than half a year to finally see the redesigned C4, they got to try out the new car's engine and transmission some 18 months before. Designated L83, the 1982 Corvette's 350 V-8 used refined versions of the tubular-header exhaust system and computer command control equipment that first appeared along with the LG4 California V-8 in 1980. And like 1981's L81, the L83 also used weight-saving magnesium valve covers.

Topping it all off was the new Cross-Fire Injection system. To many, the setup looked very much like the rare twin-carb option used on some Z/28™ 203 Camaros in 1969. But those weren't carburetors beneath that cool-looking air cleaner. They were two computer-controlled Rochester throttle-body fuel injectors mounted diagonally on an aluminum cross-ram intake manifold. Making this throttle body injection (TBI) system work was an electronic control module (ECM) that was capable of dealing with up to 80 variable (ignition timing, air/fuel mixture, idle speed, etc.) adjustments per second to maximize performance and efficiency, and offer something fuelie fans had been missing since 1965: instant throttle response.

Enhancing overall response even further was the new 700-R4 four-speed automatic transmission, which was electronically linked to the ECM. Shifts and the torque converter's lockup clutch feature were precisely controlled by the ECM, depending on varying speed and load data inputs. This power team combo clearly was as high as high-tech had ever been beneath a fiberglass hood to that point. Yet even with all that techno-wizardry, the 1982 small block still only produced 10 more horsepower than 1981's L81.

To both showcase all this new technology and mark the end of the Shark era, Chevrolet, in 1982, put together another special model, the Collector Edition, which, in Dave McLellan's words, was "a unique combination of color, equipment, and innovation [resulting in] one of the most comprehensive packages ever offered to the Corvette buyer."

At about $18,000, a base Corvette in 1982 was already expensive enough. But carrying a full load of features like 1978's Limited Edition model, the

C3 1968–1982

Corvette sport coupe production for 1982 was 18,648. Base Price was $18,290.07. Another 6,759 Collector Edition hatchbacks also were built that year.

Priced at $22,537.59, the 1982 Collector Edition hatchback features such exclusive touches as special paint and bodyside decals, unique alloy wheels, and a lavish interior.

137

The 1982 Collector Edition Corvette was the only C3 model to feature a working glass hardback.

Collector Edition became the first Corvette to crack the $20,000 realm. The exact suggested price was $22,537.59. Helping hike that bottom line up were exclusive turbine alloy wheels wearing white-letter P255/60R15 Goodyear Eagle GT rubber, glass roof panels done in unique bronze tinting, a rear window defogger, a power antenna, and special identification inside and out.

Also included was unique paint—a silver-beige finish accented by graduated gray decals and accent striping. That exclusive color carried over inside, where silver-beige leather was found on the seats and door panels. Leather wrapping also went onto the steering wheel, and luxurious, extra-deep pile carpeting covered the floor. Bringing up the rear was another harbinger, a frameless window hinged to technically expand Corvette body style choices to two—coupe and hatchback—for the first time since the convertible departed after 1975.

A suitable closing chapter for the Shark tale, the Collector Edition was offered for 1982 only. As for the old St. Louis plant, it was boarded up in the fall of 1987 after building 13 million Chevrolet cars and trucks—including about a half million Corvettes.

> A suitable closing chapter for the Shark tale, the Collector Edition was offered for 1982 only.

C3 1968–1982

Leather/vinyl or cloth/vinyl upholstery was standard inside the 1982 Corvette. Popular options included power door locks and driver's seat, cruise control, rear window defogger, and an AM/FM cassette stereo.

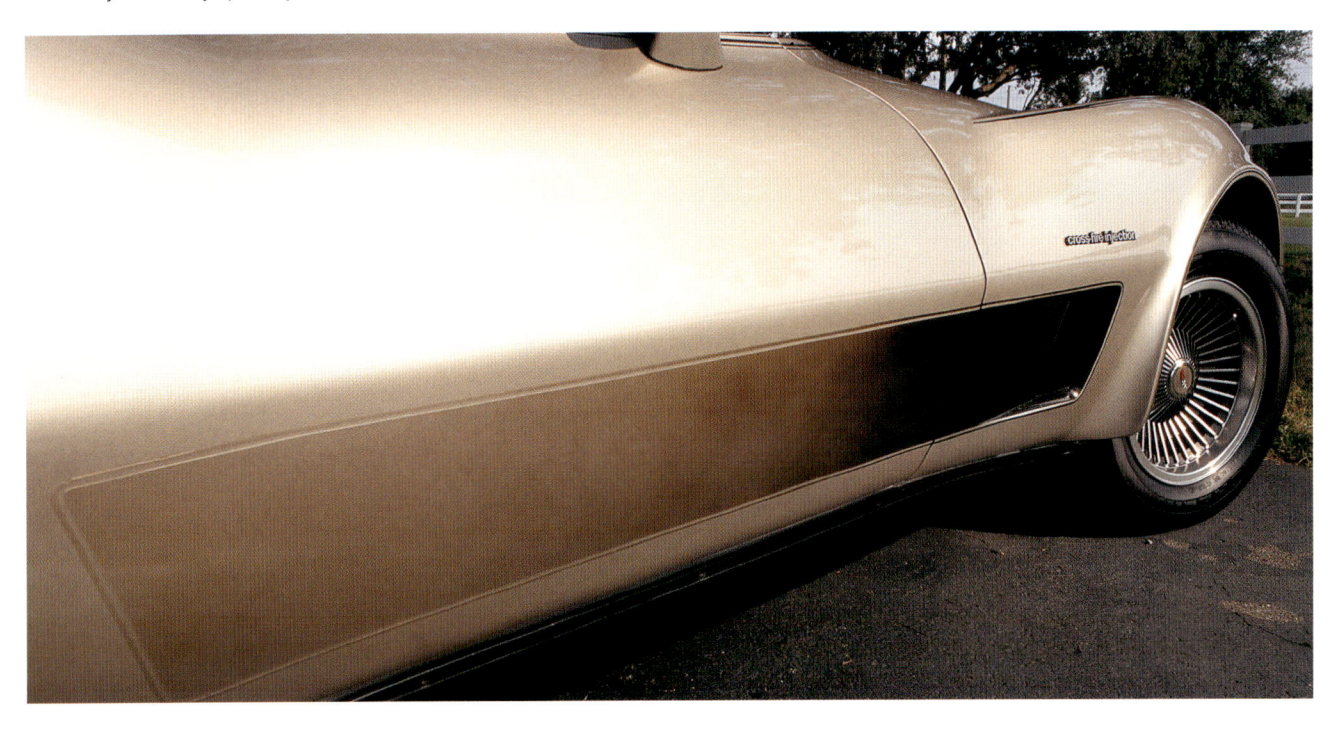

The Collector Edition's silver-beige paint was accented with graduating decals on the hood and down the body sides, and silver-beige leather was all the rage on the interior. *Mike Mueller*

139

CHAPTER FOUR

04 Better Late than Never

140

C4 1984–1994

By 1982, the Shark had grown old beyond its years. Its bulging body had been around for a decade and a half, but its foundation dated all the way back to 1963. Obviously, a change was long overdue.

Various factors contributed to the C3's marathon run, not the least of which involved its record sales successes. Annual production went up each year from 1970 to 1977, surpassing 40,000 for the first time in 1976 and staying above that level through 1981 despite a national recession. Why fix something that plainly wasn't broken?

- C4 generation was Corvette's second longest (13 model runs)
- Tuned-port injection replaces Cross-Fire Injection (1985)
- Convertible returns to lineup after 11-year hiatus (1986)
- Corvette paces Indianapolis 500 (1986)
- 35th Anniversary model offered (1988)
- ZR-1 introduced (1990)
- All Corvettes treated to ZR-1-style taillamp (1991)
- 5.7-liter Gen II V-8 (300-horsepower LT1) debuts (1992)
- 1 millionth Corvette built July 2, 1992
- Chief Engineer Dave McLellan retires, replaced by David Hill (1992)
- 40th Anniversary model offered (1993)
- Chevy General Manager Jim Perkins announces GM has approved development of the 1997 C5 Corvette, June 29, 1993
- National Corvette Museum opens in Bowling Green, Kentucky, September 2, 1994
- Corvette paces Indianapolis 500 (1995)
- LT4 V-8 offered one year only (1996)
- Collector Edition offered (1996)
- Grand Sport offered (1996)
- Zora Arkus-Duntov dies, April 21, 1996

ALTHOUGH GM EXECS WERE PLEASED as punch about all the dough the Shark brought home each year, there was an understanding among the Corvette team that the old, tired St. Louis assembly line had seen its best days. The Missouri plant could roll out a lot of C3 Corvettes, but it never would be able to handle production of a new-and-improved next-generation model. A substantially modernized C4 Corvette couldn't appear until a larger, substantially modernized facility opened for business, and that didn't occur until the Bowling Green line got rolling in June 1981.

The fourth-generation Corvette, like its predecessor, then showed up a little late for its own coming-out party. Dating back to 1977, the C4 project originally involved a 1982 model-year debut. But Bowling Green officials were far too busy ironing out bugs in 1981 to worry about retooling the line for fourth-generation assembly, and the new car was far from ready then. Dave McLellan didn't actually get upper-office approval for his beloved baby until April 22, 1980, and by that stage a 1983 introduction was proposed. Additional stumbling blocks pushed that unveiling back further, resulting in the first dealer showings coming in March 1983. Though Chevrolet officials could've used a midyear "1983-1/2" label, General Manager Robert Stempel chose instead to forego that year entirely, leaving the first C4 to emerge as a 1984 model. No 1983 Corvettes were sold to the public.

Appearing as it did in the spring of 1983, the first C4 then enjoyed an extended production run, allowing it to approach 1979's sales record. The 51,547 Corvettes built in Kentucky for 1984 stand second only to the 53,807 models born alongside the Mississippi River during John DeLorean's lengthened 1979 run.

Dave McLellan had been chief engineer for four years by the time the Corvette reached its all-time high in 1979. But the cars he was engineering were still Zora Duntov's, and much the same could've been said about early redesign proposals. Duntov was retired, yet he still dropped by McLellan's offices in Warren, Michigan, regularly during the

Above: Wind-tunnel testing helped make the 1984 C4 one of the most aerodynamic Corvettes to date.

Opposite: The new C4 Corvette was introduced in the spring of 1983 as a 1984 model. No 1983 models, though built in prototype and pilot forms, were released to the public. *Mike Mueller*

CHAPTER FOUR

The final C4 image takes shape in clay. Again, note the model year; original plans did not call for a 1983 introduction, but designers and engineers were unable to meet that deadline.

Below: On March 1, 1990, a ZR1 (foreground) and L98 (background) set endurance records on Firestone Tire's 7.7-mile test track at Fort Stockton, Texas.

> "Performance, luxury, comfort—the new Corvette is a paragon of all the things which the only true American sports car has come to stand for," bragged the car's chief engineer in February 1983.

late 1970s. The father not only couldn't walk away from his baby, he also couldn't quite forget his one dream that never came to be.

GM execs could shoot down Duntov's midengine ideal again and again, but it just wouldn't die, nor would his intriguing XP-882 platform. Late in 1976, Ed Cole's favored Wankel engine was pulled from the XP-882 chassis and replaced by a Yankee-friendly V-8, resulting in the critically acclaimed Aerovette, a gorgeous concoction that *Road & Track* described as "The 1980 Corvette."

A traditional "body drop" still occurred during C4 production, only this time the frame was already incorporated within that shell.

The Corvette team continued dabbling with midengine experiments even though it was reasonably clear that GM execs wouldn't change their minds, at least not then. So it was that when McLellan and crew went before the GM approval board in April 1980, their C4 proposal involved a traditional engine placement, leaving Duntov disappointed yet again. Zora's dream, however, was revived in 1985 when Chevy's Advanced Engineering team began work on two other middies—the Corvette Indy and CERV III—once more leading the automotive press to proclaim the coming of a truly new era. And once more the powerplant remained rooted up front when the next all-new Corvette showed up in 1997.

Duntov's hopes might have been dashed when the C4 emerged with a conventional drivetrain layout, but Dave McLellan almost couldn't say enough about how happy he was with the results. "Performance, luxury, comfort—the new Corvette is a paragon of all the things which the only true American sports car has come to stand for," bragged the car's chief engineer in February 1983. "Even in base suspension configuration, the new Corvette is a sports car absolutely superior to any production vehicle in its part of the market. To find even a peer to this vehicle, you have to look at cars produced in extremely limited numbers and at prices traditionally two or more times that of the Corvette. This car will be at home and respected on the interstate, the autobahn, or any highway in the world."

"In terms of technical innovation, we've taken a great leap forward with this new model," added Development Engineer Fred Schaafsma. Most innovative was the chassis, which featured rack-and-pinion steering, transverse mono-leaf springs front and rear, Girlock disc brakes, revised suspension locations at both ends, a five-link independent rear suspension (in place of the old three-link design), and various lightweight components.

McLellan had released some of the C4's weight-saving components early, with the first— an upgraded differential housing and its frame member mount—coming in 1980. Switching these parts to aluminum reportedly shaved off 300 pounds from that year's Shark. The 1984 Corvette's mono-leaf springs also were made of fiberglass, a weight-conscious idea first tried as part of the 1981 C3 platform.

Corvette customers also had been tipped off to the C4's drivetrain, which McLellan chose to unveil in 1982. For 1984, the L83 Cross-Fire Injection V-8 put out 5 more horsepower (205) and could've been mated to a four-speed manual transmission, a warmly welcomed addition after buyers were forced to stick with the 700-R4 four-speed automatic in 1982. As innovative as its automatic running mate, this new four-speed, supplied by Doug Nash, incorporated computer-controlled overdrive operation in its top three gears—thus its "4+3" moniker.

Along with the C4's technical innovation, its small block also exhibited unprecedented high style as designer Jerry Palmer opted to not let beauty run only skin deep. "We made sure the engine compartment would be just as pleasant to look at as the rest of the car," he said early in 1983. "I remember lengthy discussions concerning the eventual color of the high-tension cable leading to the spark plugs. We even asked Delco for a new black-and-gray battery, so it would go with the rest of the hardware." Black was the color of choice for the L83's cylinder block and heads.

At the time, Palmer was head of GM Design Staff's Chevy 3 studio, which had been home to both Camaro (F-body) and Corvette (Y-body) styling development since 1974. It was the Design

CHAPTER FOUR

The 350-cubic-inch L83 Cross-Fire Injection V-8 carried over from 1982 but produced 205 horsepower instead of 200.

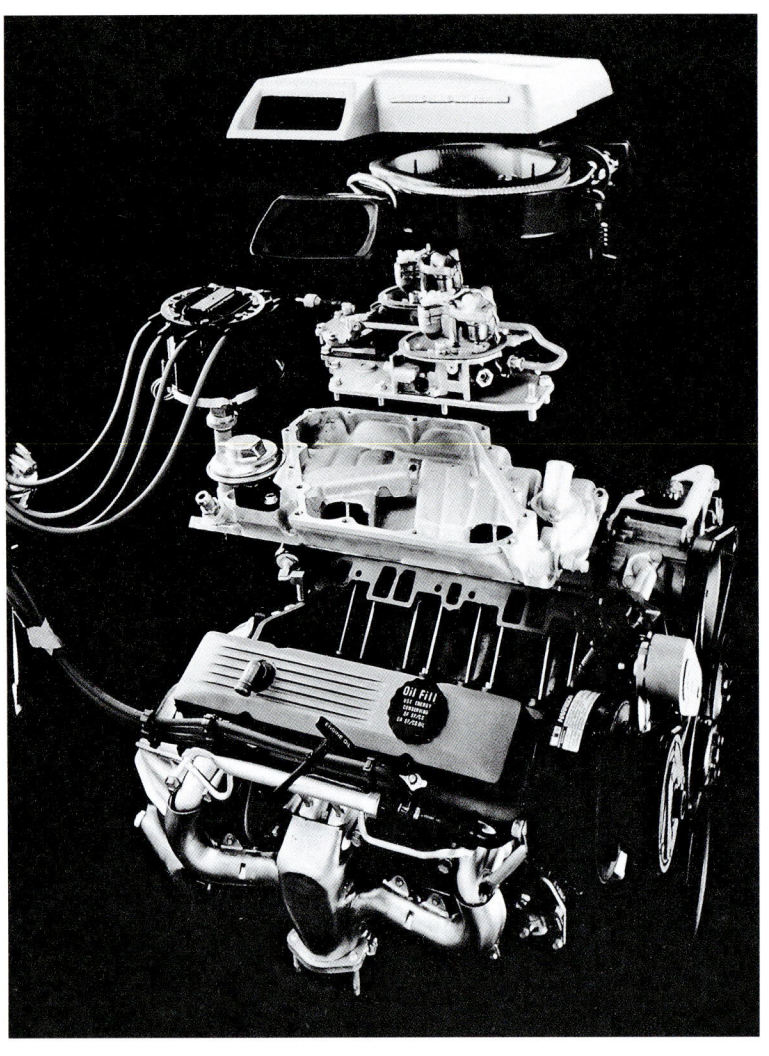

Below: Wind-tunnel testing helped make the 1984 C4 one of the most aerodynamic Corvettes to date.

Staff group that promoted Duntov's ideal, while Chevrolet Engineering people preferred to stick with the status quo, leaving Palmer squared off against his close coworker and friend, Dave McLellan.

"In the early Seventies, we were thinking along the lines of a midengine sports car as our next Corvette," explained Palmer in February 1983. "Work on such a car accelerated when the rotary engine was being viewed as a promising powerplant for a sport machine." Palmer's Chevy 3 team proposed a smaller midengine Corvette powered by a V-6, and they even built a running mule using a Porsche 914 platform as a base.

Then a funny thing happened on the way to the drawing board. In the summer of 1977, Porsche unveiled its new 928 with a front-mounted V-8, a conventional rear-driver that came about reportedly after engineers and designers in Germany had evaluated a 1976 Corvette. Chevrolet Engineering officials had already concluded that a V-6 would never do between fiberglass fenders; after Porsche's Corvette copycat appeared, the deal was done. No way the C4 would rely on anything other than a V-8 delivering torque to the rear wheels. Palmer then complied with a suitable body to match the conventional chassis layout engineered by McLellan.

On the C4's clean side, Chevy 3's totally fresh restyle incorporated improved function into its form by way of a large clamshell hood that allowed easier engine access. Created by John Cafaro, then new to Palmer's studio, this idea

144

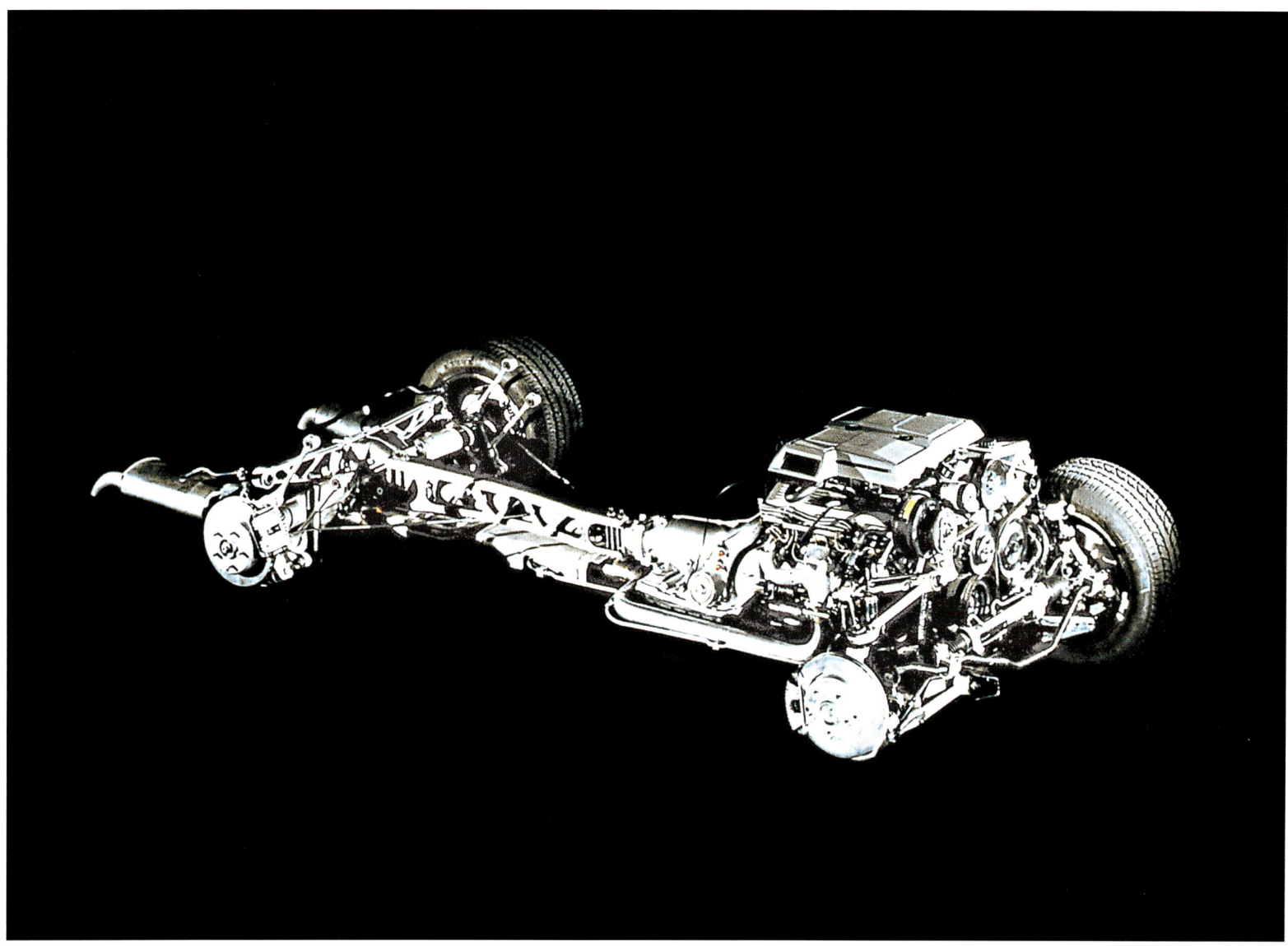

Below: New for the C4 powertrain was a "backbone" layout that tied the engine/trans combo together more rigidly with the differential via an aluminum C-section beam. Weight-saving aluminum was also used throughout the front and rear suspension.

was made possible by Cafaro's addition of a parting line running completely around the car, separating the body into upper and lower halves. In front, this parting groove served as a break between the lower fenders and the forward-hinged clamshell. Overall, the groove made for easier body construction by doing away with the pesky, often-unsightly bonding seams required previously. A rub molding installed into the groove hid the joint between upper and lower C4 body panels.

On top of the C4 was a one-piece lift roof panel; in back was a hatchback rear window that at the time represented the largest piece of compound glass ever fitted to an American car. Front glass sloped back 65 degrees, making it the most rakish windshield in Detroit history. Wind tunnel tests of this low, sleek shape resulted in a superb drag coefficient of 0.34.

While overall width increased 2 inches, wheelbase (96.2 inches) went down 2, length dropped 8.8, and height fell by 1.1. The lowered stance in turn lowered the C4's center of gravity, and that aspect, working in concert with a revised engine location (moved rearward) and widened front and rear tracks, translated into noticeably improved handling. While promotional people claimed the car's added width made for more interior room, some of those gains were negated (mostly in the footwells) by the wider transmission tunnel required to make room for the L83's exhaust system.

Overall looks were certainly fresh, yet still honored the past. "I really believe we've designed a car without compromises," said Palmer. "But we've managed to retain Corvette identity. The first time that people see this car, they're going to know what it is. They're going to say, 'Hey! That's a new Corvette!'"

Dave McLellan introduced his pride and joy to the press at the Riverside International Raceway in California in December 1982. The first C4 wearing a 1984 vehicle identification number (VIN) rolled out of its Kentucky home on January 6, 1983. Before that, 43 Corvettes wearing 1983 VINs were built: 33 pilot cars and 10 prototypes. All were supposedly scrapped, but one miraculously later showed up in the National Corvette Museum in Bowling Green. Reportedly, those 43 cars cost Chevrolet about $500,000 apiece.

The 1984 run began at serial number 00002, and the first 69 C4s built were engineering cars used for testing. The officially tagged 00001 car was actually put together after 00070 and donated to the National Council of Corvette Clubs to be raffled off as part of a fund raiser to help fight spina bifida.

Although most engineering cars eventually found private homes, serial number 00071 was the first 1984 Corvette sold to the public through traditional channels. It and the following 678 cars were shipped from Bowling Green to California in February 1983 to take part in the model's first dealer introductions, held March 24. Buyers elsewhere across America got their initial looks on April 21. The fourth-generation Corvette had arrived, and with it, another era in American sports car luxury.

CHAPTER FOUR

1984

In the mad rush to proclaim the C5 as the best 'Vette yet, few witnesses in 1997 seemed to recall a similar sensation nearly 15 years earlier. Like the all-new fifth-generation Corvette, the first C4 garnered *Motor Trend*'s prestigious "Car of the Year" award, and rightly so. Okay, the C5's trophy didn't come until the convertible version was introduced in 1998, but that's only because the Targa-topped sport coupe appeared too late to make *Motor Trend*'s 1997 balloting.

While honoring the C4, *Motor Trend* claimed it "has the highest EQ [excitement quotient] of

> The injected L98 was hot, for sure, but it also looked so damned cool.

Below: Various C4 features were released early during the third-generation run, including the 1984 Corvette's engine and transmission, which debuted in 1982. That year's Collector Edition also featured the breed's first hatchback roof, which remained a standard item on all 1984 Corvettes.

1984

Model availability	sport coupe with rear hatchback glass
Construction	fiberglass body on steel skeleton chassis, perimeter rail frame
Wheelbase	96.2 inches
Length	176.5 inches
Width	71 inches
Height	46.7 inches
Shipping weight	3,088 pounds
Tread (front/rear, in inches)	59.6/60.4
Tires	Goodyear Eagle GT P255/50VR-16
Brakes	11.5-inch discs
Wheels	alloy, 16x8.5 inches front, 16x9.5 inches rear
Fuel tank	20 gallons
Front suspension	independent short and long arms, transverse fiberglass leaf spring, stabilizer bar, tubular shock absorbers
Rear suspension	independent five-link layout with upper and lower control arms, tie rods, half shafts, and transverse fiberglass leaf spring; stabilizer bar; tubular shocks
Steering	rack and pinion
Engine	205-horsepower 350-ci V-8 (L83) with Cross-Fire Injection
Transmission	Turbo Hydra-matic 700-R4 automatic transmission
Opt. Transmission (no cost)	four-speed manual

C4 1984–1996

> Like the all-new fifth-generation Corvette, the first C4 garnered *Motor Trend*'s prestigious "Car of the Year" award, and rightly so.

anything to come out of an American factory. Ever. Its handling goes beyond mere competence; call it superb, call it leading edge, call it world class. *We certainly do.*"

Early paperwork had specified 15-inch alloy wheels wearing P215/65ZR rubber for the 1984 Corvette, but these were superseded in regular production by P255/50VR-16 tires on truly large (for the time) wheels measuring 16x8.5 inches in front, 16x9.5 in back. Initially listed as an individual option (RPO QZD), these wheels and tires also were included in the new Z51 performance handling

All four Corvette generations tour a test track in 1984.

Lights, camera, C4 action! The fourth-generation Corvette began appearing in dealer showrooms in early January 1984.

CHAPTER FOUR

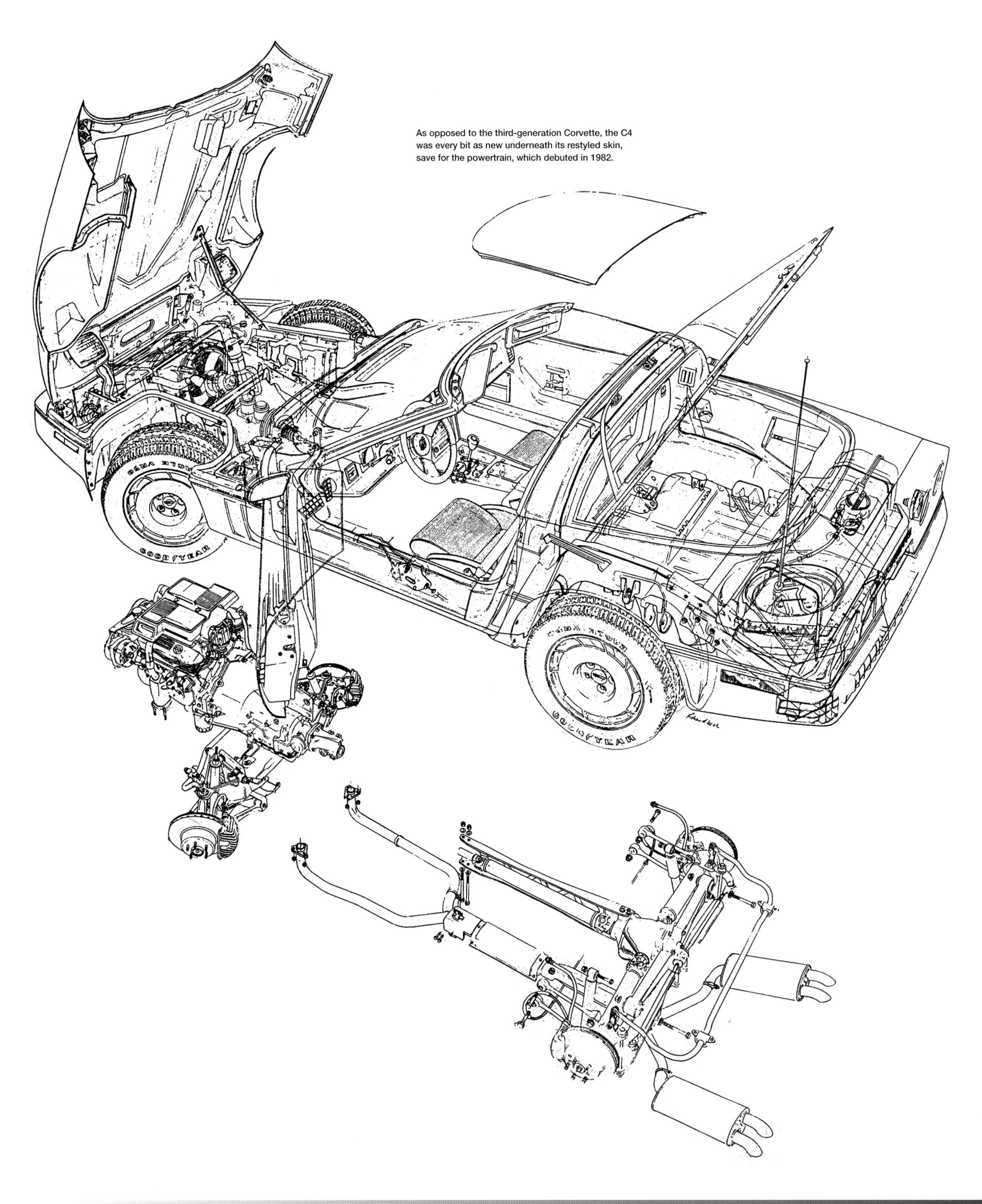

As opposed to the third-generation Corvette, the C4 was every bit as new underneath its restyled skin, save for the powertrain, which debuted in 1982.

C4 1984–1996

liquid-crystal graphic displays in place of a conventional speedometer and tach. While some critics soon would knock this so-called "arcade" design, *Motor Trend*'s Kevin Smith apparently liked what he saw early on. "The light show can get fairly spectacular," he wrote, "and while we still fail to see any advantage over old-fashioned round dials, the Corvette's is the first digital panel that we can at least imagine coming to accept." Some compliment.

1985

The second-edition C4 served as a showcase for Chevrolet's first electronic fuel injection (EFI) system, called tuned-port injection (TPI). A major step up compared to the TBI setup used in 1982 and 1984, TPI was kind to both the environment and the pocketbook, but at the same time was downright mean when it came to making muscle. Bringing back serious performance was the goal this time around, and the TPI-equipped L98 V-8 did not disappoint. Its 230 newfound horses inspired *Car and Driver* to call the new L98 Corvette the "king of the road, reborn." Torque was 330 ft-lbs, compared to 290 for 1984's L83.

"This engine is stronger than dirt," began *Car and Driver*'s Rich Ceppos in praise of the L98. "The white golf shirts that the Corvette guys handed out with a grin at Chevrolet's 1985 press preview say it all. Stenciled over the left breast are the Corvette crossed-flags insignia and the words 'Life Begins at 150.' Yes, that's *miles per*, and yes, for the first time in more than a decade, Chevy's

The 1984 Corvette was the first car developed in GM's wind-tunnel facility.

package, which featured heavy-duty springs, shocks, and bushings; stiffened stabilizer bars; quicker steering; an engine oil cooler; and a second cooling fan that pushed air into the radiator, which incorporated aluminum fins and plastic reservoirs. The oil cooler could've been ordered separately (RPO KC4), as could the heavy-duty Delco-Bilstein shock absorbers (RPO FG3).

Certainly new inside was the C4's electronic instrumentation, which featured digital readouts and

> While honoring the C4, *Motor Trend* claimed it "has the highest EQ [excitement quotient] of anything to come out of an American factory. Ever."

The 1984 Corvette wheels featured black accents in their centers, and this style carried over into 1985. Wheel centers for 1986 were left with a natural aluminum finish. *Mike Mueller*

1985

Model availability	sport coupe with rear hatchback glass
Construction	fiberglass body on steel skeleton chassis, perimeter rail frame
Wheelbase	96.2 inches
Length	176.5
Width	71 inches
Height	46.4 inches
Shipping weight	3,088 pounds
Tread (front/rear, in inches)	59.6/60.4
Tires	Goodyear Eagle GT P255/50VR-16
Brakes	11.5-inch discs
Wheels	alloy, 16x8.5 inches front, 16x9.5 inches rear
Fuel tank	20 gallons
Front suspension	independent short and long arms, transverse fiberglass leaf spring, stabilizer bar, tubular shock absorbers
Rear suspension	independent five-link layout with upper and lower control arms, tie rods, half shafts, and transverse fiberglass leaf spring; stabilizer bar; tubular shocks
Steering	rack and pinion
Engine	230-horsepower 350-ci V-8 (L98) with tuned-port injection
Transmission	Turbo Hydra-matic 700-R4 automatic transmission
Opt. transmission (no cost)	four-speed manual

149

CHAPTER FOUR

A 1985 coupe sits next to the first GTP Corvette.

plastic bi-seater will hit the magic one-five-oh mark. How sweet it is."

"No need to speculate, the car is definitely quicker," added Road & Track. "And what numbers! With the automatic transmission [L98], we ran off 0 to 60 mph in 6.2 seconds and the quarter-mile in 14.6 sec, both of which represent a good second off our previous times."

Now a true fuel-injection system (in some minds, the L83's twin throttle-body injectors were little more than force-feeding carburetors), TPI featured eight individual Bosch injectors, one at each intake port. Up front was a mass airflow sensor that directed ambient atmosphere into a large-capacity aluminum plenum. From there, the flow was separated into eight precisely tuned (thus the name) intake runners that curved somewhat elegantly down to the ports in the cylinder heads. These tubular runners helped ram the air stack into the combustion chamber then the intake valve opened. And a pair of high-pressure fuel rails, one stretched along each bank of intake ports, supplied the go-juice to those eight injectors.

The injected L98 was hot, for sure, but it also looked so damned cool. According to Rich Ceppos, the TPI equipment "makes the scenery under the clamshell hood as pretty as anything you'll find in a Porsche or BMW."

Few other changes were made for 1985. Springs were softened by 26 percent in front, 25 in back, to answer customer complaints concerning the 1984 Corvette's harsh ride. The Z51 package made for even rougher seat-of-the-pants responses, so its springs were undone too, by 16 percent in front, 25 in back. Thicker stabilizer bars were added to the Z51 option to compensate for these weaker transverse leafs. New to that year's Z51 deal was the V08 heavy-duty cooling equipment.

Inside, revisions included less colorful speedometer and tachometer graphics and larger digits for the liquid-crystal displays. The sport seat option, which was cloth in 1984, was done in leather beginning midyear.

1986

Big news in 1986 involved the standard application of a Bosch-supplied anti-lock brake system (ABS) and the return of a Corvette convertible, absent since 1975. Center high-mount stoplights (CHMS) also appeared high on the coupe's roof, at the top of the convertible's rear fascia. Wheels changed

150

C4 1984–1996

Few changes set the 1985 Corvette apart from its forerunner. Spring rates were softened front and rear to appease drivers who had complained about the first C4's rough ride. Production for the 1985 Corvette remained high at 39,729 with a base price of $24,403. Car & Driver named it America's fastest production car.

slightly as their center caps' black finish was dropped, leaving a natural metal tone.

L98 enhancements included aluminum cylinder heads, though not in all cases. Early faulty castings delayed the installation of weight-saving heads until 1986 production was well under way. Beginning midyear as well, the reborn convertible was fitted with aluminum heads from start to finish. But the majority of coupes that year were delivered with typical cast-iron heads. Aluminum-head L98 V-8s were rated at 235 horsepower, 5 more than their iron-headed counterparts.

The new convertible became the second Corvette to pace the Indianapolis 500, and Chevrolet that year was more than proud that its two-seater again (like its 1978 forerunner) needed no special engineering modifications to help bring Rick Mears, Danny Sullivan, Michael Andretti, and the rest up to speed. Save for the safety-conscious strobe lights, five-point harness, and onboard fire system, the yellow, 230-horsepower ragtop that led the pace lap on May 31, 1986, was basically identical to all other '86 Corvette convertibles sold to the public.

Retired U.S. Air Force Brigadier General Chuck Yeager, a man familiar with setting the pace, was the celebrity driver in 1986. Yeager, of course, was the first man to surpass the speed of sound, that achievement coming in October 1947. Flying much lower and slower, Bobby Rahal won the Indy 500 in May 1986.

After the race, civilian drivers were typically offered pace car replicas. But this time, Chevrolet considered every Corvette convertible built in 1986, 7,315 in all, to be a street-going Indy pacer, regardless of color. No special limited-edition package was offered; it simply was left up to the buyer to add a dealer-offered commemorative decal to the doors of his or her '86 drop top. Most didn't.

The Z51 Corvette remained the hottest thing rolling, but came only in coupe form. Even with its integral X-member reinforcement, the convertible chassis was no match for the beefed, bone-chattering Z51 suspension. The 1986 Z51 package also included a heavy-duty radiator (RPO V01) and boost fan (B4P), along with all the hot parts of previous years.

Corvette Indy

In 1985, Chevrolet's Advanced Engineering team hooked up with Lotus in Hethel, England, to build two experimental Corvettes, one targeted for the

CHAPTER FOUR

auto show stage, the other for the test track. The former was the Corvette Indy; the latter was CERV III.

The Corvette Indy showed up first, at the Detroit auto show in January 1986. Its name came from its power source: the 2.65-liter Ilmore Indy Car twin-turbo V-8. This mighty mite was mounted transversely behind a severely cramped cockpit. It existed there, however, only for display purposes—Chevy people didn't call this nonrunning mockup the "pushmobile" for nothing. Warren's latest midengine wonder didn't actually work until the Chevrolet-Ilmore V-8 was traded for the 32-valve, DOHC LT5 V-8 then being developed by Lotus for the upcoming ZR-1. A Toronado transmission was mated to the LT5 to allow for its transverse location, and the package was offered to journalists to drive at Lotus in 1988.

Futuristic simply wasn't a big enough word in this sexy two-seater's case. Sensationally sleek on the outside, the Corvette Indy, in many minds, predicted the shape of things to come, although that super-low-sloping canopy would've never made do in real-world applications. Its advanced

Below: A Corvette paced the Indianapolis 500 for the second time in 1986. Veteran fighter ace and test pilot Chuck Yeager drove the car on race day in May that year.

1986

Model availability	sport coupe (with rear hatchback glass) and convertible
Construction	fiberglass body on steel skeleton chassis, perimeter rail frame
Wheelbase	96.2 inches
Length	176.5 inches
Width	71 inches
Height	46.4 inches
Shipping weight	3,086 (coupe)
Tread (front/rear, in inches)	59.6/60.4
Tires	Goodyear Eagle GT P255/50VR-16
Brakes	11.5-inch discs with ABS
Wheels	alloy, 16x9.5 inches
Fuel tank	20 gallons
Front suspension	independent short and long arms, transverse fiberglass leaf spring, stabilizer bar, tubular shock absorbers
Rear suspension	independent five-link layout with upper and lower control arms, tie rods, half shafts, and transverse fiberglass leaf spring; stabilizer bar; tubular shocks
Steering	rack and pinion
Engine	230-horsepower 350-ci V-8 (L98) with tuned-port injection and cast-iron cylinder heads (235 horsepower with aluminum heads)
Transmission	Turbo Hydra-matic 700-R4 automatic transmission
Opt. transmission (no cost)	four-speed manual

C4 1984–1996

Corvette enthusiasts rejoiced to see the first Corvette convertible in nine years.

Below: The Corvette Indy showcar debuted publicly at the Detroit auto show in January 1986. Its name came from its power source: a 2.65-liter Ilmore Indy Car twin-turbo V8.

153

CHAPTER FOUR

A long list of paint choices greeted Corvette customers in 1986, including four two-tone finishes. Bright red was most popular that year, followed by black and white.

mechanicals, however, were certainly feasible. Both its drive-by-wire throttle control and its monocoque chassis' central backbone (in this case done in Kevlar and carbon fiber) reappeared later as part of the C5's makeup.

Additional innovations included all-wheel drive with traction control, four-wheel steering, anti-lock brakes, and a definitely advanced active suspension system designed by Lotus. The latter did away with conventional springs, shocks, and stabilizer bars, relying instead on computer-controlled hydraulics at each wheel to instantly adjust ride and handling characteristics per changing road conditions. Some witnesses felt this feature could eventually reach production but not until, as Road & Track's Joe Rusz explained it, such "futuristic concepts can be assimilated into the world of mass production. The prototype active suspension system costs perhaps $100,000," he continued. "Mass produced, it could cost only a few hundred dollars."

Still others were sure that the Corvette Indy signified that Zora Duntov's dream of a midengine redesign wasn't dead. By 1993, claimed some publications, the next new Corvette would also feature a midship-mounted V-8, perhaps additionally fitted with twin turbos like the original Corvette Indy. Wishful thinking again.

1987

Identifying a new Corvette in 1987 was made somewhat easy by a revised wheel finish. For 1984 and 1985, the wheel's center cap and radial slots were painted black. In 1986, only the slots wore black paint. In 1987, both cap and slots were treated to argent gray coloring.

C4 1984–1996

1987

Model availability	sport coupe (with rear hatchback glass) and convertible
Construction	fiberglass body on steel skeleton chassis, perimeter rail frame
Wheelbase	96.2 inches
Length	176.5 inches
Width	71 inches
Height	46.7 inches, coupe; 46.4 inches, convertible
Shipping weight	3,086 pounds (coupe)
Tread (front/rear, in inches)	59.6/60.4
Tires	Goodyear Eagle GT P245/60VR-16
Brakes	11.5-inch discs with ABS
Wheels	alloy, 16x9.5 inches
Fuel tank	20 gallons
Front suspension	independent short and long arms, transverse fiberglass leaf spring, stabilizer bar, tubular shock absorbers
Rear suspension	independent five-link layout with upper and lower control arms, tie rods, half shafts, and transverse fiberglass leaf spring; stabilizer bar; tubular shocks
Steering	rack and pinion
Engine	240-horsepower 350-ci V-8 (L98) with tuned-port injection
Transmission	four-speed manual or Turbo Hydra-matic automatic

For 1984 and 1985, the wheel's center cap and radial slots were painted black. In 1986, only the slots wore black paint. In 1987, both cap and slots were treated to argent gray coloring.

Production totals for the 1987 Corvette coupe and convertible were 20,007 and 10,625.

155

CHAPTER FOUR

Car & Driver named the 1987 Corvette one of the 10 best automobiles in the world in its January 1987 issue.

New beneath the C4 clamshell were 240 horses, which resulted from a switch to friction-reducing roller lifters. Additional performance enhancements included a finned power-steering cooler and convertible-style structural reinforcements for the Z51 package, then limited to four-speed manual installations in coupes.

Another chassis upgrade came by way of RPO Z52, introduced for 1987 with all the Z51 stuff save for its raucous springs. Called the sport handing package, the Z52 deal relied on the softer standard springing, and thus was available on both coupes and convertibles. There was no transmission limitation, either.

A second new-for-1986 option briefly came and went due to technical difficulties. A low-tire-pressure indicator (RPO UJ6) appeared early that year wearing a hefty $325 price tag. But some of the units were able to trigger the warning alarms in other UJ6-equipped Corvettes if they wandered too close.

Reportedly, 46 UJ6 installations were made before the faulty option was withdrawn and sent back to the drawing board. It would return in fine shape in 1989.

An existing option, electronic air conditioning (RPO C68), became available in coupes and convertibles in 1987. Originally announced as a late 1985 option but not released until the following year, C68 had been limited to coupes in 1986.

1988
All-new six-slot wheels appeared in 1988, still wearing P255/50ZR-16 rubber. Technical improvements included better brakes and a revised front suspension. The latter was redesigned to reduce unwanted steering wheel actions induced by brake torque and varying road condition inputs. The former featured new, dual-piston brake calipers and an upgraded parking brake system that relied on the rear disc pads instead of the separate small drums used previously. Larger calipers and rotors (12.9-inch fronts, 11.9-inch rears) were added to the Z51 package, as were bigger (17x9.5 inches) 12-slot wheels shod in P275/40ZR tires.

Less-restrictive mufflers (for coupes with the optional 3.07:1 performance axle only) in 1988 boosted L98 output to 245 horsepower. Deemed too loud for convertibles (and coupes with taller 2.59:1 gears), these mufflers didn't go behind some L98s that year, resulting in a repeat of 1987's 240-horse rating in those cases.

Representing the biggest news for 1988 was the Corvette's second anniversary model, the 35th special edition package, RPO Z01. Available for coupes only, this option added exclusive white paint (accented by a black roof bow), white wheels, appropriate exterior badges, and a console-mounted anniversary plaque. Inside, the steering wheel was covered in white leather, and more white cowhide graced the seats, which featured anniversary embroidery on their headrests.

1988

Model availability	sport coupe (with rear hatchback glass) and convertible
Construction	fiberglass body on steel skeleton chassis, perimeter rail frame
Wheelbase	96.2 inches
Length	176.5 inches
Width	71 inches
Height	46.7 inches, coupe; 46.4 inches, convertible
Shipping weight	3,229 pounds
Tread (front/rear, in inches)	59.6/60.4
Tires	Goodyear Eagle GT P255/50ZR-16
Brakes	11.5-inch discs with ABS
Wheels	alloy, 16x9.5 inches
Fuel tank	20 gallons
Front suspension	independent short and long arms, transverse fiberglass leaf spring, stabilizer bar, tubular shock absorbers
Rear suspension	independent five-link layout with upper and lower control arms, tie rods, half shafts, and transverse fiberglass leaf spring; stabilizer bar; tubular shocks
Steering	rack and pinion
Engine	240-horsepower (or 245 horses, depending on mufflers) 350-ci V-8 (L98) with tuned-port injection
Transmission	four-speed manual or Turbo Hydra-matic automatic

> In 1980, the International Motor Sports Association (IMSA) created a new racing class for Grand Touring Prototype (GTP) cars, truly exotic machines able to run well beyond 200 miles per hour.

Various options (electronic air conditioning, Z52 suspension, etc.) were thrown in as part of the Z01 deal, explaining its hefty asking price of $4,795. Z01 production was 2,050.

Corvette GTP

In 1980, the International Motor Sports Association (IMSA) created a new racing class for Grand Touring Prototype (GTP) cars, truly exotic machines able to run well beyond 200 miles per hour. Corvettes by then were no strangers to IMSA events, but in lower, slower, production-based classes. Competing at such a high level in front of far more fans would surely provide far more publicity for Chevrolet's two-seater, or so thought some company officials. Thus came the Corvette GTP, one of the wildest vehicles to tour an IMSA track during the 1980s.

The Bowtie banner was first carried onto the GTP battlefield by the Chevy-powered Lola T-600, a midengine coupe that first ran in May 1981. Inspired by the T-600's successes, Chevrolet people—no longer afraid to admit to

Below: New six-slot wheels appeared for 1988 coupes and convertibles. Brakes and the front suspension also were improved. Bowling Green produced 15,382 Corvette coupes and 7,407 convertibles in 1988.

CHAPTER FOUR

The steering wheel of the 1988 35th Anniversary Corvette was wrapped in white leather. *Mike Mueller*

The Corvette's optional manual transmission from 1984 to 1988 was the so-called 4+3 gearbox, with computer-controlled overdrives for the top three gears. Installations of the 4+3 transmission in 1988 totaled 4,282. *Mike Mueller*

Wheels on the 35th Anniversary Corvette were painted white too. *Mike Mueller*

Z01 interiors featured leather upholstery, a console-mounted anniversary plaque, and this commemorative embroidery on the seats' headrests. *Mike Mueller*

participating in racing—aspired to morph the powerful package into a competition vehicle that fans would directly associate with Chevy's hottest street-going product. The hush-hush world of racing for Chevrolet was long gone.

About the same time, performance product promotion chief Vince Piggins had engine builder Ryan Falconer working on a competition V-6 that Piggins first planned to take racing at Indy. But plans for the Brickyard fell flat, leaving Falconer to suggest that Chevrolet put his turbocharged V-6 to work on the IMSA circuit. The idea then became to plant the V-6 into a modified T-600 chassis and wrap it up with bodywork resembling that of the existing Corvette.

Jerry Palmer's GM design studio supplied the body, which looked somewhat like a Corvette up front, but from the doors back was a pure prototype racer. Lola delivered the 10th of 12 T-600 platforms built, and the vehicle became Chevrolet's first Corvette GTP. Created for show purposes only, the racer-in-name-alone debuted at the Detroit Grand Prix Expo in June 1983. Lotus handled chassis refinement and testing, and one Corvette GTP was later fitted with the British firm's innovative active suspension.

Lola's one-and-only T-710 chassis served as a base for the Hendrick Motorsports GM Goodwrench Corvette GTP, the first of the breed to actually race. Powered by a 3.4-liter turbo V-6, the Hendrick car debuted at Road America in Elkhart Lake, Wisconsin, on August 25, 1985, completing 69 of 125 laps to finish 33rd. Later honors included a lap-speed record at Daytona in December that year and a victory, in record time, at Road Atlanta in April 1986.

Although Hendrick put in the bulk of the Corvette GTP's track time, other teams also contributed. After Hendrick Motorsports took delivery of its first GTP car in May 1984, Lee Racing brought home one of its own later that December. Peerless Automotive accepted a V-8-powered Corvette GTP in May 1988, by which point the breed's time in the sun was waning. The No. 76 Peerless car, driven by Jacques Villeneuve and Scott Goodyear, was the last Corvette GTP to race, finishing 11th at Watkins Glen in July 1989.

While track records and awesome speeds were relatively plentiful during the Corvette GTP's short, happy career, victories weren't.

1989

The 17-inch 12-slot wheels included in the Z51 and Z52 packages in 1988 were bolted on to all Corvettes for 1989. Standard power was again either 240 or 245 horses, depending on the exhaust system installed.

New behind the L98 for 1989 was a Zahnradfabrik Friedrichshafen (ZF) six-speed manual, a no-cost option that replaced the oft-maligned 4+3 Doug Nash gearbox. The ZF transmission incorporated computer-aided gear selection (CAGS), an innovation that also attracted detractors. According to *Road & Track*'s Douglas Kott, "The best thing about [CAGS] is that it goes unnoticed most of the time."

The most noticeable difference between the 1984 Corvette (right) and the 1989 rendition (left) involved redesigned wheels. A center high-mount stoplight (CHMS) was also added to the coupe's roof (barely visible here) in 1986. *Mike Mueller*

CHAPTER FOUR

1989

Model availability	sport coupe (with rear hatchback glass) and convertible
Construction	fiberglass body on steel skeleton chassis, perimeter rail frame
Wheelbase	96.2 inches
Length	176.5 inches
Width	71 inches
Height	46.7 inches, coupe; 46.4 inches, convertible
Shipping weight	3,229 pounds (coupe), 3,269 pounds (convertible)
Tread (front/rear, in inches)	59.6/60.4
Tires	Goodyear Eagle GT P275/40VR-17
Brakes	11.5-inch discs with ABS
Wheels	alloy, 17x9.5 inches
Fuel tank	20 gallons
Front suspension	independent short and long arms, transverse fiberglass leaf spring, stabilizer bar, tubular shock absorbers
Rear suspension	independent five-link layout with upper and lower control arms, tie rods, half shafts, and transverse fiberglass leaf spring; stabilizer bar; tubular shocks
Steering	rack and pinion
Engine	240-horsepower (or 245 horses, depending on mufflers) 350-ci V-8 (L98) with tuned-port injection
Transmission	four-speed manual or Turbo Hydra-matic automatic
Optional transmission	six-speed manual

Chevrolet built 56 street-legal Corvette Challenge cars that first year, featuring engines specially assembled in Flint and delivered in sealed fashion to Bowling Green for installation.

This wheel first appeared as part of the Z51 and Z52 options packages in 1988. In 1989, it became the standard Corvette rim. *Mike Mueller*

In loafing-rpm situations, CAGS activated a shifter detent to bypass second and third. Below 19 miles per hour, a driver was forced to shift from first to fourth, with the idea to improve fuel economy by limiting engine revs when they weren't needed. Like Kott, more than one owner complained about this nuisance, but avoiding it was simply a matter of surpassing 19 miles per hour in first or blipping the throttle lightly while attempting a low-speed shift into second.

A selective ride and handling package (RPO FX3), priced at $1,695, appeared on 1989's options list, while the Z52 suspension disappeared. Available only on Z51 Corvettes, the electronic FX3 system adjusted shock damping per suspension firmness levels specified by a console-mounted switch. Three modes were available: touring, sport, or competition. Once again, the Z51 option was offered only for manual-transmission coupes. Furthermore, optional sport leather seats (RPO AQ9) were installed only in Z51 cars.

Corvette Challenge racers

Corvettes dominated SCCA Showroom Stock racing during the early 1980s, so much so that SCCA officials in 1987 basically asked Chevrolet to take its two-seater home after running four straight years without a loss.

The 1989 Corvette differed little from its predecessor. One of the most remarked-upon changes was a new Zahnradfabrik Friedrichshafen (ZF) six-speed manual transmission for the L98, replacing the old 4+3 Doug Nash gearbox.

The Corvette Challenge series then followed, organized by Canadian race promoter John Powell and fully supported by Chevrolet. Powell's plan called for pitting 50 identical L98 Corvettes against each other in competition for a $1 million purse. Ten Corvette Challenge events were run in 1988; another 12 followed in 1989 before the popular series ran its course. Bill Cooper claimed the first purse; Stu Hayner copped the championship in 1989.

Chevrolet built 56 street-legal Corvette Challenge cars that first year, featuring engines specially assembled in Flint and delivered in sealed fashion to Bowling Green for installation. Fifty of these cars were then shipped to Protofab in Wixom, Michigan, to be fitted with roll bars and other competition-conscious equipment.

Another 60 Challenge cars followed in 1989, all equipped with standard engines. However, 30 of these were soon refitted with higher-horsepower V-8s especially created by CPC's Flint Engine facility. After assembly, these engines were sent to Specialized Vehicles Inc. (SVI), in Troy, Michigan, where each was sealed after being equalized for power output. Cars and SVI engines then met at Powell Development America, in Wixom, Michigan, where roll bars and other safety gear were installed and the original V-8s were swapped for their warmed-up counterparts. Reportedly, those original engines were later returned to each car owner at season's end.

Although there was no Corvette Challenge after 1989, Chevrolet still offered a special race package in 1990 intended for the SCCA World Challenge series. Listed under merchandising code R9G, the deal involved various heavy-duty deviations from typical build specifications. Reportedly, 23 R9G Corvettes were built for 1990.

1990

A modernized interior appeared in 1990 with an instrument panel that toned down at least some of the snickers heard concerning the arcade dash. While digital graphics remained for the speedometer, all other readouts went back to analog. New, too, were a supplemental inflatable restraint (SIR) air bag system on the driver's side and a glove box on the passenger's.

The standard ABS system was improved, a more efficient radiator was installed, and optional leather sports seats became available for all 1990 models. That radiator worked so well there was no longer a need for the optional boost fan (RPO B24) offered from 1986 to 1989. Additional changes on the 1990 options list included widening FX3's scope (it was no longer limited to Z51 cars) and the introduction of a 200-watt premium Delco-Bose stereo system (RPO U1F) priced at $1,219. The U1F stereo featured a compact disc player. Only one stereo option (without a CD system) was listed in 1989.

L98 output went up 5 horses (again depending on the exhausts installed) to 245 and 250 horsepower. More cam, more compression, and

161

CHAPTER FOUR

1990

Model availability	sport coupe (with rear hatchback glass) and convertible
Construction	fiberglass body on steel skeleton chassis, perimeter rail frame
Wheelbase	96.2 inches
Length	176.5 inches
Width	71 inches
Height	46.7 inches, coupe; 46.4 inches, convertible
Shipping weight	3,223 pounds (coupe), 3,263 pounds (convertible)
Tread (front/rear, in inches)	59.6/60.4
Tires	Goodyear Eagle GT P275/40VR-17
Brakes	11.5-inch discs with ABS
Wheels	alloy, 17x9.5 inches
Fuel tank	20 gallons
Front suspension	independent short and long arms, transverse fiberglass leaf spring, stabilizer bar, tubular shock absorbers
Rear suspension	independent five-link layout with upper and lower control arms, tie rods, half shafts, and transverse fiberglass leaf spring; stabilizer bar; tubular shocks
Steering	rack and pinion
Engine	245-horsepower (or 250 horses, depending on mufflers) 350-ci V-8 (L98) with tuned-port injection
Transmission	four-speed manual or Turbo Hydra-matic automatic
Optional transmission	six-speed manual

a switch from mass-air to speed-density controls helped free up those extra ponies. These two ratings then rolled over into 1991, the year the L98 finally retired.

CERV III

Chevrolet Chief Engineer Don Runkle and Lotus engineering boss Tony Rudd first discussed building the midengine Corvette Indy in June 1985, in part to showcase the new Indy racing V-8 created for Chevy by Ilmore. Home to the Corvette Indy's futuristic carbon-fiber shell was the Chevy 3 studio, where, as Jerry Palmer told *Road & Track* in 1990, "there was a lot of clay flying around" as the "extremely emotional program" (more Palmer words) progressed.

Lotus handled the innovative mechanicals, while Ken Baker's Advanced Vehicle engineering (AVE) group was tasked with integrating things. AVE engineer Dick Balsley was the main man behind the Corvette Indy's development, as well as that of the CERV III to follow.

Three Corvette Indy cars were built, the first the nonfunctional Ilmore-equipped model displayed

An interior restyle represented the major change for 1990. Coupe production that year was 16,016.

As Corvette moved out of the 80s and into the 90s, the interior looked more and more like a jet fighter cockpit. Gauges were both analog and digital.

at the Detroit auto show in 1986. A running counterpart, painted white and fitted with an LT5 V-8, was completed at Lotus late in 1986 and served as a pilot car of sorts for a more-refined runner finished in January 1987. A third working midengine Corvette also was up and running by early 1988, and that exceptional machine was given the CERV designation, which at that time stood for "corporate experimental research vehicle."

While CERV III shared all of the Corvette Indy's advanced technologies (all-wheel drive, four-wheel steering, etc.), its overall makeup was redone to make it more practical and less experimental. Initial plans called for secrecy—the public wasn't meant to see this engineering exercise, which was intended to morph the Indy ideal into a reality-friendly prototype that perhaps could become the latest best 'Vette yet, and the various changes made to accomplish this goal explained why

CERV III took a little longer to complete compared to its three forerunners. But when GM President Robert Stempel set eyes on the car, he immediately ordered it onto an auto show stage. So it was that CERV III was introduced to show-goers in Detroit in January 1990. By then, it had clocked about 2,000 miles of testing time.

First and foremost among CERV III upgrades was a modified, roomier (both taller and wider) greenhouse complemented with Lamborghini-style swing-up doors that incorporated power windows. Remember, the Corvette Indy cockpit was no place for full-sized drivers, thanks to its windshield and side glass sloping in so sharply that, according to *Road & Track*'s Joe Rusz, "only a Munchkin can sit [inside] without scraping his head." Like its Indy predecessors, CERV III also was crowned by a removable roof but a predictably larger panel in this case.

Further concessions to real-world realities involved opening up the wheelhouses for 3.5 inches of wheel travel and reshaping the rocker panels to accommodate fuel cells. The nose was reshaped to meet federal guidelines concerning bumper height and headlight locations, and the atter became hideaway units instead of exposed beneath clear covers.

Power was supplied by a muscled-up LT5 fed by two intercooled Garrett T3 turbochargers. Output was an outrageous 650 horsepower and 655 ft-lbs of torque. Compression dropped to 8.5:1 to deal with the turbos' boost, and beefed connecting rods and Mahle pistons were added to further handle the extra strain.

All that power was transmitted by chain from the engine's torque converter to a compound transmission consisting of a strengthened Turbo Hydra-matic three-speed and a custom-built

CHAPTER FOUR

two-speed box. After that came a bevel-gear differential that turned the horses around a corner, redirecting them both fore and aft to two more differentials, both Posi-Traction units, that supplied the final drive to front and rear wheels. The middle differential relied on computer controls to deliver the power where it was needed most: if the back wheels slipped, the front wheels automatically compensated.

CERV III's top end was estimated at 225 miles per hour. To bring that much performance back to Earth, Dick Balsley specified a unique ABS brake system incorporating two huge disc rotors at each wheel. Done in carbon fiber, these discs and their pads were supplied by AP Racing, a firm known for its Formula 1 stopping systems.

Suspension pieces included titanium coil springs at the corners, but they were only installed to keep CERV III off the ground when not running. Once the turbocharged LT5 took over, Lotus' hydraulic active suspension handled all supportive chores.

The CERV III (second from right) debuted at the 1990 Detroit auto show. It was basically a modified Corvette Indy platform featuring a more practical cockpit and hideaway lights.

CERV III

Body style and construction	midengine two-seater with Kevlar/carbon-fiber/Nomex shell
Drag coefficient	0.277
Wheelbase	97.6 inches
Chassis	carbon-fiber monocoque with central backbone
Length	193.6 inches
Width	80 inches
Height	45.2 inches
Curb weight	3,400 pounds
Tread	(front/rear, in inches) 66.1/66.1
Tires	275/40ZR-17, front; 315/35ZR-17, rear
Brakes	twin discs at each wheel with ABS
Wheels	cast magnesium; 17x9.5 inches, front; 17x11 inches, rear
Fuel tank	23.3 gallons
Suspension	upper and lower A-arms with coil springs and hydraulic actuators, front and rear
Steering	power-assisted rack and pinion at both ends
Engine	transverse-mounted 32-valve DOHC LT5 V-8 with twin turbos and intercoolers
Transmission	custom two-speed automatic and three-speed Turbo Hydra-matic in tandem

Restyled wheels and new fender louvers announced the 1991 Corvette's arrival. Coupe production that year was 14,467.

1991

Model availability	sport coupe (with rear hatchback glass) and convertible
Construction	fiberglass body on steel skeleton chassis, perimeter rail frame
Wheelbase	96.2 inches
Length	178.6 inches
Width	71 inches
Height	46.7 inches, coupe; 46.4 inches, convertible
Shipping weight	3,223 pounds (coupe), 3,263 pounds (convertible)
Tread (front/rear, in inches)	59.6/60.4
Tires	Goodyear Eagle GT P275/40VR-17
Brakes	11.5-inch discs with ABS
Wheels	alloy, 17x9.5 inches
Fuel tank	20 gallons
Front suspension	independent short and long arms, transverse fiberglass leaf spring, stabilizer bar, tubular shock absorbers
Rear suspension	independent five-link layout with upper and lower control arms, tie rods, half shafts, and transverse fiberglass leaf spring; stabilizer bar; tubular shocks
Steering	rack and pinion
Engine	245-horsepower (or 250 horses, depending on mufflers) 350-ci V-8 (L98) with tuned-port injection
Transmission	four-speed manual or Turbo Hydra-matic automatic
Optional transmission	six-speed manual

Clearly advanced inside and out, CERV III certainly would have made one helluva real-world Corvette. But all that innovation didn't come cheap. According to Don Runkle, the car would have carried an out-of-this-world price tag in the $300,000 to $400,000 range. As he told *Road & Track*'s John Lamm in 1990, "We did a business case on the CERV III, but the intent was not really to do it for production. Rather, we put production specifications on the design because it's more interesting to build cars like that than to do an Indy. That car is a sculpture and doesn't need any reality to it."

1991

The 1991 C4 shell experienced a major makeover. First, the car's nose was revised with more rounded corners and wraparound parking lights. Fender louvers increased from two to four and were horizontal instead of vertical; the black rub strip was replaced by a wider, body-colored molding; and wheels were redesigned. But most noticeable was the new tail, which borrowed the look from the ZR-1 but was not as wide. All Corvettes that year looked alike at a glance, with only the center high-mount

CHAPTER FOUR

Above: Chevrolet called the 1992 LT1 its Gen II V-8. Born in 1955, the original Gen I block ran for 37 years.

Right: Doug Tuner's #1 Dieline-sponsored Corvette was the one to beat in the 1991 SCCA World Challenge series.

166

stoplights setting them apart: this extra brake light appeared on the ZR-1's roof but showed up on the restyled rear fascia of L98-powered coupes.

A finned power-steering cooler became standard, as did a power outlet for cell phones and such. Less-restrictive mufflers were installed for improved performance, but advertised L98 output carried over unchanged from 1990.

Nearly all options rolled over, too, with the most prominent exception involving RPO Z51, which was traded in 1990 for the Z07 adjustable suspension package. Priced at an intimidating $2,155, the Z07 deal was similar to the old Z51 package in that it added beefed-up brakes, stiffened underpinnings, and an oil cooler for the L98. The electronic FX3 system also was incorporated. But therein lay the differences. Z51 Corvettes fitted with FX3 in 1989 and 1990 relied on relaxed standard springing to allow ride adjustments, running from soft to firm. The Z07 option, on the other hand, kept the Z51 stiff suspension parts to make for three truly firm adjustable modes. Like the Z51 option, the uncompromising Z07 deal was limited to manual-transmission coupes, though apparently at least one automatic example was built. Z07 production was only 733 in 1991.

1992

Model availability	sport coupe (with rear hatchback glass) and convertible
Construction	fiberglass body on steel skeleton chassis, perimeter rail frame
Wheelbase	96.2 inches
Length	178.6 inches
Width	71 inches
Height	46.3 inches, coupe; 47.3 inches, convertible
Shipping weight	3,223 pounds (coupe), 3,269 pounds (convertible)
Tread (front/rear, in inches)	57.7/59.1
Tires	Goodyear GS-C P275/40ZR-17
Brakes	11.5-inch discs with ABS
Wheels	alloy, 17x9.5 inches
Fuel tank	20 gallons
Front suspension	independent short and long arms, transverse fiberglass leaf spring, stabilizer bar, tubular shock absorbers
Rear suspension	independent five-link layout with upper and lower control arms, tie rods, half shafts, and transverse fiberglass leaf spring; stabilizer bar; tubular shocks
Steering	rack and pinion
Engine	300-horsepower 350-ci (5.7-liter) V-8 (LT1) with multi-port injection
Transmission	four-speed manual or Turbo Hydra-matic automatic
Optional transmission	six-speed manual

The base price for the 1991 convertible was $38,770.

CHAPTER FOUR

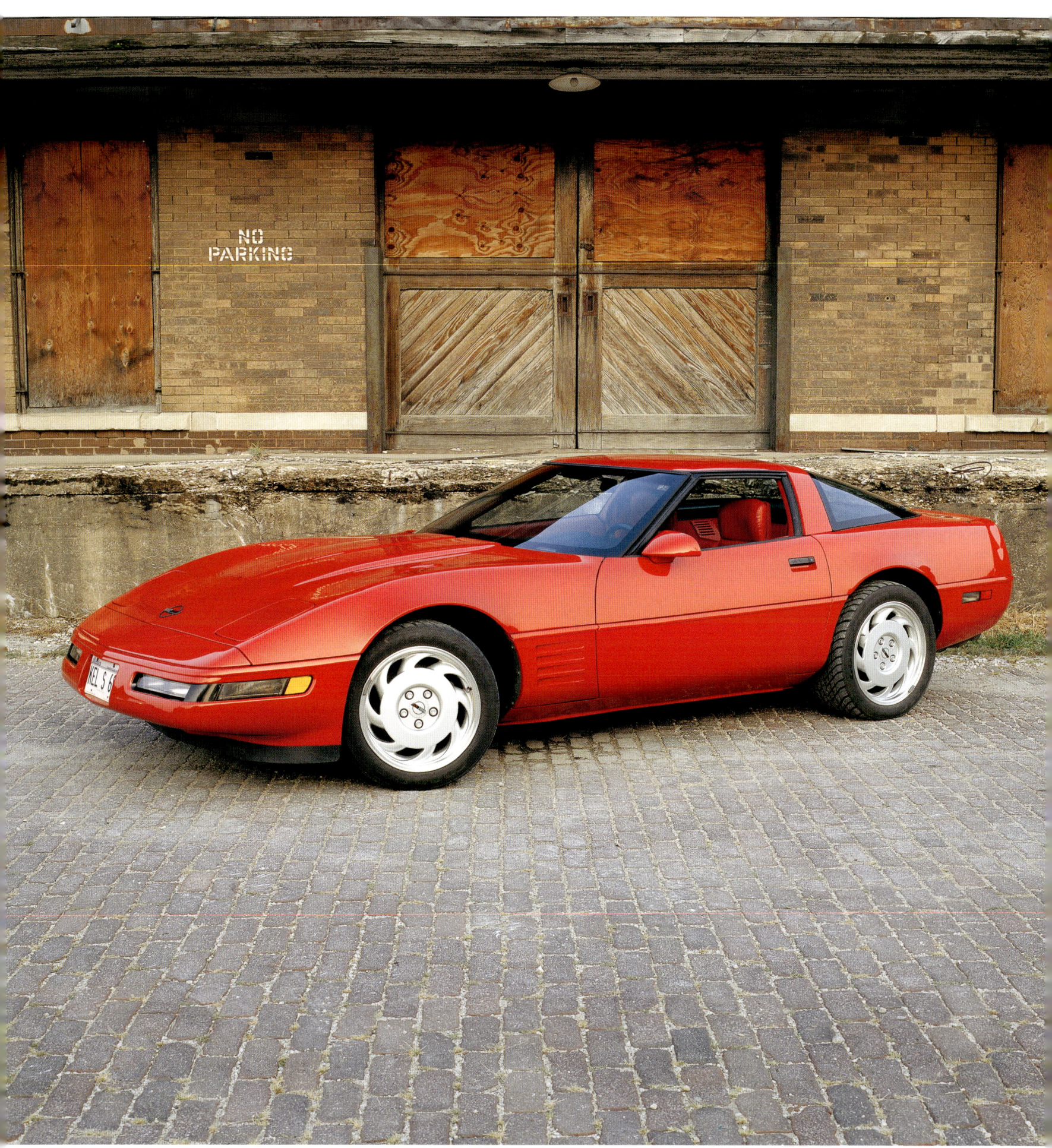

No visible changes marked the arrival of the 1992 Corvette. Coupe production that year was 14,604. *Mike Mueller*

C4 1984–1996

> "We want to continue to offer a car that people can own as their only car and still do all the things they want to do—go to the golf course or go on a week's vacation."

1992

Chevrolet's mouse motor turned 35 in 1990, making it rather young by human standards but certainly venerable in automotive engineering terms. As the 1990s dawned, most anyone could see that the time had come to update the tried-and-true small block, and it was then left to GM Powertrain Assistant Chief Engineer Anil Kulkarni to get the job done.

Kulkarni was issued a somewhat-intimidating list of goals, among which were increased reliability, reduced noise, and minimized external dimensions. A higher, flatter, longer torque curve was also specified, as was better fuel economy coupled with more performance. Making 50 more horses was what Kulkarni's bosses had in mind.

Kulkarni's resulting design met all those expectations and then some, and represented such a marked improvement that Chevrolet officials chose to announce the coming of a second-generation small block. And to honor the impressive arrival of the aptly named "Gen II" V-8, those same folks also opted to let history repeat itself, naming this small block "LT1," after the famous LT-1 built from 1970 to 1972.

According to Dave McLellan, the LT1 earned its revered title because of its strength—maximum output was 300 horsepower at 5,000 rpm. Kulkarni had been instructed to better the L98's output by 20 percent and that's exactly what he had done—while also increasing fuel economy by 1 mile per gallon. These numbers aside, it was the seat of your pants that told the true tale. The new LT1 punch literally represented a rebirth for the Corvette, with published road test results running

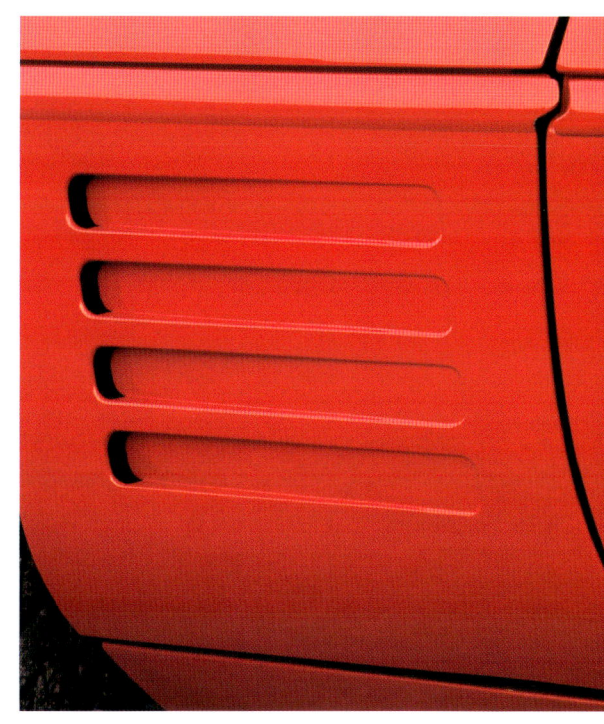

Above: The 1992 Corvette's fender louver layout first appeared in 1991. *Mike Mueller*

Below: Topless Corvette production in 1992 was 5,875. Folding roof color choices were blue (with white), beige, black, and white.

This 1992 coupe served as a paint trial car, in this case testing a shade called Melon Metallic, which didn't make the cut that year. This is not the same color as 1994's rare Copper Metallic. *Mike Mueller*

> As the 1990s dawned, most anyone could see that the time had come to update the tried-and-true small block, and it was then left to GM Powertrain Assistant Chief Engineer Anil Kulkarni to get the job done.

as low as 4.92 seconds for the 0–60 run and 13.7 seconds (topping out at 103.5 miles per hour) for the quarter-mile.

Chevy's new small block truly was born again. Very little interchanged between LT1 and L98, with carryovers including block height, bore spacing, and displacement. Most everything else was drawn up on a clean sheet of paper, beginning with the iron cylinder block and aluminum heads, which were recast to incorporate a revised reverse-flow cooling system. Gen I coolant typically had been pumped into the block first, then to the heads. But, as Kulkarni explained, cooler heads and warmer cylinder bores are key to maximizing both performance and fuel economy. Thus, the Gen II design sent the coolant first to the heads, then into the block and back to the pump.

That pump also incorporated the crossover passage (used to deliver coolant from head to head) formerly cast into the Gen I's intake manifold, an improvement that essentially allowed engineers to work with more space inside the Gen II's intake while maximizing the all-important air/fuel flow. Removing the coolant crossover from the intake also helped translate into a lower engine silhouette, a major priority considering the Corvette's low, low hoodline. Overall, the LT1 measured almost 3.5 inches shorter than its L98 forerunner.

Working in concert with that low-rise, cast-aluminum intake to reduce engine height was a new induction layout, a multi-port fuel-injection (MPFI) system that did away with the TPI's exposed long-tube runners. MPFI equipment also delivered the coals to the latest, greatest small block's fire as efficiently as ever. According to Kulkarni, the exact matching of all components—from the low-restriction air snorkel to the bigger, better throttle body; from the short-runner, one-piece intake to the more-precise AC Rochester Multec injectors—was the key to the LT1's newfound performance.

Cylinder heads were massaged for better breathing, compression was boosted up to 10.5:1

> According to Dave McLellan, the LT1 earned its revered title because of its strength—maximum output was 300 horsepower at 5,000 rpm.

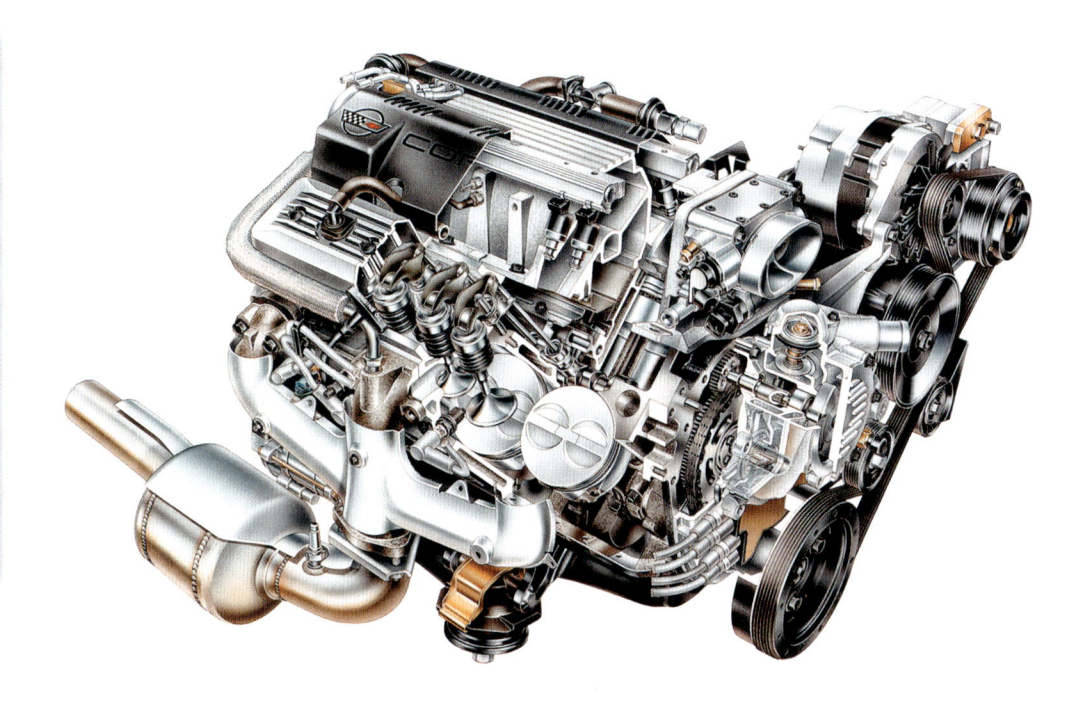

Chevrolet engineers dusted off the LT-1 tag, deleted the hyphen, and used it to announce the rebirth of Corvette performance in 1992. Output for the 1992 5.7-liter LT1 small block was 300 horsepower.

(from 9.5:1), and a more-aggressive roller camshaft was installed. Lift increased from the L98 cam's 0.415/0.430 (intake/exhaust) to 0.451/0.450 for the lumpier LT1 unit. Valve specs rolled over from the L98: 1.94-inch intakes, 1.50 exhausts.

The L98's high-energy ignition (HEI) system was exchanged for the new Optispark distributor, located up front where it was driven off the camshaft. This unit used optical signals to govern ignition timing. Light shining through a stainless-steel shutter created 360 pulses per crank revolution, and the pulses were translated into electronic signals sent to the powertrain control module (PCM), which then determined spark firing. The most-precise small-block ignition ever was the end result, though some bugs were encountered early on.

Chevrolet engineers filled the LT1 with Mobil 1 synthetic oil and recommended it for continued use throughout each 1992 Corvette's life. Those engineers also determined that synthetic lubricants don't require special cooling, and thus the KC4 oil cooler option was deleted.

Introduced for 1992 were new Goodyear GS-C tires with directional, asymmetrical tread. Acceleration slip regulation (ASR) traction control also became standard that year. Supplied by Bosch, the ASR system could be deactivated by a dash-mounted switch. In operation, it would combine various controls (spark retardation, throttle shutdown, and braking) to avert wheel spin, and a driver knew when it was working by the way the accelerator pushed back on his or her right foot.

Another bit of important Corvette news came in November 1992 as 49-year-old David C. Hill was made the car's third chief engineer, replacing Dave McLellan, who had retired a couple of months prior. Like Ed Cole, Hill came to Chevrolet from Cadillac, where he had spent most of his time after joining GM following his 1965 graduation from Michigan Technological University. He finished a master's degree in engineering at the University of Michigan in 1969 and became chief engineer for the Allante project a dozen years later. By 1992, he was Cadillac's engineering program manager; then came the opening at Chevrolet.

Hill's new job represented a suitable assignment considering his past experience with sports cars. Not long after going to work at GM, he had purchased a 1948 MG TC and restored it. Next, he took a Lotus Super 7 racing in SCCA competition, winning two national events in 1972. By then, his daily driver (as well as his first new car) was a 1970 Corvette coupe, which he toured the country in, piling up 7,500 miles over 13 vacation days.

The Corvette's future fell right into his lap in November 1992, and early on he even considered reviving Duntov's everlasting dream. As he told *Automobile* magazine's Rich Ceppos in 1994, "I myself was not at first satisfied that we shouldn't do a midengined car, and I spent some energy on that when I arrived. But cars like the Acura NSX don't have anywhere near the utility of a Corvette. We want to continue to offer a car that people can own as their only car and still do all the things they want to do—go to the golf course or go on a week's vacation."

Of that last aspect, Hill clearly knew from whence he spoke.

Sting Ray III

Early in 1989, Chevy 3 studio's John Cafaro thought he deserved the honor of drawing up a suitably sensational shell for the Corvette's next new generation, the C5. But GM design chief Charles Jordan didn't quite agree.

At the time, Chevy 3 was busy completing the latest next-generation Camaro. Once that job was done, thought Cafaro, he could give the C5 Y-car project his full attention. But a week before Chevy 3 was to release the new F-body design in January, Chuck Jordan ordered a laundry list of changes inspired by a wild Camaro show car built earlier by John Schinella's Advanced Concepts Center (ACC) in Newbury Park, a northwest suburb of Los Angeles.

Jordan didn't think much of Chevy 3's C5 proposals, anyway, and he told Cafaro if he couldn't do better, Schinella's West Coast boys could. Then he ordered Schinella to sculpt a "California Corvette." Not long afterward, Jordan contacted ACC again, specifying a running car, not just a clay mockup.

Schinella's creation appeared in the spring of 1990. Sweet and sassy, this Black Cherry convertible, nicknamed "the purple car," featured a functional trunk and exposed headlights up front, both features not seen on regular-production Corvettes since 1962. The former showed up on the new C5 convertible in 1998, and the latter reappeared on the C6 in 2005. Another purple car feature, a rear-mounted transmission, also carried over into C5 production.

Coil-over shock absorbers at all four corners suspended this cool convertible, known officially as the Sting Ray III. Wheels were massive: 18x10 in front, 19x10 in back. Mounted on those flashy three-spoke rims were 285/35ZR tires at the nose, 305/35ZR at the tail. Overall, the Sting Ray III was 2 inches shorter than a C4 Corvette, but its wheels were farther apart. Wheelbase was 103 inches, up nearly 7 inches compared to the C4.

Although Cafaro and others back east weren't all that impressed with the Sting Ray III, it was a big hit with the public when it was unveiled at the

CHAPTER FOUR

Detroit International Auto Show in January 1992. Most loved its fresh face. "ACC wrapped all this engineering within an exterior shape that's exciting and aggressive," wrote *Road & Track*'s John Lamm. "With its rear haunches, it looks like a big cat ready to leap. They fitted a soft top that allows the car to look as nice closed as open. And they gave the Sting Ray III those sexy eyes."

Lamm also was impressed with estimates claiming a production Sting Ray III would sell in the $20,000 to $25,000 range. "Come on, Chevrolet!" he continued. "Keep the Corvette out of the technostratosphere and get it back on the asphalt where it belongs: as a Sting Ray III roadster more young Americans—and Europeans and Japanese—can afford."

These pleas, of course, fell on deaf ears. And in the end, John Cafaro did do the bulk of the C5 sketching.

1993

Five years after rolling out its white Z01 coupes, Chevrolet returned with its third celebratory Corvette, this one marking 40 years. Listed under RPO Z25, the 40th anniversary package was available for coupes, convertibles, and the fire-breathing ZR-1. This widened scope, combined

1993

Model availability	sport coupe (with rear hatchback glass) and convertible
Construction	fiberglass body on steel skeleton chassis, perimeter rail frame
Wheelbase	96.2 inches
Length	178.6 inches
Width	70.7 inches
Height	46.3 inches, coupe; 47.3 inches, convertible
Shipping weight	3,333 (coupe), 3,383 (convertible)
Tread (front/rear, in inches)	57.7/59.1
Tires	Goodyear GS-C P255/45ZR-17, front; P285/40ZR-17, rear
Brakes	12-inch discs with ABS
Wheels	alloy; 17x8.5 inches, front; 17x9.5 inches, rear
Fuel tank	20 gallons
Front suspension	independent short and long arms, transverse fiberglass leaf spring, stabilizer bar, tubular shock absorbers
Rear suspension	independent five-link layout with upper and lower control arms, tie rods, half shafts, and transverse fiberglass leaf spring; stabilizer bar; tubular shocks
Steering	rack and pinion
Engine	300-horsepower 350-ci (5.7-liter) V-8 (LT1) with multi-port fuel injection
Transmission	four-speed manual or Turbo Hydra-matic automatic
Optional transmission	six-speed manual

Yet another anniversary package, RPO Z25, appeared in 1993 to help celebrate the Corvette's 40th birthday. Ruby Red paint was exclusive to the Z25 cars, which were offered as LT1 coupes, ZR-1 coupes, or convertibles. The production breakdown was 4,204 coupes, 2,043 convertibles, and 245 ZR-1s.

with an easier-on-the-wallet price ($1,455 compared to $4,795), helped Chevrolet more than triples sales of its latest birthday present; 2,050 35th Anniversary cars were built in 1988, while the Z25 total for 1993 was 6,749, including 4,333 coupes, 2,171 convertibles, and 245 ZR-1s.

Like 1988's Z01, the Z25 Corvette was treated to an exclusive finish: Ruby Red Metallic. Special chrome emblems on the hood and fuel filler door, color-keyed wheel centers, and matching Ruby Red leather seats were also included. Headrests on those seats featured "40th" embroidery, and the same logo appeared in badge form on each fender. All leather seats in 1993 featured anniversary logos (cloth seats did not) even if Z25 wasn't ordered, meaning all ZR-1 coupes featured this stitching inside because all ZR-1s featured leather interiors.

Standard appearance features carried over from 1992 save for a freshened appearance for the wheels from a revised machining process. Front wheel width also decreased 1 inch for 1993, and tire size correspondingly dropped: in place of 1992's P275/40ZR-17 front rubber were P255/45ZR-17 rollers. Rear tires, meanwhile, got bigger: P285/40ZR-17, compared to 1992's P275/40ZR-17. Z07-equipped Corvettes that year used 17x9.5 wheels and P275/40ZR-17 tires all the way around.

Spring rates were lowered slightly, and various minor upgrades were made beneath the hood. A two-piece heat shield (as opposed to the previous one-piece unit) and polyester (instead of magnesium) valve covers helped quiet the LT1 V-8 down in 1993. Inside, revised cam specs on the exhaust side bumped maximum torque from 330 ft-lbs to 340. Horsepower remained at 300.

New for 1993 was GM's first passive keyless entry (PKE) system, a neat techno trick that used a mini transmitter in the key fob to automatically lock or unlock the doors whenever the fob moved into or out of range. Simply walking up to the car with the keys in your pocket instantly disarmed the theft deterrent system, popped the locks, and turned on interior lights. Walking away reversed the process with a little abbreviated beep from the horn to remind you that your CDs were safe.

1994

Appearance features again rolled over, with the only noticeable updates for 1994 involving two new paints: Admiral Blue and Copper Metallic. In the latter's case, compromised finish quality limited applications to only 116 1994 Corvettes, creating an instantly recognized rarity.

Interior upgrades were relatively plentiful: a new passenger-side air bag (that deleted the glove box), an express-down power window on the driver side, revised upholstery and door panels, a redesigned two-spoke steering wheel, and new instrument graphics that went from white to tangerine after dark. All seats in 1994 were leather, as cloth upholstery was dropped, and base and optional sport styles were again available.

Among technical changes was a new transmission, the 4L60-E four-speed automatic, that did the previous 4L60 one better by incorporating electronic controls for improved shifts and seamless operation. Shifting the 4L60-E from park also required depressing the brake pedal, another first.

New powdered-metal connecting rods replaced the forged rods inside 1994's LT1 small block, but most notable that year was a switch to a markedly improved sequential fuel-injection system that used a mass airflow (MAF) sensor in place of 1993's speed-density system. Working in concert with a more powerful ignition (that improved cold-starting), sequential fuel injected (SFI) equipment in turn enhanced throttle response and idle quality, and lowered emissions. Advertised output remained unchanged.

Commemorative wheel centers were included in the Z25 deal. Mike Mueller

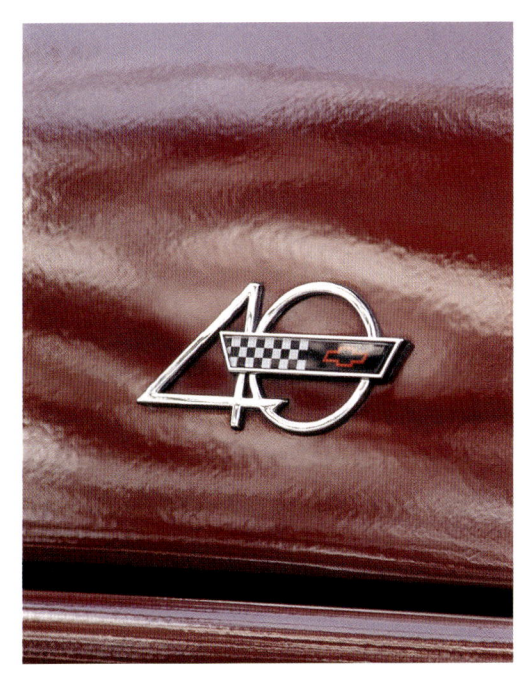

Appropriate fender badges graced the 40th Anniversary Corvette. Additional anniversary embroidery appeared inside the seats' headrests. Mike Mueller

1994

Model availability	sport coupe (with rear hatchback glass) and convertible
Construction	fiberglass body on steel skeleton chassis, perimeter rail frame
Wheelbase	96.2 inches
Length	178.5 inches
Width	70.7 inches
Height	46.3 inches, coupe; 47.3 inches, convertible
Shipping weight	3,317 pounds (coupe), 3,358 pounds (convertible)
Tread (front/rear, in inches)	57.7/59.1
Tires	Goodyear GS-C P255/45ZR-17, front; P285/40ZR-17, rear
Brakes	12-inch discs with ABS
Wheels	alloy; 17x8.5 inches, front; 17x9.5 inches, rear
Fuel tank	20 gallons
Front suspension	independent short and long arms, transverse fiberglass leaf spring, stabilizer bar, tubular shock absorbers
Rear suspension	independent five-link layout with upper and lower control arms, tie rods, half shafts, and transverse fiberglass leaf spring; stabilizer bar; tubular shocks
Steering	rack and pinion
Engine	300-horsepower 350-ci (5.7-liter) V-8 (LT1) with multi-port fuel injection
Transmission	four-speed manual or Turbo Hydra-matic automatic
Optional transmission	six-speed manual

Above: Next to nothing changed for 1994, save for color choices, which then included Admiral Blue and Copper Metallic. Copper Metallic was offered for 1994 only and was applied to a mere 116 Corvette coupes and convertibles. *Mike Mueller*

Right: The long-awaited grand opening for the National Corvette Museum, located nearly next door to the Corvette plant in Bowling Green, Kentucky, finally came on Labor Day weekend in 1994. Literally everyone was on hand, including, of course, Zora Arkus-Duntov, shown here during the VIP dinner the night before the official opening. *Mike Mueller*

174

C4 1984–1996

Interior color choices in 1994 numbered four: black, light beige, light gray, or red. *Mike Mueller*

> Among technical changes was a new transmission, the 4L60-E four-speed automatic, that did the previous 4L60 one better by incorporating electronic controls for improved shifts and seamless operation.

On the options list, the carryover FX3 selective ride and handling package was fitted with lowered spring rates to soften seat-of-the-pants responses. A new RPO, WY5, added the world's first run-flat rubber, the Goodyear GS-C Extended Mobility Tire (EMT), which could travel up to 200 miles at 55 miles per hour at absolutely zero pressure, allowing the driver to make it safely somewhere after suffering a flat. The low-pressure indicator (RPO UJ6) was mandatory with the EMT option because an airless run-flat tire looks no different from one filled to the brim. RPO WY5 was not available with the Z07 package or on the ZR-1 coupe.

1995
Revised fender vents set 1995's Corvette apart from its forerunners. Truly distinctive were the 527 Indy 500 pace car replicas built to mark the breed's third appearance (this time with Chevy General Manager Jim Perkins driving) at the Brickyard, on May 28, 1995. Listed under RPO Z4Z, that year's pace car convertible wore splashy graphics on a Dark Purple/Arctic White finish and featured special leather seats embroidered with Indianapolis 500 logos. Price for the Z4Z option was an eye-popping $2,816.

Dark Purple Metallic was a new finish that year, filling in where the deleted Copper and Black Rose metallic paints left off. Inside, stronger French seams graced the optional sport seats, and an automatic transmission fluid temperature readout was added to the instrument panel display. The 4L60-E automatic was improved (enhanced clutch controls and a lighter yet stronger torque converter) for smoother shifts and a quieter cooling fan went under the hood. Six-speed manuals were also upgraded with a high-detent operation (in place of 1994's reverse lockout) for easier operation.

New on the options list was RPO N84, made possible by the growing popularity of the EMT run-flat tires. The WY5 rubber basically rendered a traditional spare moot; thus came the N84 spare

continued on page 178

CHAPTER FOUR

Chevrolet built 15,771 coupes for 1995. The base price for the coupe was $36,785, and the base price for the convertible was $43,665. *Mike Mueller*

1995

Model availability	sport coupe (with rear hatchback glass) and convertible
Construction	fiberglass body on steel skeleton chassis, perimeter rail frame
Wheelbase	96.2 inches
Length	178.5 inches
Width	70.7 inches
Height	46.3 inches, coupe; 47.3 inches, convertible
Shipping weight	3,203 pounds (coupe), 3,360 pounds (convertible)
Tread (front/rear, in inches)	57.7/59.1
Tires	Goodyear GS-C P255/45ZR-17, front; P285/40ZR-17, rear
Brakes	13-inch discs, front; 12-inch discs, rear; with ABS
Wheels	alloy; 17x8.5 inches, front; 17x9.5 inches, rear
Fuel tank	20 gallons
Front suspension	independent short and long arms, transverse fiberglass leaf spring, stabilizer bar, tubular shock absorbers
Rear suspension	independent five-link layout with upper and lower control arms, tie rods, half shafts, and transverse fiberglass leaf spring; stabilizer bar; tubular shocks
Steering	rack and pinion
Engine	300-horsepower 350-ci (5.7-liter) V-8 (LT1) with multi-port fuel injection
Transmission	four-speed manual or Turbo Hydra-matic automatic
Optional transmission	six-speed manual

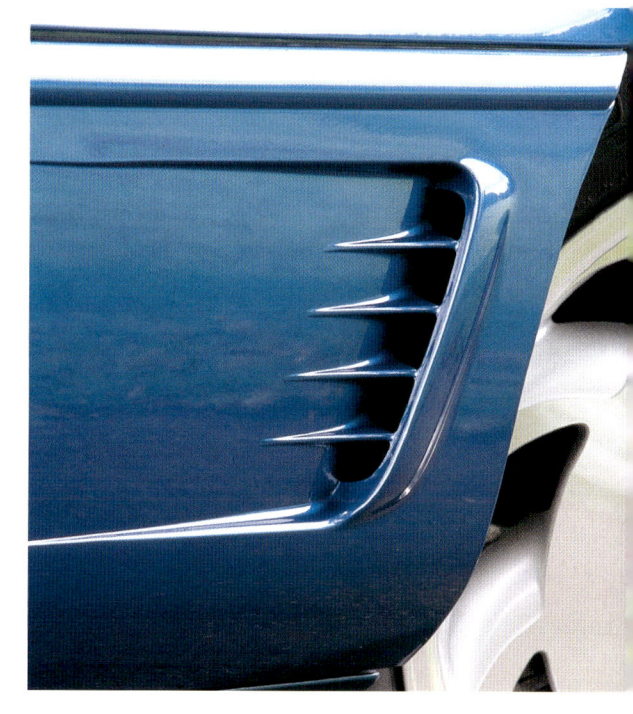

Above: Fender louvers were restyled yet again for the 1995 Corvette, which rolled on the same style wheels seen in 1994. *Mike Mueller*

Below: A Corvette became the prestigious pace car for the Indianapolis 500 for the third time in 1995. Chevrolet built 527 replica convertibles that year.

CHAPTER FOUR

Leather seats and a Delco stereo with cassette player were included in the 1995 Corvette's base price, as had been the case the previous year. *Mike Mueller*

Continued from page 175

tire delete deal, which put $100 back into the buyer's pocket. Due to the EMT design's success, the Corvette spare tire was done away with completely when the C5 debuted.

Overall ride was improved by less-stiff De Carbon gas-charged shocks and lower spring rates. Brakes, meanwhile, grew tougher as the big front discs (13x1.1 inches) previously limited to the Z07 option and ZR-1 coupe became standard on all 1995 Corvettes in place of 1994's 12x0.79 rotors. Further enhancing the 1995 Corvette's stopping performance was the new Bosch V ABS system

Reminding many of its 1982 ancestor, the 1996 Collector Edition helped honor the end of an era as the curtain closed that year on the C4 run.

1996

This was the last year for Corvette's Gen II small block, and to honorably mark its farewell tour engineers put together an upgraded version of the LT1, the 330-horsepower LT4. All LT4 V-8s delivered in 1996 were backed by six-speed manuals, while the base 300-horse LT1 was mated only to automatics.

LT4 enhancements began with aluminum cylinder heads, which featured taller ports and bigger valves: 2.00-inch intakes, 1.55-inch exhausts. Those valves had hollow stems to save weight, and they used special oval-wire springs that could handle more lift without binding. Helping

1996

Model availability	sport coupe (with rear hatchback glass) and convertible (Grand Sport coupe featured wide fender flares)
Construction	fiberglass body on steel skeleton chassis, perimeter rail frame
Wheelbase	96.2 inches
Length	178.5 inches
Width	70.7 inches
Height	46.3 inches, coupe; 47.3 inches, convertible
Shipping weight	3,298 pounds (coupe), 3,360 pounds (convertible)
Tread (front/rear, in inches)	57.7/59.1
Tires	Goodyear GS-C P255/45ZR-17, front; P285/40ZR-17, rear
Tires (Grand Sport coupe)	Goodyear Eagle GT P275/40ZR-17, front; P315/35ZR-17, rear
Brakes	13-inch discs, front; 12-inch discs, rear; with ABS
Wheels	alloy; 17x8.5 inches, front; 17x9.5 inches, rear
Wheels (Grand Sport coupe)	17x9.5 inches, front; 17x11 inches, rear
Fuel tank	20 gallons
Front suspension	independent short and long arms, transverse fiberglass leaf spring, stabilizer bar, tubular shock absorbers
Rear suspension	independent five-link layout with upper and lower control arms, tie rods, half shafts, and transverse fiberglass leaf spring; stabilizer bar; tubular shocks
Steering	rack and pinion
Engine	300-horsepower 350-ci (5.7-liter) V-8 (LT1) with multi-port fuel injection, only available with automatic transmission
Optional engine	330-horsepower 350-ci (5.7-liter) V-8 (LT4) with multi-port fuel injection (standard with Grand Sport), only available with manual transmission
Optional transmission	six-speed manual, only for LT4

Arctic White was the fourth most popular paint choice in 1996, behind Sebring Silver Metallic, Torch Red, and black. Production of 1996 Corvette coupes was 17,167. *Mike Mueller*

CHAPTER FOUR

Again, no changes marked a new Corvette's arrival in 1996. Convertible production that year was 4,369. *Mike Mueller*

C4 1984–1996

Interior color choices remained the same in 1996, with red trim standing out like a Yugo in a Grand Prix. Mike Mueller

increase the LT4's valve lift were higher-ratio (1.6:1) roller rocker arms supplied by Crane. Revised lift specs were 0.476 inch on intake, 0.479 on exhaust. Cam duration, too, was increased, from 200 degrees to 203 on the intake side and from 207 to 210 on exhaust.

Further LT4 upgrades included a freer-flowing intake (with taller ports to match the heads), a compression increase to 10.8:1, and a roller-type timing chain. The LT4's crank, cam, water-pump drive gear, and main bearing caps were beefed, and premium head gaskets were installed to deal with the extra compression. Topping things off were various high-profile red engine cover accents.

The LT4/six-speed combo was optional for the coupe, convertible, and a new third concoction, the Collector Edition. The 330-horse package came standard with another special 1996 model, the Grand Sport.

Reminding many of its 1982 ancestor, the 1996 Collector Edition helped honor the end of an era as the curtain closed that year on the C4 run. Priced at $1,250, this package (RPO Z15) included an exclusive Sebring Silver finish, chrome "Collector Edition" badges, silver-painted ZR-1 five-spoke wheels with special centers, black brake calipers

continued on page 184

All 5.7-liter LT4 V-8s wore special red trim and "Grand Sport" identification, regardless of the application. Mike Mueller

181

CHAPTER FOUR

Perforated sport seats were standard inside the Collector Edition model. *Mike Mueller*

The Grand Sport package (RPO Z16) was offered in both coupe and convertible forms in 1996. The Z16 price in coupe applications was $3,250. Z16 coupe production was 810.

Appropriate badges predictably adorned the 1996 Collector Edition Corvette. *Mike Mueller*

Collector Edition seats were typically adorned with special embroidery. *Mike Mueller*

Like the last C3 in 1982, the C4's passing in 1996 was honored with a special Collector Edition model, done in Sebring Silver Metallic paint. Listed under RPO Z15, the package cost $1,250 and was available on both coupes and convertibles. *Mike Mueller*

CHAPTER FOUR

Admiral Blue paint with a white accent and red Sebring stripes on the left front fender readily identified the Grand Sport.

Continued from page 181
with "Corvette" lettering, and perforated sport seats complemented with "Collector Edition" embroidery. Z15 Corvettes came as coupes and convertibles.

Also offered in both body styles, the Grand Sport was listed under RPO Z16, the same code used 31 years earlier for Chevrolet's first SS 396 Chevelle. The idea was to honor the lightweight race cars built by Zora Duntov late in 1962, thus the reasoning behind the red hash marks (or "Sebring stripes") on the left front fender and the white racing stripe accenting the car's exclusive Admiral Blue Metallic paint. Further enhancing the competitive image were blacked-out ZR-1 five-spoke wheels (again with special center caps) wearing big, bad tires measuring P275/40ZR in front, P315/35ZR in back. Grand Sport coupes were fitted with fender flares in back to better house all that extra rubber.

Grand Sport convertibles rolled on smaller tires (P255/45ZR in front, P285/40ZR in back) and therefore didn't require the flares.

Additional Grand Sport flair included black brake calipers with bright "Corvette" lettering, appropriate fender badges, and perforated bucket seats in black or red/black with "Grand Sport" embroidery. A unique serial number also was part of the deal that added $3,250 to a coupe's bottom line, $2,282 to a convertible's. Production was 1,000: 810 coupes, 190 convertibles.

Two new options appeared on the 1996 list, one with a familiar RPO code. From 1984 through 1990, the hottest Corvette available (discounting the ZR-1) was the Z51 with its big brakes and beefed suspension. A Z51 performance handling package returned for 1996, this time with stiff Bilstein shocks, thicker stabilizer bars, higher-rate

> Also offered in both body styles, the Grand Sport was listed under RPO Z16, the same code used 31 years earlier for Chevrolet's first SS 396 Chevelle. The idea was to honor the lightweight race cars built by Zora Duntov late in 1962.

184

springs, bigger wheels and tires, and a heavy-duty power steering cooler. Wheels were 17x9.5 ZR-1 five-spokes, front and rear, wearing P275/40ZR Goodyear GS-C tires.

The Z51 deal was limited to coupes, and when ordered for the closed Grand Sport it included the latter's larger rear wheels and tires. When included along with an automatic transmission, RPO Z51 mandated the installation of the 3.07:1 performance axle, RPO G92.

The second newly introduced option was electronic selective real-time damping (RPO F45), which picked up where 1995's FX3 package left off and even carried the same price tag, $1,695. A marked improvement, this driver-adjustable, Delco-supplied ride control system relied on sensors at each wheel and a powertrain control module to deliver data with lightning-quick speed to a central computer that in turn controlled damping rates individually for all four shock absorbers. The FX3 system adjusted the four shocks simultaneously at the same rate. Furthermore, when chassis engineers said "real time," they meant it—maximum damping alterations occurred every 10 to 15 milliseconds, which translated into an F45 Corvette adjusting to each foot of changing road at 60 miles per hour.

Advanced suspensions; a new, hotter small block; and two coveted limited-edition models—what a suitable sendoff for the last of the C4 line.

Right: **Fender badges honored the original Grand Sport Corvettes, built in 1963.** *Mike Mueller*

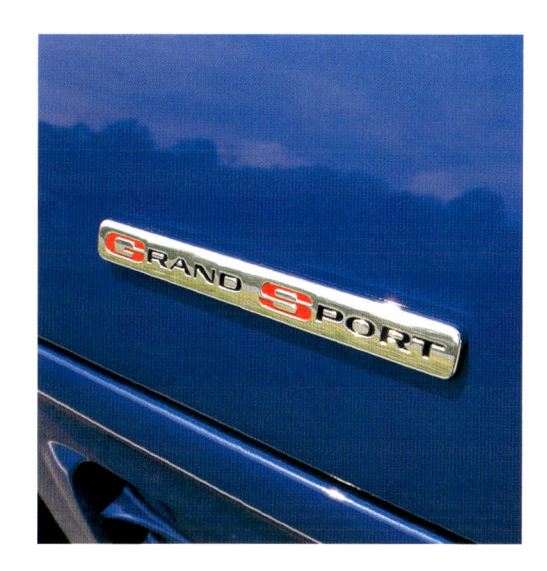

Red or a red/black combo represented the only Grand Sport interior color choices. *Mike Mueller*

CHAPTER FIVE

Long Live the King: ZR-1

ZR-1
1990–1995

General Motors' people began considering the Corvette's near future even as the most advanced example yet, the new C4, was wowing the world in 1984. Most prominent among these forward-thinkers early on was Lloyd Reuss, who that year became GM vice president and general manager of the newly formed Chevrolet-Pontiac-Canada (CPC) group.

- All ZR-1 Corvettes were coupes with high-mounted brake light on roof
- All ZR-1 Corvettes featured six-speed manual transmissions
- A ZR-1 breaks a 50-year-old 24-hour endurance speed record (1990)
- ZR-1 fender emblems added (1992)
- Dodge's 400-horsepower Viper appears to wrest away ZR-1's title as America's most powerful production car (1992)
- LT5 output boosted to 405 horsepower, putting ZR-1 back on top of Detroit's power rankings (1993)
- 40th Anniversary package (RPO Z25) offered for ZR-1 (1993)
- Last ZR-1 built at Bowling Green (1995)

A LONG-TIME POWERTRAIN ENGINEER, Reuss considered keeping Chevy's "halo vehicle" (his words) on top a personal priority, and he was fearful of the growing threat then posed by sports cars made in Japan. American automakers already were being taught a hard lesson in the compact classroom by their Far East rivals; was Chevrolet about to take an after-school beating on the high-performance playground as well?

Building more muscle was Reuss' prime concern, and he then made it Dave McLellan's. Both recognized that the boost from 205 horsepower to 230 in the works for the 1985 Corvette was nowhere near big enough. Three hundred horses was more like it, and to that end McLellan passed the ball on to Powertrain Engineering Director Russ Gee, who in turn put V-type engines' chief engineer Roy Midgley in pursuit of those ponies. Midgley at first experimented with a turbocharged V-6 before all involved concluded that Corvette customers would never settle for a six. Exhaust-boosted induction couldn't produce acceptable fuel economy (at mandated performance levels) anyway, and this inherent downside also kept an experimental twin-turbo V-8 from getting off the ground.

Further investigation led Gee toward a newly emerging power-boosting, fuel-efficient technology: multi-valve heads with overhead cams. In October 1984, he instituted a program to create new four-valve engines in four-cylinder, V-6, and V-8 forms. But CPC engineers already were buried up to their pocket protectors in multi-port fuel injection development work, leaving Gee to look outside his company for help.

In November 1984, Lotus Cars' Engineering Managing Director Tony Rudd came over from England to Warren, Michigan, for a visit, bringing with him some interesting news: at the time, Lotus was making 350 horsepower with a four-liter dual-overhead-cam (DOHC) V-8. Gee got the bright idea to try mating Lotus' four-valve heads atop the existing Chevy V-8 cylinder block. But such a proposal ended up being too wide for the C4 engine compartment, inspiring Rudd in April 1985 to suggest building an entirely new engine top to bottom, an expensive proposition to say the least. Rudd promised he could make as much as 400 horsepower from a more compact DOHC V-8 if he was allowed to construct his own block.

Gee relayed this alarming idea to Reuss that same month, and the corporate VP then relied on his own clout, as well as some powerful support from GM Board Chairman Roger Smith, to help ram Rudd's risky proposal past bean-counting killjoys. The result was a cooperative design deal between GM and Lotus to create what would soon be known as the LT5 V-8, the high-tech, all-aluminum,

Above: After falling behind the 400-horsepower Viper in Detroit's horsepower race in 1992, the 405-horse 1993 ZR-1 regained the title of America's most powerful production car. *Mike Mueller*

Opposite: The 1990 ZR-1 looked like a standard Corvette from the windshield forward, but body panels from the doors back were exclusive pieces created to widen the tail by 3 inches. *Mike Mueller*

CHAPTER FIVE

Developed by Lotus in England, the ZR-1 Corvette's 375-horsepower, 32-valve, DOHC V-8 was supplied by Mercury Marine's MerCruiser division, based in Stillwater, Oklahoma.

Plans for a 1989 introduction fell flat, leaving the ZR-1 Corvette to debut as a 1990 model. Production that first year was 3,049, and the Bowling Green plant rolled out its last ZR-1 in 1995. *Mike Mueller*

32-valve DOHC engine that eventually became the heart of the ZR-1 Corvette.

Initially labeled the "King of the Hill" by Chevrolet Chief Engineer Don Runkle, the ZR-1 was unveiled to the world in March 1989 at the Geneva auto show. The previous June, a prototype had blown past gawking American journalists (at upward of 150 miles per hour) at Riverside Raceway, leaving many magazines to report that Chevrolet would be unleashing this super-duper Corvette as a 1989 model. Indeed, that was the original plan. But lagging development efforts forced Reuss, McLellan, and the gang to make the same decision made during C4 development. While a public introduction midway into 1989 would be possible, a traditional midyear model designation was out of the question.

The official decision to designate the first ZR-1 a 1990 model came on April 11, 1989, and was announced, along with Chevrolet's best 'Vette yet, to the SAE at Selfridge Air Force Base, north of Detroit, one week later. Up until then, all references to the first ZR-1 inside and outside of GM had used a 1989 tag, including a commonly displayed color cutaway done by noted automotive artist David Kimble. Distributed by Chevrolet earlier that year, a poster made from Kimble's "1989 ZR-1" X-ray view instantly became a collectors' item.

News reports of the 1989 ZR-1's midyear arrival prior to this point had been further substantiated in September 1988, when journalists were allowed an up close and personal look, again at Riverside, at both the King of the Hill and its awesome LT5 heart. In the latter's case, the introduction was especially thorough. While Roy Midgley and his comrades from Lotus explained its many merits, two technicians nearby assembled an actual engine—then started it up with nary a hiccup, inspiring an ovation from the audience that resounded nearly as profoundly as the distinctive LT5 exhaust note. One of those highly talented (and well-rehearsed) wrenchmen was Ron Opszynski from CPC. The other was Chris Allen from MerCruiser, the firm GM contracted to build the LT5 after Lotus completed development.

> Initially labeled the "King of the Hill" by Chevrolet Chief Engineer Don Runkle, the ZR-1 was unveiled to the world in March 1989 at the Geneva auto show

In 1985, the plan was for Lotus to design the LT5 and GM to construct it. However, as Manufacturing Manager Dick Donnelly told Midgley, the corporation's engine people were far too busy to take on such a complicated, small-volume project. Russ Gee instructed Midgley to again seek outside help, and he first considered Lotus and another British firm, Coventry Climax. He even checked the John Deere tractor factory, but none of these facilities offered the required production capabilities.

Midgley next turned to the Mercury Marine division of the Brunswick Corporation in Fond du Lac, Wisconsin. He was interested in Mercury Marine's MerCruiser division, based in Stillwater, Oklahoma. For years, MerCruiser had been GM's best customer as far as special products were concerned, buying engines to go along with its state-of-the-art stern drives and inboard boat construction. The inboard industry's acknowledged leader, MerCruiser surely looked like it could handle the production of land-based powerplants.

Midgley first contacted Mercury Marine Vice President of Manufacturing Joe Anthony, who met with him in Stillwater on January 27, 1986, to ice the deal.

After lining up the many required component suppliers and ironing out tooling bugs, the Oklahoma firm completed its first three production LT5 V-8s on July 13, 1989. Its first preproduction LT5 had sputtered to life at about 7:00 P.M. on Christmas Eve 1987, a suitable gift to engineer Terry Stinson. Stinson had been working

The ZR-1 was born near the end of Dave McLellan's watch as Corvette chief engineer. Here, he stands next to VIP ceremonies at the opening of the National Corvette Museum in 1994 to pose with a "cast from life" statue of himself by New York sculptor Karen Atta, who created 20 such works of art for the museum prior to its grand opening. Mike Mueller

Countless nuts and bolts are on display at the National Corvette Museum's Design & Engineering section, including a long line of Corvette drivetrain pieces. The ZR-1 Corvette's 375-horsepower LT5 V-8 shows off its innards here. Mike Mueller

CHAPTER FIVE

A wider tail was required to house the massive tires needed to deal with the LT5's 375 horses.

feverishly, with able assistance from his Lotus counterparts, for nearly two years to prove that Midgley wasn't wrong in choosing MerCruiser, and to demonstrate to his boss, Bud Agner, that Anthony was right in signing off on this cumbersome, pressure-packed project.

Like MerCruiser, Lotus had required more than a little time to complete its assignment, as creating such a mechanical work of art was no simple task.

One stumbling block appeared right out of the gates in May 1985 after Roy Midgley noted a perceived problem with Lotus' initial design. To fit the valve sizes required to make 400 horsepower, Tony Rudd's engineers had put the new engine's bore centers 4.55 inches apart, as opposed to the existing Chevy V-8's long-running 4.40-inch measurement. Though nothing at all carried over from the traditional small block to the LT5, Midgley was adamant about at least continuing the mouse motor's time-honored "440" legacy, something Chevy's radically redesigned Gen III (LS1) small block also would do in 1997. When Rudd replied that squeezing the cylinders closer together would reduce maximum bore width, which in turn would limit valve diameters and thus output (to no more than 385 horses), Midgley couldn't have cared less—to hell with the consequences, get it done. Meanwhile, displacement, at 5.7 liters, also rolled over in familiar fashion from the conventional small block to the exotic LT5.

Lotus' first working prototype, or "Phase I" engine, was up and running on May 1, 1986. Chevrolet sent various Corvettes, with their L98 V-8s removed, to England for experimental LT5 installations. The first Phase I mule was roaring around Lotus' test track in August. Production of the more-refined Phase II engine (of which about 25 were built) began in March 1987, and one of the last of these was used in a successful 200-hour nonstop durability run staged in November that year. After passing that test with flying colors, the LT5 graduated to its Phase III stage, with production of these nearly ready-for-prime-time prototype players beginning in January 1988. Further testing and refinement of both engine and car continued throughout most of 1988, reducing the ZR-1's chances for a 1989 model year introduction as each month passed.

> "The ZR-1 is scheduled for early next year, but we won't be surprised if it's delayed a few months," added *Car & Driver*'s Csaba Csere. "One can hardly blame the Chevrolet Motor Division for wanting to make the world's first 190-mph production sports car as perfect as can be."

ZR-1 1990–1995

Lotus was ordered to destroy all 1989 ZR-1 prototypes sent to England for testing. But a few somehow managed to find their way back across the Atlantic. This wreck, shown in Mike Yager's Mid America Motorworks shop in Effingham, Illinois, was later restored and is proudly displayed today in Yager's Midwestern museum. *Mike Mueller*

Early complete-car testing, both in England and America, involved LT5 installations in stock-bodied Corvettes. But from the outset, Reuss and McLellan knew a new shape would be in order. In Reuss' case, he wanted to make sure the King of the Hill stood out from the crowd even more so than a typical Corvette. McLellan's motivation involved fewer form considerations and more functionality. Wider rear tires surely would be needed to harness those LT5 horses, meaning that bodywork in back would require modification to house all that extra tread.

McLellan's engineers first met with Goodyear tire guys in December 1985 to discuss suitable rear rubber for the ZR-1. Goodyear's response was a sticky roller measuring 1.5 inches wider than the standard Corvette issue. Bolting on a pair meant that the ZR-1's tail would have to be widened by 3 inches. Enter Studio 3 Director Jerry Palmer and his ace designer, John Cafaro.

Palmer's people recognized that add-on flares would never do. Besides looking tacky, they would require even-less-desirable tack-on modifications to the trailing edge of each door, which opened oh so close to the rear wheelhouses. Regardless of the extra expense, the only choice was to mold new doors and quarter panels to smoothly incorporate the 3 extra inches in back. The rest of the car, in Palmer's opinion, should remain stock. Although Reuss wanted more distinction, including additional nose treatments, Studio 3's simpler design won out, and widened rear body parts started showing up in Bowling Green for prototype builds in the spring of 1987.

There was no way to miss that fat tail, which was made even more distinctive by trading the C4's existing round taillights for square units. As for further identification, the ZR-1 name wasn't finalized until 1989. Like Kimble's first cutaway, early pilot cars up until then had featured LT5 emblems on the lower right corner of their rear fascias, a no-no after GM execs that year decreed that engine RPO codes could no longer be displayed on the outside of the corporation's products. Codes for performance packages, however, remained fair game. Corvette Development Manager Doug Robinson chose "ZR-1," basically because it was available and it sounded cool. That it also honored the car's heritage—RPO ZR1 had been used from 1970 to 1972—was icing on the cake.

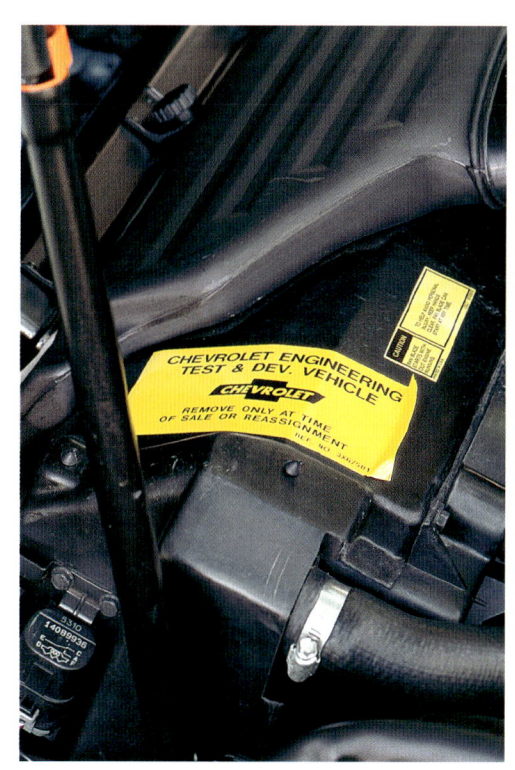

All 1989 ZR-1s were preproduction test vehicles, as this label attests. *Mike Mueller*

191

CHAPTER FIVE

Above: **This ZR-1 prototype also was restored from a crushed wreck. Reportedly 84 1989 ZR-1 Corvettes were built for testing and press review purposes, but none were meant for public consumption.** *Mike Mueller*

Opposite: **Note the British license plate and prototype "LT5" badge on the tail of this 1989 ZR-1.** *Mike Mueller*

ZR-1 1990–1995

193

CHAPTER FIVE

A few of those LT5-badged 1989 models are rolling around today. Reportedly, at least 15 of these preproduction vehicles were built, with perhaps all ending up in Lotus' hands for testing, after which time they were supposed to be destroyed per instructions from Warren. But apparently "destroy" isn't defined the same way in England as it is in the United States. A couple of severely crushed 1989 ZR-1s were salvaged by American collectors with relative ease and painstakingly restored during the 1990s, to the utmost dismay of GM officials who previously were sure that these unauthorized machines had been scattered to the winds.

Identifying an official ZR-1 in 1990 was easy enough thanks to its 3-inch-wider rear end, which also incorporated four exclusive, rectangular exhaust trumpets. Even though the rest of the Corvette line had ZR-1-style tailamps in 1991, the ZR-1 still stood out, however mildly, due to its roof-mounted extra brake light, and continued doing so until its demise in 1995. All other Corvettes during that span incorporated this light at the top center of their rear fascias.

Along with the exclusive LT5 V-8, all ZR-1s built from 1990 to 1995 featured CAGS-controlled ZF six-speed manual transmissions, and all were coupes. A few stock-bodied convertible ZR-1s were tested early on, followed by the experimental DR-1 drop top in 1990. A topless ZR-1 proponent from the beginning, Don Runkle that year put Chevy's Advanced Engineering team to work developing his idea, which was transformed into reality by the American Sunroof Company (ASC), a Detroit-area firm known for its convertible conversions. ASC, however, built only the single DR-1 and then teamed up with Advanced Engineering and Studio 3 in 1991 to create the chopped-windshield ZR-1 Spyder, a cool custom convertible assembled on the Bowling Green line. Unfortunately, no regular-production ZR-1 ragtops ever left that line.

The ZR-1 coupe was a fully loaded machine. Options for the car's first year numbered only two: electronic air conditioning and dual removable roof panels. Standard features included leather power seats, a specially laminated "solar" windshield, a Delco-Bose CD stereo, a low-tire-pressure warning light, the electronic FX3 selective ride and handling package, and the heavy-duty Z51 suspension. That latter was exclusive to the ZR-1 deal, with softened springs and stabilizer bars compared to the optional Z51 equipment found beneath conventional 1990 Corvettes.

Last, but by no means least, were those Goodyear Eagle Gatorback tires. Mounted on 9.5-inch-wide alloy wheels, the fronts were P275/40ZR-17 units, same as those found on garden-variety L98 Corvettes. But mandating that big butt were huge P315/35 z-rated Goodyears on big 11-inch-wide steam rollers, making the ZR-1 a king of the road in more ways than one.

1990

Model availability	coupe
Construction	integral perimeter frame birdcage body in reinforced composite, perimeter-rail frame
Wheelbase	96.2 inches
Length	176.5 inches
Width	74 inches
Height	46.7 inches
Curb weight	3,465 pounds
Tread (front/rear, in inches)	59.6/61.9
Tires	P275/40R-17, front; P315/35ZR-17, rear
Brakes	power-assisted four-wheel vented discs with aluminum calipers and ABS
Brake dimensions	12.9 inches, front; 11.9 inches, rear
Wheels	aluminum-alloy; 17x9.5 inches, front; 17x11 inches, rear
Fuel tank	20 gallons
Front suspension	independent short- and long-arm (SLA) upper and lower control arms and steering knuckles, transverse spring, stabilizer bar, tubular shock absorbers
Rear suspension	independent five-link with U-jointed half shafts and forged-aluminum control links and knuckles, camber adjustment, transverse spring, steel tie rods and stabilizer bar, tubular shocks
Steering	power-assisted rack and pinion, 15.73 1 ratio
Engine	5.7-liter all-aluminum 32-valve DOHC LT5 V-8, designed by Lotus, built by MerCruiser
Induction	Rochester electronic port injection
Bore and stroke	3.90x3.66 inches
Compression	11.0 1
Output	375 horsepower at 6,200 rpm, 370 ft-lbs torque at 4,500 rpm
Transmission	aluminum-case ML9 six-speed manual with CAGS
Production	3,049
RPO ZR1 price	$27,016

1990

Minor press mentions of the upcoming new king of the Corvette realm first appeared late in 1986, and journalist Rich Ceppos was among the earliest to discuss this mystery machine at any length (but with few real facts) in *Car and Driver*'s June 1987 issue. *Automobile* followed with a one-page report, simply and appropriately titled "King of the Hill," in its September 1987 edition and correctly identified the car's "Lotus-designed LT5, a 32-valve, 5.7-liter V-8." The report included a Barry Penfound spy photo of an LT5 test mule wearing a bulging hood, which was "thought to be temporarily necessary to clear the sixteen aluminum intake runners" of the innovative multi-valve engine.

Another spy shot, this one squeezed off by veteran undercover lensman Jim Dunne, appeared in the February 1988 issue of *Road & Track*. "Could this be the 400-bhp, aluminum-block, aluminum-head, four-cam V-8 Corvette ZR-1 that General Motors Vice President Lloyd Reuss announced at the Specialty Equipment Manufacturers Association (SEMA) trade show in Las Vegas this past November?" asked *R&T*'s Ron Sessions. "Could be and is."

Another Dunne photo showed up in the May 1988 copy of *Car and Driver* with the announcement: "Coming soon: the highest-performance production car on the planet." All ZR-1 details were correctly reported by C/D's Csaba Csere, save for GM's early 400-horse prognostication and those horses' expected delivery date.

"The ZR-1 is scheduled for early next year, but we won't be surprised if it's delayed a few months," added Csere. "One can hardly blame the Chevrolet Motor Division for wanting to make the world's first 190-mph production sports car as perfect as can be."

When the ZR-1 finally made it into public hands in 1990, it offered 375 horsepower and 370 ft-lbs of torque—nothing to sneeze at; after all, it still qualified as America's most powerful production car. The Lotus-engineered LT5's externally ribbed aluminum block featured forged-aluminum cylinder liners with Nikasil coating and was beefed up on the bottom end by a bolt-on aluminum girdle that incorporated cast-in nodular-iron main bearing caps. The crank and connecting rods were forged steel, the pistons aluminum. Compression was 11.0:1.

The ZR-1 RPO code was first used in 1970 for a rarely seen race-ready LT-1 Corvette. The code was then dusted off in 1990 for what some inside Chevrolet called the "King of the Hill." *Mike Mueller*

Valve sizes in the twin-cam heads were 1.54 inches for the two intakes, 1.39 for the exhausts. Lift was 0.39 inches on intake and exhaust. Duration was 252 degrees for the eight primary intakes, 272 degrees for their secondary counterparts, and 252 degrees for all 16 exhaust valves. Cam drive was by duplex roller chain, while sending voltage to the LT5's centrally located spark plugs was the job of a crank-controlled direct-fire system.

A two-phase electronic injection system shot the juice to those primary intake valves only during calm moments; mashing the pedal to the metal put the other intake valve in each combustion chamber to work after about 3,500 rpm or so. A key-operated switch on the console controlled these two phases. Keeping the key turned to the right on "full" allowed all of the LT5's 16 intake valves to do their stuff as designed; turning it to the left (identified as "reduced" on pilot cars, then "normal" in production) left only the less-aggressive primary intakes and their injectors in action for more economic operation. Putting the key in your pocket after leaving your ZR-1 behind in its half-power mode discouraged any extracurricular activity by the likes of, say, parking attendants—thus the "valet key" nickname.

More than one uninitiated ZR-1 driver early on failed to turn that key to the right, leaving them disappointed with performance as flat as a pancake. But those who didn't suffer brain fade found the results thrilling to say the least.

> They didn't call it King of the Hill for nothing.

"The phrase 'in any gear, at any speed' might have been coined for the LT5," began a *Motor Trend* report. "Power flows from the ZR-1 in a Niagara-like rush that makes the slick-shifting six-speed seem ridiculously redundant."

Early estimates claimed a top end of about 170 miles per hour, and a *Car and Driver* test reported 0–60 in 4.6 seconds and the quarter-mile in 12.9 clicks at 111 miles per hour. Such wild, world-class

Installation of huge 17x11 rear wheels mandated a wider tail. Mounted on those rims were truly fat P315/35ZR tires. *Mike Mueller*

CHAPTER FIVE

All ZR-1s were equipped with a six-speed manual transmission. *Mike Mueller*

performance, however, came at a price. RPO ZR1, the special performance package, alone cost a whopping $27,016, making it far and away the most expensive option in Detroit history. Throw in a 1990 sport coupe's base price of $31,979, and it became clear that not just anyone would be standing in line to be the first on their block with a King of the Hill Corvette.

Nonetheless, that line was a long one by the time the most eagerly awaited Corvette of all time appeared in showrooms in 1990. So hot was demand that more than one dealer was asking six digits right out of the box—and getting it. Most who rolled out that much dough for a '90 ZR-1 were sure they had themselves a piece of history, and indeed many of these high-priced high rollers were stashed away in climate-controlled storage quicker than you could say "museum piece."

But the boys in Bowling Green just had to rain on the planned parade toward automotive immortality. High demand led to a relatively hefty supply of ZR-1s that first year, 3,049 to be exact. To say the market was flooded is akin to calling a Corvette simply "a car." Yet, within a few years, a '90 ZR-1 could've been had for a typical used car price sometimes running lower than $30,000.

Though it failed as a collector classic, the ZR-1 never let anyone down when it came to rolling up numbers on a speedometer. Truly historic were the figures established by a 1990 ZR-1 specially prepared by Corvette Development Manager John Heinricy and veteran Corvette racer Tommy Morrison.

In March 1990, this Morrison Motorsports machine took to Firestone's 7.7-mile test track in Stockton, Texas, to challenge a 50-year-old endurance record. In 1940, Ab Jenkins' Mormon Meteor III had run for 24 hours straight on the Bonneville Salt Flats, averaging 161.180 miles per hour. For years, many challengers tried to eclipse this endurance standard, but all failed. All save for the ZR-1. After 24 hours, the LT5-powered Morrison Corvette had averaged more than 175 miles per hour for 4,200 miles—a new record. In all, the car established three new world endurance records (over varying distances) and 12 class standards.

They didn't call it King of the Hill for nothing.

1991

Changes to the ZR-1 were few during its five-year stay, and nearly all were minor, beginning with a revised valet parking key for 1991 that automatically defaulted to the low-power mode when the ignition was switched off. A second revision involved customer preferences. In 1990, only 124 ZR-1 buyers had opted to stick with the standard manually controlled air conditioner, so Chevrolet made the far-more-popular electronic climate control system standard for 1991. And like all Corvettes that year, the second-edition ZR-1 also featured a new wheel design.

1991

Model availability	coupe
Construction	integral perimeter frame birdcage body in reinforced composite, perimeter-rail frame
Wheelbase	96.2 inches
Length	176.5 inches
Width	74 inches
Height	46.7 inches
Curb weight	3,465 pounds
Tread (front/rear, in inches)	59.6/61.9
Tires	P275/40R-17, front; P315/35ZR-17, rear
Brakes	power-assisted four-wheel vented discs with aluminum calipers and ABS
Brake dimensions	12.9 inches, front; 11.9 inches, rear
Wheels	aluminum-alloy; 17x9.5 inches, front; 17x11 inches, rear
Fuel tank	20 gallons
Front suspension	independent short- and long-arm (SLA) upper and lower control arms and steering knuckles, transverse spring, stabilizer bar, tubular shock absorbers
Rear suspension	independent five-link with U-jointed half shafts and forged-aluminum control links and knuckles, camber adjustment, transverse spring, steel tie rods and stabilizer bar, tubular shocks
Steering	power-assisted rack and pinion, 15.73:1 ratio
Engine	5.7-liter all-aluminum 32-valve DOHC LT5 V-8, designed by Lotus, built by MerCruiser Induction Rochester electronic port injection
Bore and stroke	3.90x3.66 inches
Compression	11.0 1
Output	375 horsepower at 6,200 rpm, 370 ft-lbs torque at 4,500 rpm
Transmission	aluminum-case ML9 six-speed manual with CAGS
Production	2,044
RPO ZR1 price	$31,683

Various special events are held at the National Corvette Museum each year. Reunions for models like the venerable ZR-1 are common, like this one from May 1991. *Mike Mueller*

CHAPTER FIVE

Like all 1991 Corvettes, the second-edition ZR-1 was fitted with restyled wheels. ZR-1 production for 1991 was 2,044, down from 3,049 the year before. *Mike Mueller*

ZR-1 1990–1995

> "The new small block is so good it almost makes you wonder why anyone would ante up $30,000 or so for 75 rarely used extra horsepower and 40 lb-ft made by the ZR-1's unchanged LT5," wrote *Motor Trend*'s Mac Demere after driving the new LT1.

Demand dropped off considerably after 2,044 ZR-1s hit the streets that year. Accountants didn't help matters by hiking the option price to $31,683, bringing the latest bottom line to $64,138.

1992

Goodyear GS-C rubber became standard on all 1992 Corvettes, including the ZR-1, as did ASR traction control. And along with those new tires came slightly reduced spring rates front and rear. In back, 1991's quad exhaust tips were traded for two large rectangular trumpets.

Specific ZR-1 changes included appropriate emblems added to the front fenders and a temporary demotion to the options list for the formerly standard six-way power passenger seat. It became part of the basic package again in 1993.

Further dimming the ZR-1 attraction was the introduction of the new LT1 Gen II V-8. The LT1's 300 horsepower, working in concert with the overall package's lighter weight, instantly transformed the regular-issue 1992 Corvette into a machine boasting much of the ZR-1's punch at about half the cost. "The new small block is so good it almost makes you wonder why anyone would ante up $30,000 or so for 75 rarely used extra horsepower and 40 lb-ft made by the ZR-1's unchanged LT5," wrote *Motor Trend*'s Mac Demere after driving the new LT1.

RPO ZR1's asking price carried over from 1991, but an increase in the Corvette coupe's bottom line boosted the total sticker to $65,318. Production, meanwhile, dropped to 502.

1992

Model availability	coupe
Construction	integral perimeter frame birdcage body in reinforced composite, perimeter-rail frame
Wheelbase	96.2 inches
Length	178.5 inches
Width	73.1 inches
Height	46.3 inches
Curb weight	3,465 pounds
Tread (front/rear, in inches)	57.7/60.6
Tires	Goodyear GS-C; P275/40R-17, front; P315/35ZR-17, rear
Brakes	power-assisted four-wheel vented discs with aluminum calipers and ABS
Brake dimensions	12.9 inches, front; 11.9 inches, rear
Wheels	aluminum-alloy; 17x9.5 inches, front; 17x11 inches, rear
Fuel tank	20 gallons
Front suspension	independent short- and long-arm (SLA) upper and lower control arms and steering knuckles, transverse spring, stabilizer bar, tubular shock absorbers
Rear suspension	independent five-link with U-jointed half shafts and forged aluminum control links and knuckles, camber adjustment, transverse spring, steel tie rods and stabilizer bar, tubular shocks
Steering	power-assisted rack and pinion, 15.73:1 ratio
Engine	5.7-liter all-aluminum 32-valve DOHC LT5 V-8, designed by Lotus, built by MerCruiser (four-bolt main bearing caps replace previous two-bolt mains)
Induction	Rochester electronic port injection
Bore and stroke	3.90x3.66 inches
Compression	11.0:1
Output	375 horsepower at 6,200 rpm, 370 ft-lbs torque at 4,500 rpm
Transmission	aluminum-case ML9 six-speed manual with CAGS
Production	502
RPO ZR1 price	$31,683

Specific changes from the 1991 ZR-1 included appropriate emblems added to the front fenders and a temporary demotion to the options list for the formerly standard six-way power passenger seat, which would return in 1993.

CHAPTER FIVE

1993

During 1990 and 1991, the ZR-1 stood tall as this country's most powerful production car. Then along came Dodge's Viper with its 400- horsepower V-10 in 1992. No problem: Lotus engineers simply went back to their drawing boards and found 30 more ponies for the LT5, making the ZR-1 the king of Detroit's high-performance hill once again in 1993.

This jump up to 405 horsepower came about more or less by massaging the heads for improved flow. And with the aluminum engine making so much more power (torque increased, too, to 385 ft-lbs), more beef was designed in as pistons were strengthened, and the earlier block's two-bolt main bearing caps were replaced by tougher four-bolt pieces. Also new were a Mobil 1 synthetic oil requirement, platinum-tipped spark plugs, and an electrical linear exhaust gas recirculation (EGR) system that reduced emissions.

> The ZR-1's days became numbered after November 23, 1993, when Mercury Marine ceased production of LT5 V-8s.

1993

Model availability	standard coupe or 40th Anniversary (RPO Z25) coupe
Construction	integral perimeter frame birdcage body in reinforced composite, perimeter-rail frame
Wheelbase	96.2 inches
Length	178.5 inches
Width	73.1 inches
Height	46.3 inches
Curb weight	3,465 pounds
Tread (front/rear, in inches)	57.7/60.6
Tires	Goodyear GS-C; P275/40R-17, front; P315/35ZR-17, rear
Brakes	power-assisted four-wheel vented discs with aluminum calipers and ABS
Brake dimensions	12.9 inches, front; 11.9 inches, rear
Wheels	aluminum-alloy; 17x9.5 inches, front; 17x11 inches, rear
Fuel tank	20 gallons
Front suspension	independent short- and long-arm (SLA) upper and lower control arms and steering knuckles, transverse spring, stabilizer bar, tubular shock absorbers
Rear suspension	independent five-link with U-jointed half shafts and forged-aluminum control links and knuckles, camber adjustment, transverse spring, steel tie rods and stabilizer bar, tubular shocks
Steering	power-assisted rack and pinion, 15.73:1 ratio
Engine	5.7-liter all-aluminum 32-valve DOHC LT5 V-8, designed by Lotus, built by MerCruiser
Induction	Rochester electronic port injection
Bore and stroke	3.90x3.66 inches
Compression	11.0:1
Output	405 horsepower at 5,800 rpm, 385 ft-lbs torque at 4,800 rpm
Transmission	aluminum-case ML9 six-speed manual with CAGS
Production	448 (245 Z25 anniversary models)
RPO ZR1 price	$31,683

After falling behind the 400-horsepower Viper in Detroit's horsepower race in 1992, the 405-horse 1993 ZR-1 regained the title of America's most powerful production car. *Mike Mueller*

200

ZR-1 1990–1995

Above: The 40th Anniversary package, RPO Z25, was available for all Corvettes in 1993, including the ZR-1 coupe. Of the 448 1993 ZR-1s built, 245 featured the Z25 package and its Ruby Red paint. *Mike Mueller*

Far left: Exclusive fender badges identified a 40th Anniversary ZR-1 coupe in 1993. *Mike Mueller*

Left: Special embroidery on the seats was included on all Z25 models in 1993. *Mike Mueller*

201

CHAPTER FIVE

Remaining ZR-1 standard features rolled over from 1992. But prominently new on the options list was RPO Z25, the same 40th Anniversary package offered in 1993 for LT1 Corvette coupes and convertibles. Of the 448 ZR-1s built that year, 245 were Ruby Red Metallic Z25 models, which added another $1,455 to the already steep base price of $66,278, which included the same amount charged for RPO ZR1 during the two previous years

The ZR-1's days became numbered after November 23, 1993, when Mercury Marine ceased production of LT5 V-8s. Remaining supplies were shipped to Bowling Green to fulfill predetermined production runs planned for 1994 and 1995. As in 1993, the ZR-1 totals for those last two years numbered 448 cars each.

1994

While the handwriting was clearly on the wall, at least the ZR-1 was treated to exclusive five-spoke wheels in 1994. New, too, was the passenger-side air bag added to all 1994 Corvettes. RPO ZR1's cost this time decreased, to $31,258, but a typical increase in the coupe's base price again translated into a total rise to $67,443.

> From start to finish, the ZR-1 coupe was a fully loaded machine.

1994

Model availability	coupe
Construction	integral perimeter frame birdcage body in reinforced composite, perimeter-rail frame
Wheelbase	96.2 inches
Length	178.5 inches
Width	73.1 inches
Height	46.3 inches
Curb weight	3,465 pounds
Tread (front/rear, in inches)	57.7/60.6
Tires	Goodyear GS-C; P275/40R-17, front; P315/35ZR-17 rear
Brakes	power-assisted four-wheel vented discs with aluminum calipers and ABS
Brake dimensions	12.9 inches, front; 11.9 inches, rear
Wheels	aluminum-alloy five-spoke (exclusive to ZR-1); 17x9.5 inches, front; 17x11 inches, rear
Fuel tank	20 gallons
Front suspension	independent short- and long-arm (SLA) upper and lower control arms and steering knuckles, transverse spring, stabilizer bar, tubular shock absorbers
Rear suspension	independent five-link with U-jointed half shafts and forged-aluminum control links and knuckles, camber adjustment, transverse spring, steel tie rods and stabilizer bar, tubular shocks
Steering	power-assisted rack and pinion, 15.73:1 ratio
Engine	5.7-liter all-aluminum 32-valve DOHC LT5 V-8, designed by Lotus, built by MerCruiser
Induction	Rochester electronic port injection
Bore and stroke	3.90x3.66 inches
Compression	11.0:1
Output	405 horsepower at 5,800 rpm, 385 ft-lbs torque at 4,800 rpm
Transmission	aluminum-case ML9 six-speed manual with CAGS
Production	448

Little changed for the 1994 ZR-1, and its short, happy life was soon to end on April 28, 1995. The production tally for that final year was again 448. *Mike Mueller*

ZR-1 1990–1995

The same exclusive wheels rolled over into 1995 for the last ZR-1 Corvette. *Mike Mueller*

1995

Model availability	coupe
Construction	integral perimeter frame birdcage body in reinforced composite, perimeter-rail frame
Wheelbase	96.2 inches
Length	178.5 inches
Width	73.1 inches
Height	46.3 inches
Curb weight	3,465 pounds
Tread (front/rear, in inches)	57.7/60.6
Tires	Goodyear GS-C; P275/40R-17, front; P315/35ZR-17, rear
Brakes	power-assisted four-wheel vented discs with aluminum calipers and ABS
Brake dimensions	12.9 inches, front; 11.9 inches, rear
Wheels	aluminum-alloy five-spoke (exclusive to ZR-1); 17x9.5 inches, front; 17x11 inches, rear
Fuel tank	20 gallons
Front suspension	independent short- and long-arm (SLA) upper and lower control arms and steering knuckles, transverse spring, stabilizer bar, tubular shock absorbers
Rear suspension	independent five-link with U-jointed half shafts and forged-aluminum control links and knuckles, camber adjustment, transverse spring, steel tie rods and stabilizer bar, tubular shocks
Steering	power-assisted rack and pinion, 15.73:1 ratio
Engine	5.7-liter all-aluminum 32-valve DOHC LT5 V-8, designed by Lotus, built by MerCruiser
Induction	Rochester electronic port injection
Bore and stroke	3.90x3.66 inches
Compression	11.0:1
Output	405 horsepower at 5,800 rpm, 385 ft-lbs torque at 4,800 rpm
Transmission	aluminum-case ML9 six-speed manual with CAGS
Production	448
RPO ZR1 price	$31,258

1995

At 1:12 P.M. on Friday, April 28, 1995, Chevrolet's Bowling Green assembly line came to a halt to honor the passing of a legend. Even though it actually had been built the previous Monday, the 6,939th and last ZR-1 Corvette was paraded off the line that afternoon with plant manager Wil Cooksey and UAW member Billy Jackson on board. Chevrolet General Manager Jim Perkins, with checkered flag in hand, watched over the roll-off, then took the wheel of the Torch Red coupe and drove it across the street to the National Corvette Museum with retired Chief Engineer Dave McLellan along for the ride. Even further closure came that evening at a farewell dinner when Roy Midgley announced his own retirement.

One main reason behind the ZR-1's short but happy life was fairly obvious: the base LT1 Corvette had taken a serious bite out of its big brother's customer base. But Perkins also mentioned a slumping supercar market and the prohibitive costs required to make the ZR-1 meet 1996's tightened emissions standards.

Midgley had visited Lotus in September 1991 to inform his British counterparts that 1995 would be the ZR-1's final year, this after explaining that updating the car's emissions aspects would cost more than $1 million. Reportedly, Lotus engineers had begun work in 1991 on an improved LT5 that would've made between 450 and 475 horsepower, all for naught.

Further nails went into the coffin in 1992 when the car's strongest supporter, Lloyd Reuss, was removed from the GM president's seat given to him by Board Chairman Robert Stempel two years before. Replacing Reuss was John "Jack" Smith,

CHAPTER FIVE

Chevrolet officials knew as early as September 1991 that 1995 would be the ZR-1's final year. The high cost of keeping the LT5 V-8 emissions legal for 1996 was the main reason given for the cancellation. *Mike Mueller*

ZR-1 1990–1995

Above: The 1995 LT5 looked no different than its 1990 forerunner, but beneath all that aluminum were 405 horses instead of 375. Notice the yellow test vehicle sticker on the radiator. *Mike Mueller*

who contrarily found no soft spot in his heart for such a costly niche-market product as the ZR-1.

And so the axe fell on the killer Corvette that cost $68,603 that last year. Whether it was a good deal or not was argued right up to the end, as some critics still couldn't put the supreme C4 down.

"Forget about the detractors or the sales numbers," began an October 17, 1994, *AutoWeek* report on the vehicle's downhill slide. "With rocket-sled acceleration, mastiff grip, and right-now brakes, the ZR-1 has the largest performance envelope of any mass-produced American car ever built. It also has the amenities and acceleration of a modern passenger car and none of the temperament of a race car. With a slight tailwind, it will go 180 miles an hour, but it will also trudge through gridlock without complaint, A/C on kill and CD player on loud. We may never see the likes of it again."

Long live the King.

Long live the King.

CHAPTER SIX

06 50 Years Young

C5 1997–2004

It's highly unlikely that any future Corvette generation will manage to stay on the scene as long as the C3 did. The C4 lineage came close, falling short by two model runs, but only because it was forced to fill in for a couple more years than planned while the radically redesigned fifth-generation Corvette remained stuck in the works. Like the first Shark in 1968 and Dave McLellan's pride and joy in 1984, the C5 Corvette took its sweet time going from drawing board to the street. Make that seriously sweet.

- Chevrolet's Gen III small-block V-8 introduced along with C5 Corvette (1997)
- Convertible reintroduced to lineup after one-year hiatus (1998)
- Corvette trunk (last seen in 1962) reintroduced with C5 convertible (1998)
- Corvette paces the Indianapolis 500 (1998)
- First time three body styles (coupe, convertible, hardtop) offered (1999)
- C5-R racing Corvette introduced (1999)
- Standard LS1 V-8 output boosted from 345 horsepower to 350 (2001)
- Z06 (powered by new LS6 V-8) introduced (2001)
- LS6 V-8 output goes from 385 horsepower to 405 (2002)
- Z06 becomes first Corvette (discounting ZR-1) with a base price beyond $50,000 (2002)
- A 2003 50th Anniversary coupe paces the Indianapolis 500 (2002)
- Commemorative Edition package introduced for all Corvette models to honor C5-R racing program (2004)

Above: At $45,900, the convertible was the most expensive model in 2000. Base prices for the coupe and hardtop were $39,475 and $38,900, respectively.

Opposite: The 2002 base coupe differed little from its predecessors.

DATING BACK TO 1988, C5 development initially called for an all-new Corvette to debut for 1993 to help mark the two-seat breed's 40th anniversary. But various cash crunches and delayed decisions pushed the introductory date back repeatedly. In May 1989, the unveiling was rescheduled for 1994 after the engineering budget for the year was slashed, and then the goal changed again to 1995 three months later. In October 1990, the coming out was reslated for 1996, and by October 1992, it seemed that no one at GM could predict when the C5 really would show.

The prime motivating factor behind the C5's long ride to market involved the severe financial difficulties GM experienced in the early 1990s. Awash in cash in 1988, Detroit's corporate giant found itself losing a record $2 billion two years later, and that was followed by a staggering $24.2 billion worth of red ink in 1992.

Muddying the waters further was the destabilization of corporate leadership that occurred after Chairman of the Board Roger Smith reached mandatory retirement in August 1990. New Chairman Robert Stempel inherited a financial nightmare, and then complicated matters by filling his old president's seat with Lloyd Reuss, a man other board members felt wasn't right for the job. They weren't mistaken. Reuss was shown the door within two years, replaced by Jack Smith, who was eventually responsible for finally giving the C5 project a green light.

The C5 design team also experienced lineup changes during that time. Dave McLellan retired in 1992 and was replaced that November by former Cadillac engineer David Hill. Stepping down at the same time was the man who had tight-fistedly controlled the C5's early styling developments, GM Design Vice President Charles Jordan. His replacement, Wayne Cherry, was chosen over Jerry Palmer, head of Chevy 3 (the Corvette's studio home since 1974) and the man responsible for the C4 restyle. But while Cherry was officially made design department head in September 1992, his responsibilities involved more business than art—it was Palmer who actually remained in charge of GM studios and thus directly influenced the designs drawn up within. Palmer also oversaw John Cafaro,

207

CHAPTER SIX

David Hill, shown here with the C5 coupe at the National Corvette Museum in 1997, became the Corvette's third chief engineer in 1992. He retired in 2006. *Mike Mueller*

First considered for the Q-Corvette in 1957, a rear-mounted transmission helped balance the C5's weight on all four wheels. Notice the transverse mufflers behind the rear wheels in the upper illustration.

the designer who ended up riding herd over C5 styling during its evolution.

Jordan first instructed Cafaro to begin sketching a C5 image in a Chevy 3 studio basement in August 1988. But Cafaro's earliest efforts didn't do it for the "Chrome Cobra," as Jordan was known around design department halls. GM's silver-haired exec turned to John Schinella's ACC in California for other options, leading to ACC's creation of its sexy Sting Ray III in 1990. Jordan also ordered Tom Peters in GM's Advanced 4 studio to make it a three-way design contest, all this without informing Cafaro, who, needless to say, wasn't sorry to see Cherry take over in 1992.

In April 1991, Jordan gave his blessing to one of Peters' designs but also directed all involved to blend in many of Cafaro's ideas. Almost a year later, Cafaro came back with his stunning "Black Car," which was based on the C4's birdcage chassis but looked completely new and exciting on the top side. Chevrolet General Manager Jim Perkins was

208

C5 1997–2004

A C5 prototype awaits a display position outside the National Corvette Museum in 2001. *Mike Mueller*

> "If, as they say, God is in the details, then this is the first holy Corvette," gushed *Car and Driver*'s Csaba Csere in reverent honor of the first C5.

Full frontal imagery was a hotly debated topic throughout the C5 design process.

a big fan of the Black Car, which indeed predicted much of the eventual C5 look.

Perkins also was responsible for shepherding C5 chassis development through GM's economic quagmire, first creatively charging early work to the existing C4 budget. An official C5 development budget still wasn't in place late in 1992 when Perkins managed to siphon off $1 million for a running test mule, the CERV IV. Unlike its three exotic forerunners, CERV IV was created using as many actually proposed production features as possible. Though its body was a bastardized C4 shell, underneath was a backbone chassis incorporating a rear-mounted transmission—both ideas that were C5 goals from the outset. An all-new aluminum engine, the proposed Gen III small-block V-8, wasn't ready yet, so power came from another C4 carryover.

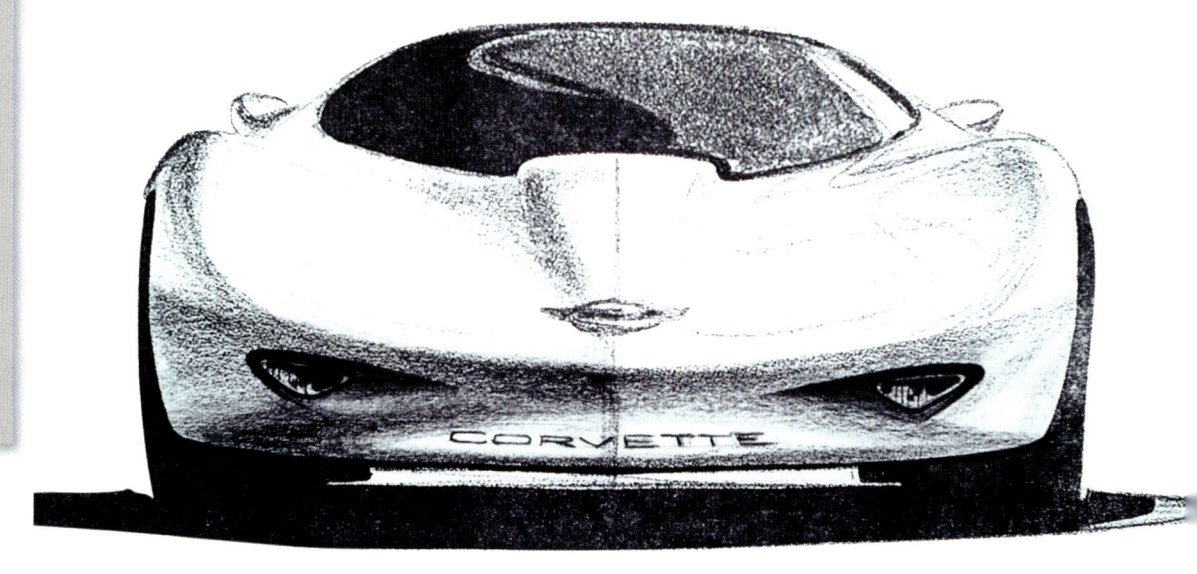

Perkins' rolling proposal was clandestinely completed in May 1993, and it was followed by another in January 1994. Called CERV IVb (making its forerunner CERV IVa), the second machine was fitted with an iron-block Gen III V-8 as an aluminum cylinder block had yet to make the grade. Both CERV IV models were followed by more-advanced test cars called "alpha" and "beta" machines. Alphas wore true C5 bodies, and betas demonstrated various improvements dictated by alpha testing.

GM President Jack Smith gave his approval to the project in June 1993. Soon afterward, David Hill decided to concentrate development work on a Targa-top coupe model even though the superstrong C5 chassis was designed with convertible applications first in mind. But little work on a topless body had been completed by August 1993, and if Hill and crew were going to have a new fifth-generation Corvette ready for the latest deadline, 1997, they needed to complete the coupe already well on its way. So it was that a convertible C5 followed its closed running mate one year late, and that duo was then joined by another variety—a hardtop with a trunk—in 1999, making

CHAPTER SIX

C5 test cars were labeled "alpha" and "beta." About 30 betas were built, all painted white, after initial bugs were ironed out in alpha tests. Beta tests included enduring heat in the Australian outback and surviving northern Canada winters.

it the first time in Corvette history that buyers could choose from three different body styles.

When finally introduced at Detroit's North American International Auto Show on January 6, 1997, the first C5 stood as Chevrolet's first truly all-new Corvette. Previous next-generation ground breakers in 1963, 1968, and 1984 had brought along at least a little something from their pasts, with carryover powertrains showing up most prominently. Even those Polo White two-seaters in 1953 incorporated many components already familiar to Chevrolet passenger-car buyers.

Next to nothing old or borrowed showed up on the 1997 Corvette coupe, and total components were cut by a third compared to the 1996 model, meaning the C4 had an amazing 1,500 more parts than the C5. This reduction came about as part of a plan to minimize the shakes, rattles, and rolls that inspired common complaints during the C4 era. Fewer pieces also simplified production and maintenance. As Corvette Quality Engineering Manager Rod Michaelson explained, "The 1,500 parts eliminated equates to 1,500 opportunities for something to go wrong that aren't there any more."

As for the remaining components that made up the C5, they inspired nothing but raves wherever they went.

1997

"If, as they say, God is in the details, then this is the first holy Corvette," gushed *Car and Driver*'s Csaba Csere in reverent honor of the first C5. "The '97 Corvette is a home run in every way," added *Automobile*'s David E. Davis Jr. "Like no Corvette before, [the C5] now possesses the sort of smoothness and refinement that, if it were a scotch, could only be attributed to decades of little lessons learned about distillation and years of quiet aging," concluded a *Road & Track* review.

1997

Model availability	coupe
Construction	fiberglass body on steel frame with central backbone frame and hydroformed perimeter rails; transmission located at rear axle
Wheelbase	104.5 inches
Length	179.7 inches
Width	73.6 inches
Height	47.7 inches
Curb weight	3,229 pounds (automatic transmission), 3,218 pounds (manual)
Tread (front/rear, in inches)	62/62.1
Tires	Goodyear Extended Mobility; P245/45ZR-17, front; P275/40ZR-18, rear
Brakes	power-assisted four-wheel discs with Bosch ABS
Brake dimensions	11.9 inches, front and rear
Wheels	cast aluminum; 17x8.5 inches, front; 18x9.5 inches, rear
Fuel tank	19.1 gallons
Front suspension	short- and long-arm double wishbone (forged aluminum, top; cast aluminum, bottom), transverse leaf spring, stabilizer bar, monotube shock absorbers
Rear suspension	short- and long-arm double wishbone (cast-aluminum control arms, top and bottom), transverse leaf spring, stabilizer bar, monotube shock absorbers
Steering	speed-sensitive, power-assisted rack and pinion
Engine	5.7-liter OHV V-8 (LS1) with aluminum cylinder block and heads, sequential fuel injection
Bore and stroke	3.90x3.62 inches
Output	345 horsepower at 5,600 rpm; 350 ft-lbs at 4,400 rpm
Transmission	four-speed 4L60-E automatic transmission
Optional transmission	six-speed manual

The early C5 clay model looked too much like a Camaro at the tail.

Chevrolet people, too, weren't ashamed about bragging a little—or a lot. Describing the C5 as "the best 'Vette yet," David Hill claimed, "You won't find a car in Corvette's price range that provides the same level of quality, power, ride, handling, and refinement." Indeed, offering nearly 170 miles per hour for about $40,000, the 1997 Corvette ranked right up there with the fastest street-legal production cars for the dollar ever built in America. But sizzling speed wasn't the car's sole attraction.

"We designed the [C5] with a synchronous mindset," added Interior Designer Jon Albert. "We focused on individual goals, such as improved performance, reduced mass, and increased reliability, within the overall framework of the whole car. We evaluated and balanced each change so as to optimize the total car."

So many changes worked together to create that total car. An innovative frame with a rigid center tunnel and hydroformed perimeter rails made the C5's foundation 4.5 times stiffer than the C4's. At the time a new process, hydroforming uses extreme water pressure to literally blow up round steel tubes into desired shapes to exacting specifications. The various welding operations required to build C4 foundations always produced tiny differences from section to section, from frame to frame, meaning you couldn't quite count on suspension precision since overall consistent physical geometry couldn't be guaranteed.

Along with improving both handling and ride, the battleship-strong hydroformed chassis also did away with many of the squeaky gremlins inherent in earlier Corvettes. The C5's rear-mounted transmission improved ride and handling, too. This long-discussed idea not only helped bring weight distribution closer to the preferred 50/50 balance, it also freed up space beneath the passenger compartment, meaning both occupants had more room down there in the footwells to stretch out and ride comfortably. Entry and exit was enhanced as well, thanks to those strengthened frame rails, which traded excess mass for a lower sill height, down 3.7 inches.

Don't forget the new 345-horse Gen III all-aluminum small-block V-8, better brakes and tires, and a more sophisticated suspension. Clearly, no stones were left unturned during the C5 design process.

"The fifth-generation Corvette is a refined Corvette, in all the right ways," boasted Chief Engineer David Hill in 1997. "It's more user-friendly, it's easier to get in and out of, and it's more ergonomic. It has greater visibility; it's more comfortable and more functional. It provides more sports car for the money than anything in its market segment. It'll pull nearly 1 g, and it starts and stops quicker than you can blink."

Built at GM's engine plant in Romulus, Michigan, the C5's LS1 V-8 at the same time honored previous small blocks and left the past behind.

The first C5, similar to the 1997 coupe on the right, rolled off the Bowling Green assembly line on October 1, 1996. *Mike Mueller*

The only things the LS1 and LT1 V-8s shared were numbers: both featured the traditional 4.40-inch bore center measurement, and both also displaced 5.7 liters. LS1 output in 1997 was 345 horsepower.

CHAPTER SIX

"Based on a timeless design by former Chief Engineer Ed Cole, the 'Gen III' marks a bright new chapter in the highly respected lineage that GM small blocks have established in more than 40 years," claimed a Chevrolet press release in 1996.

Like all small blocks before it, the LS1 was a traditional pushrod, 16-valve V-8, and it also shared the same time-honored "440" cylinder block layout (measuring 4.40 inches from bore center to bore center) and the familiar 5.7-liter displacement label. "After all," said LS1 Engine Program Manager John Juriga, "some things are sacred." Next to nothing carried over from there, though, as the Gen III represented a real redesign, not just a modernization.

First off, the LS1 relied on revised bore/stroke parameters. Compared to the Gen II, the Gen III V-8's bore (3.90 inches) was less and its stroke (3.62 inches) was more, this adjustment made to allow more cooling space between skinnier cylinders. The Gen II's reverse-flow cooling system was traded for a conventional layout as Gen III coolant was once again pumped into the block first, then to the heads. But easily the most notable innovation was the LS1's lightweight all-aluminum construction, a first for a regular-production Chevy small block.

A rigid central tunnel served as a base for the C5 frame. Its perimeter rails were created with water pressure in a process known as "hydroforming."

> "The fifth-generation Corvette is a refined Corvette, in all the right ways," boasted Chief Engineer David Hill in 1997.

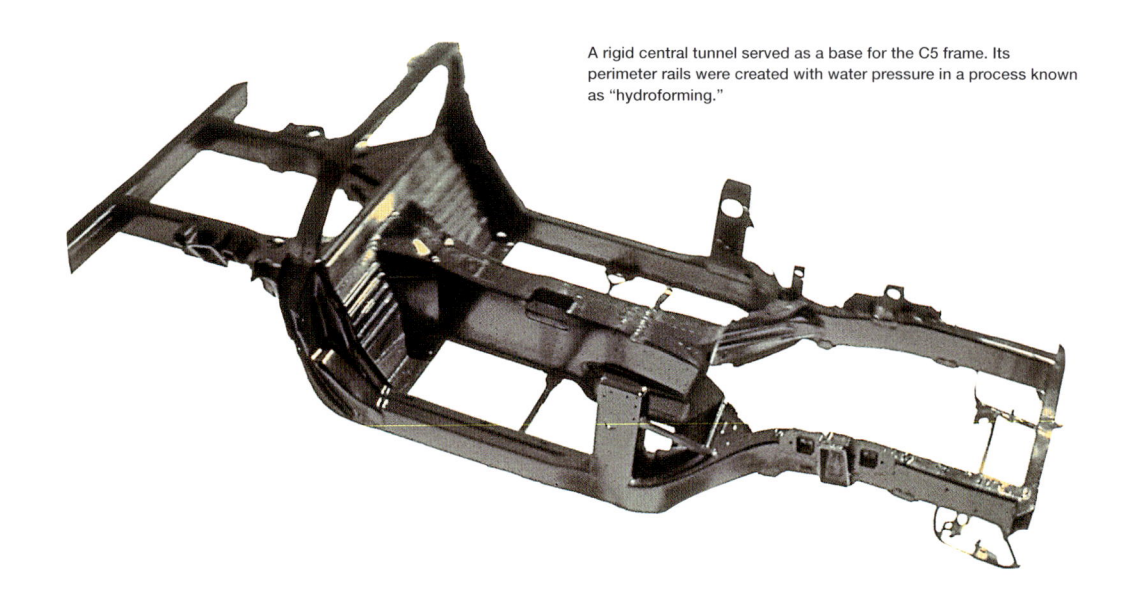

C5 Corvette production started up at the Bowling Green plant late in 1996.

Gen III development dated back to late 1991. Anil Kulkarni, the man behind the Gen II LT1, was originally in charge of this project, but he stepped aside, giving up control to long-time small-block engineer Ed Koerner. The Gen III program really got cooking in 1993, with testing in early C5 prototypes beginning the following summer using mostly iron-block Gen III pre-runners. Most prototypes were refitted with all-aluminum counterparts by late 1995. Chevrolet then officially introduced the Gen III V-8 to the press in June 1996.

Everything about the LS1 was finely engineered to the limit, beginning with the cast-aluminum cylinder block with its centrifugally cast gray-iron cylinder liners. At only 107 pounds, this aluminum block weighed 53 pounds less than its iron Gen II predecessor, and the entire engine was 88 pounds lighter. Yet at the same time, the LS1 was much stronger, thanks to extensive external stiffening ribs and its deep-skirt construction. Unlike typical V-8 cylinder blocks that end at the crankshaft's centerline, the LS1 block extended below the main bearing caps, encasing the crank in a girdle of aluminum. This extended skirt also made it possible to cross-bolt the main bearing caps for additional rigidity. LS1 bearing caps features six bolts: four in the conventional vertical location on each side of the crank and one each running through the skirt horizontally into each side of the cap.

Cutting-edge cylinder head design was one of the main keys to the LS1's success. Flow wizard Ron Sperry, who had joined GM Powertrain in late 1987 and had produced the LT1 and LT4 heads, was the man responsible for the LS1 heads, which incorporated replicated ports. Previous small-block ports were located in two closely squeezed siamesed pairs on the intake side, resulting in widely varying internal structures as those passages bent and turned differently with differing volumes, producing varying flow characteristics in the process. Keeping flow rates constant from cylinder to cylinder is vital to maximizing performance, and that's exactly what the replicated-port design accomplished.

Additional features included roller rocker arms and a roller-lifter cam. Intake valves measured 2.00 inches, exhaust valves 1.55. Compression was 10.2:1. While the LS1's sequential electronic port fuel injection was nothing new, its drive-by-wire electronic throttle control (ETC) was, at least as far as GM gasoline engines were concerned.

The LS1 got its spark from a distributorless ignition system featuring one coil per cylinder. Atop

Comfort and convenience issues weren't overlooked by C5 designers. Getting behind the wheel (and out again) was easier in 1997, due primarily to lowered doorsills.

every cylinder, mounted beneath a stylish plastic shield on each valve cover, were eight individual coils and coil-driver assemblies. Once the spark plugs did their job, spent gases were hauled off by hydroformed tubular exhaust manifolds. Bringing up the LS1's bottom was a bat wing oil pan, a shallow unit created to allow ample road clearance.

An electronic four-speed automatic transmission was standard behind the LS1, with a Borg-Warner six-speed manual available on the C5 options list. Optional carryovers from 1996 included electronic selective real-time damping (RPO F45) and the Z51 performance handling package.

1998

Although an optional removable hardtop was available beginning in 1956, all Corvettes built before 1963 were convertibles. Then along came the stunning Sting Ray and its sexy coupe shell. Faced with a choice between hardtop and convertible, most Corvette buyers still favored the open-air style until 1969, after which time closed-body sales finally took over. Chevrolet product planners eventually gave up on the topless Corvette after 1975.

Although a drop-top model did return in 1986, convertible popularity has never been as strong as it was during the early Sting Ray years. Open-air C4s numbered only 4,369 in 1996, making up about 20 percent of the total run. But such low numbers didn't deter David Hill and the gang in the least; they knew that a Corvette convertible had to make a second comeback.

The C5 convertible showed up in 1998, just in time to help the latest Corvette cop *Motor Trend*'s coveted Car of the Year trophy. "Like the coupe, this is the best 'Vette yet," bragged Hill again. Calling the new convertible "the most desirable 'Vette since 1967," *Car and Driver*'s Csaba Csere announced that "not even the glorious ZR-1 models had as many of our staffers muttering about owning a Corvette as does this new roadster."

New was the key word. No other convertible in Corvette history could stand up to the 1998 rendition in the way it stood up to the real-world realities of skin-to-the-wind touring, this due to the effort Hill's engineers put into making the C5 chassis as rigid as hell, with or without a roof. "It was critical that we didn't just take the coupe and chop off the top to make a convertible," said Hill. "Corvette's structure has been designed to achieve world-class open-car stability and strength."

Sawing the roof off a Corvette, off any car, has always represented a compromise, as a coupe's

CHAPTER SIX

1998

Model availability	coupe and convertible
Construction	fiberglass body on steel frame with central backbone frame and hydroformed perimeter rails; transmission located at rear axle
Wheelbase	104.5 inches
Length	179.7 inches
Width	73.6 inches
Height	47.7 inches
Curb weight	3,245 pounds (coupe), 3,246 pounds (convertible)
Tread (front/rear, in inches)	62/62.1
Tires	Goodyear Extended Mobility; P245/45ZR-17, front; P275/40ZR-18, rear
Brakes	power-assisted four-wheel discs with Bosch ABS
Brake dimensions	11.9 inches, front and rear
Wheels	cast aluminum; 17x8.5 inches, front; 18x9.5 inches, rear
Fuel tank	19.1 gallons
Front suspension	short- and long-arm double wishbone (forged aluminum, top; cast aluminum, bottom), transverse leaf spring, stabilizer bar, monotube shock absorbers
Rear suspension	short- and long-arm double wishbone (cast-aluminum control arms, top and bottom), transverse leaf spring, stabilizer bar, monotube shock absorbers
Steering	speed-sensitive, power-assisted rack and pinion
Engine	5.7-liter OHV V-8 (LS1) with aluminum cylinder block and heads, sequential fuel injection
Bore and stroke	3.90x3.62 inches
Output	345 horsepower at 5,600 rpm; 350 ft-lbs at 4,400 rpm
Transmission	four-speed 4L60-E automatic transmission
Optional transmission	six-speed manual

The 1998 C5 convertible took *Motor Trend's* Car of the Year Award for 1998. The base price was $44,425, and production topped out at 11,849. *Mike Mueller*

top has always played a major role in the structural scheme of things. Removing that structure has always meant additional bracing had to go back in, most commonly in the cowl area. Such bracing helps, but no convertible has ever measured up as solidly as its full-roofed counterparts. Cowl shakes, steering wheel vibrations, and general roughness, rattles, and rumbles have been standard Corvette convertible features from the get-go.

On top of all that, extra bracing had always added more than a few unwanted pounds. And due to this weight handicap, drop-top muscle machines have all been slower than their full-roofed alter egos. Inherent realities of convertible ownership have always included less performance and more annoyance. Not so after 1998.

Beneath that smooth skin was a skeleton that simply refused to bend or flex. Cowl shake and other common convertible maladies were virtually eliminated—without a single extra brace or frame member, or any additional weight. The C5 convertible, at 3,246 pounds, amazingly weighed only 1 pound more than its coupe running mate. It was also 114 pounds lighter than its C4 convertible forerunner.

According to engineers, the new C5 convertible at most only measured 10 percent less torsionally rigid compared to the coupe with its Targa top on and latched down. The topless C5 was so rigid it could be equipped with the tough Z51 suspension package, an option that wasn't offered to C4 convertible buyers. The less-solid C4 frame and hard-as-nails Z51 handling equipment just didn't mix.

Neither did the C5 and a roll bar. A rollover hoop was considered for the '98 convertible, but, according to Hill, observed customer indifference and design complications ruled it out of the C5 plan.

Also ruled out was a power-operated top. Like the C4, the C5's folding roof required two pairs of hands, not just one finger, to put it in its place. As Hill explained, the reasoning behind the choice was simple: keep costs down and available storage space up. As it was, the C5's top folded up so easily it was almost a shame to complain about the power outage.

Simple, lightweight operation, along with a tight seal when up and an equally tight fit when down, were targets of Chevrolet designers. They penciled out what they wanted and then turned to Dura Convertible Systems, Inc., to make it reality. The result was an articulating five-bow pressurized top, pressurized in that when it unfolded, a special linkage in the rearmost fifth bow both squeezed itself down tight against the body and pushed the first four bows upward toward the windshield header, where traditional latches made the seal. Pressure in back was so strong no latching mechanism was required.

The new top also proved exceptionally aerodynamic. The '98 C5 drop top's drag coefficient, at 0.32, was only 0.03 higher than the

Unlike previous topless Corvettes, the C5 convertible didn't require excess reinforcement to stay strong without a roof in place. Based on a chassis that was 450 percent stiffer than its C4 predecessor, the 1998 convertible weighed only 1 pound more than its coupe counterpart.

Below: Chevy officials loved to brag about the 1998 C5 convertible's ability to carry two golf bags in its trunk. Storage space was 11.2 cubic feet with the top down, 13.9 with the roof unfolded overhead.

wind-cheating coupe at a cost of only 0.2 inches of headroom compared to the full-roofed C5.

When down, the C5 top stowed beneath a nicely styled tonneau, reminiscent of high-flying Corvette racing days gone by. More important, the tonneau's double-hump headrest fairings allowed designers that much more room below to hide the top. Once folded, the compact C5 top rested in one helluva tight area, all the better to preserve precious storage space in the new convertible's trunk.

Thirty-six years after a Corvette last brought up its rear with real cargo space, the '98 C5 convertible appeared with a deck lid. The C5's dual-compartment gas tank, deletion of a spare tire (by making extended-mobility run-flat tires standard), and the compact, nonpower aspects of that folding top all worked in concert to make a trunk possible. The C5 trunk measured 11.2 cubic feet in top-down mode, 13.9 with roof unfolded and in place. Try to imagine two bags of golf clubs nestled in there—that's the suitable analogy Chevrolet's promotional people used to push the point of just how spacious that trunk was.

Among other standard convertible features in 1998 was neat stuff like speed-sensitive steering, the run-flats' tire-pressure monitoring system, a monstrous Bose stereo system, and a driver's power seat. A new option available for the '98 Corvette convertible or coupe was RPO JL4, the active-handling chassis control system. Active handling relied on a system of sensors to read steering inputs, yaw rates, and lateral g forces to better stabilize the car in emergency situations by selectively activating either the ABS or traction control gear. And it could've been modulated to work without rear-wheel oversteer control for more experienced drivers able to feel their way through the twisties better than the rest of us mere mortals. Called "Bondurant-in-a-box" (in reference to driving school guru and veteran Corvette racer Bob Bondurant) by *Sports Car International*, the well-received active-handling equipment was priced at $500. JL4 installations in 1998 numbered 5,356.

A second new option made the $1,695 selective real-time damping package look like a bargain. Originally designed for 1997 export models, lightweight magnesium wheels, supplied

CHAPTER SIX

Corvette Brand Manager Dick Almond. "Part of our potential customer base really wants a simpler, more elemental, yet high-performance machine," he said in August 1998. "The new hardtop is the ultimate hot sports car, yet it will carry the lowest base price in the Corvette family. Those factors combined should make the consumer appeal for Corvette even greater."

Like the convertible, this hot new hardtop featured a trunk, basically because it was more or less a drop-top model with a roof bolted and molded in place of its much softer, folding counterpart. And, along with being the cheapest of the three 1999 body styles, it was also the lightest, weighing about 80 pounds less than the coupe. At the same time, it was the stiffest of the trio, measuring some 12 percent more rigid than the coupe with its Targa top latched in place. As for performance, according to *Motor Trend*, the slightly lightened 1999 hardtop could run 0–60 in 4.8 seconds. Quarter-mile performance was 13.3 clicks at 108.6 miles per hour.

Plans early on called for cutting costs even further by, among other things, making smaller wheels and tires, manual door locks, and cloth seats standard. Such frugal ideas fortunately fell from grace, allowing black leather seats to

A 1998 convertible became the fourth Corvette to pace the Indianapolis 500. Chevrolet sold 1,163 Indy pace car replicas that year.

> "The new hardtop is the ultimate hot sports car, yet it will carry the lowest base price in the Corvette family. Those factors combined should make the consumer appeal for Corvette even greater."

by Speedline in Italy, appeared in 1998 wearing a wallet-wilting price tag of $3,000. These bronze-toned wheels each weighed 8 pounds less than the stock C5 rim. Whether or not those weight savings were worth the cost was up to the individual owner—2,029 apparently thought so in 1998.

1999

Originally conceived as a low-buck alternative to its flashier Targa-top and convertible counterparts, a C5 hardtop debuted for 1999 priced $400 less than that year's sport coupe. The logic behind its introduction was simple, at least according to

1999

Model availability	coupe, hardtop, and convertible
Construction	fiberglass body on steel frame with central backbone frame and hydroformed perimeter rails; transmission located at rear axle
Wheelbase	104.5 inches
Length	179.7 inches
Width	73.6 inches
Height	47.8 inches (coupe), 47.7 inches (convertible), 47.9 inches (hardtop)
Curb weight	3,245 pounds (coupe), 3,246 pounds (convertible), 3,153 pounds (hardtop)
Tread (front/rear, in inches)	62/62.1
Tires	Goodyear Extended Mobility; P245/45ZR-17, front; P275/40ZR-18, rear
Brakes	power-assisted four-wheel discs with Bosch ABS
Brake dimensions	12.6 inches, front; 11.8 inches, rear
Wheels	cast aluminum; 17x8.5 inches, front; 18x9.5 inches, rear
Fuel tank	19.1 gallons
Front suspension	short- and long-arm double wishbone (forged aluminum, top; cast aluminum, bottom), transverse leaf spring, stabilizer bar, monotube shock absorbers
Rear suspension	short- and long-arm double wishbone (cast-aluminum control arms, top and bottom), transverse leaf spring, stabilizer bar, monotube shock absorbers
Steering	speed-sensitive, power-assisted rack and pinion
Engine	5.7-liter OHV V-8 (LS1) with aluminum cylinder block and heads, sequential fuel injection
Bore and stroke	3.90x3.62 inches
Output	345 horsepower at 5,600 rpm; 350 ft-lbs at 4,400 rpm
Transmission	four-speed 4L60-E automatic transmission
Optional transmission	six-speed manual

C5 1997–2004

Chevrolet announced its factory-backed C5-R Corvette racing team late in 1998. The C5-R first raced in February 1999 at the Rolex 24 at Daytona Beach, finishing third in the GT2 class. On February 3, 2001, a C5-R became the overall winner at that year's Rolex 24.

eventually become part of the base hardtop package. Other interior color choices weren't offered, and options like sports seats, F45 suspension, and four-speed automatic transmission weren't available, either. Along with the six-speed manual gearbox, the beefy Z51 suspension was included in the standard deal, while exterior paint choices were limited to five of the eight shades listed that year: Arctic White, Light Pewter Metallic, Torch Red, Nassau Blue Metallic, and black.

C5 coupes and convertibles for 1999 also could've been painted Sebring Silver Metallic, Navy Blue Metallic, and Magnetic Red Metallic—the latter finish being an extra-cost ($500) option due to its special clear-coat application. Other than new doorsill trim, improved steering gear, and kinder, gentler, next-generation air bags, all other features represented 1998 rollovers, while new options included a power telescopic steering column (RPO N37) and Twilight Sentinel (T82), which automatically turned the headlights on and off. Both N37 and T82 weren't available on the hardtop model.

Another new option, RPO UV6, was initially limited to coupe and convertible customers early in the year, and then extended to their hardtop comrades later on. UV6 consisted of a heads-up display (HUD) system that projected instrument readouts onto the windshield in front of the driver, making him or her feel much like a jet fighter pilot in the process. This equipment still rates every bit as cool as it is functional.

C5-R racer

Though nearing 50, the Corvette was still running around racetracks like an Olympic sprinter in his prime in 1999, led by the latest in a long line of proud competition cars, the C5-R. But unlike so many of its forerunners, this Corvette racer hit the track with full factory backing. This time around Chevrolet clearly was "in racing."

Obviously a lot has changed since Zora Duntov kept himself busy trying to sneak race-ready Corvettes out of GM's back door. Those following in Duntov's big shoes today are more than willing to both build race cars and brag about them openly.

"The racing program we have created reinforces and underscores our commitment to the Corvette and its magnificent heritage," explained the breed's marketing director, Rick Baldick, in 2003. "All of us feel a responsibility to preserve and enhance the car's image. The racing program is designed to help us fulfill that responsibility."

"It was Zora Duntov's vision in 1953 that Corvette should lead the way with race-ready parts and designs," added GM Racing Group Manager Joe Negri. "I think he would be proud of what has been achieved."

Proud indeed. Announced in 1998, the C5-R went into production the following year using much of the car's existing macho machinery.

217

CHAPTER SIX

"We designed the competition engine using as many production parts and processes as possible," said GM Racing engine specialist Ron Sperry. "Powertrain engineers had some prototype block configurations they were studying for future products, and they made some of these available to us for the C5-R engine development program." A modified LS1 V-8, the C5-R's powerplant was bored and stroked to just a tad short of 7.0 liters. Output was 600 horsepower, more than enough muscle to allow speeds well in excess of 200 miles per hour. Compression was 12.5:1 and oiling was by a typical racing-style dry-sump system. Custom Delphi engine management software controlled this monster mill's electronic fuel injection.

A black-and-silver-painted C5-R made its racing debut early in 1999 at Daytona's Rolex 24, where ace driver Ron Fellows managed an impressive third-place finish. Let loose on unsuspecting GTS competitors in IMSA's American Le Mans Series (ALMS) that year, the C5-R cars quickly built a winning reputation like none ever witnessed during the Corvette's fabled half-century history. An impressive 2001 season kicked off with an overall win at the Rolex 24, "probably the most significant victory in the history of the marque," in GM Racing chief Herb Fishel's opinion. An equally historic 1-2 finish in class followed that summer at Le Mans, and the C5-R team repeated that result in France in 2002. In between, Corvettes ran away with the 2001 ALMS Manufacturer's Championship, winning six of eight races. Two more ALMS titles followed in 2002 and 2003.

2000

Model availability	coupe, hardtop, and convertible
Construction	fiberglass body on steel frame with central backbone frame and hydroformed perimeter rails; transmission located at rear axle
Wheelbase	104.5 inches
Length	179.7 inches
Width	73.6 inches
Height	47.7 inches (coupe), 47.7 inches (convertible), 47.8 inches (hardtop)
Curb weight	3,246 pounds (coupe), 3,248 pounds (convertible), 3,173 pounds (hardtop)
Tread (front/rear, in inches)	62.1/62.2
Tires	Goodyear Extended Mobility; P245/45ZR-17, front; P275/40ZR-18, rear
Brakes	power-assisted four-wheel discs with Bosh ABS
Brake dimensions	12.6 inches, front; 11.8 inches, rear
Wheels	cast aluminum; 17x8.5 inches, front; 18x9.5 inches, rear
Fuel tank	18.5 gallons
Front suspension	short- and long-arm double wishbone (forged aluminum, top; cast aluminum, bottom), transverse leaf spring, stabilizer bar, monotube shock absorbers
Rear suspension	short- and long-arm double wishbone (cast-aluminum control arms, top and bottom), transverse leaf spring, stabilizer bar, monotube shock absorbers
Steering	speed-sensitive, power-assisted rack and pinion
Engine	5.7-liter OHV V-8 (LS1) with aluminum cylinder block and heads, sequential fuel injection
Bore and stroke	3.90x3.62 inches
Output	345 horsepower at 5,600 rpm; 350 ft-lbs at 4,400 rpm
Transmission	four-speed 4L60-E automatic transmission
Optional transmission	six-speed manual

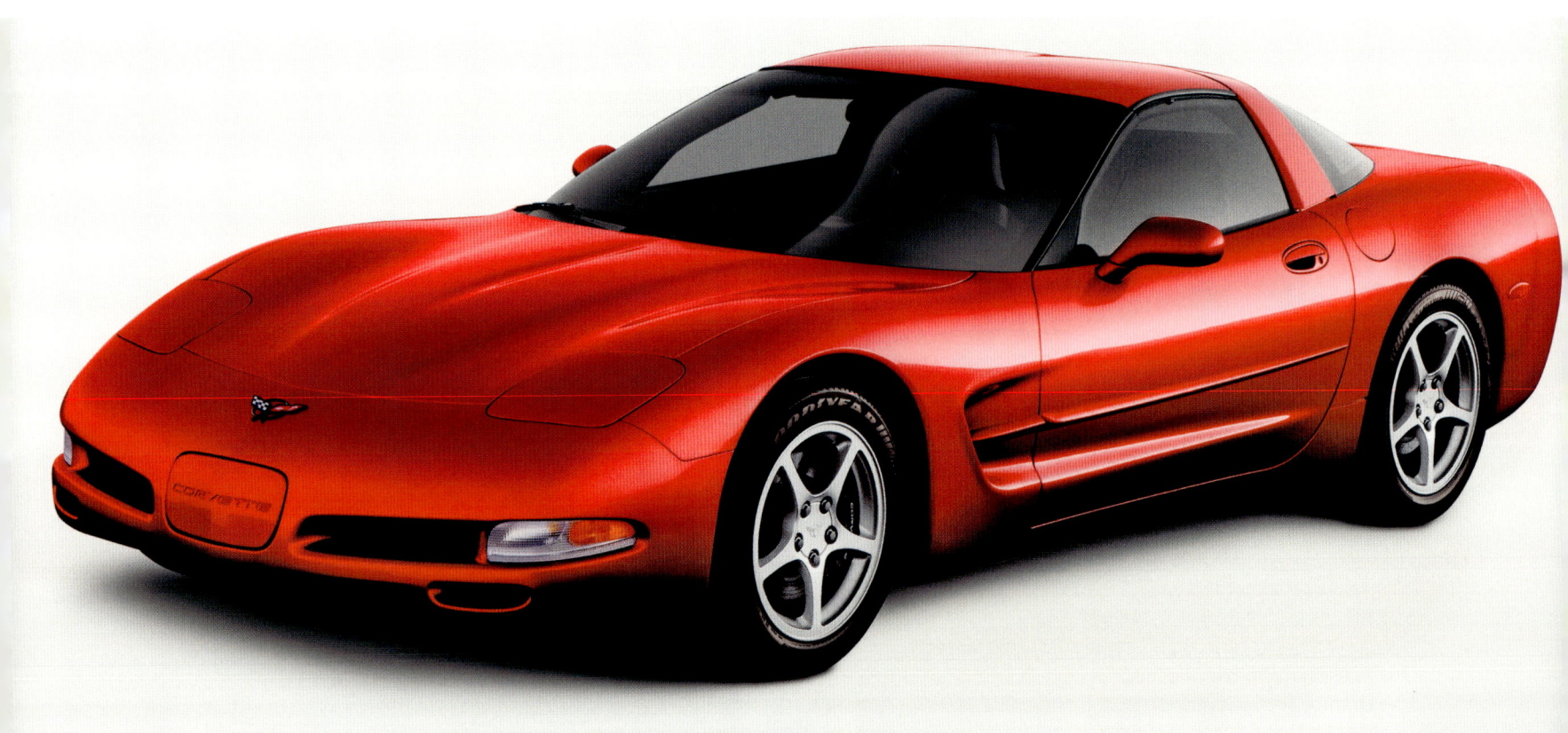

At $45,900, the convertible was the most expensive model in 2000. Base prices for the coupe and hardtop were $39,475 and $38,900, respectively.

Priced at $47,500, the Z06 hardtop was the most expensive Corvette offered in 2001. Production was 5,773—all equipped with the new LS6 V-8, backed by a six-speed manual transmission. Five colors were available for the 2001 Z06: Quicksilver Metallic, Speedway White, Torch Red, black, and the extra-cost ($600) Millennium Yellow.

"Somewhere, Zora Duntov is smiling, as are more than a million Corvette owners worldwide," said Baldick after the milestone Daytona win in 2001. "This victory signals the Corvette heritage is stronger than ever."

2000

C5 features carried over nearly unchanged again into 2000, with the most notable difference coming at the corners. A new thin-spoke forged-aluminum wheel wearing a painted finish was now standard. Exchanging paint for polish was possible by checking off RPO QF5, priced at $895. More than 15,000 customers opted for these bright wheels, and this strong demand inspired Chevrolet officials to turn to an outside supplier for another standard rim to preserve its own supply then waiting to be polished. Announced in January 2000, this second standard wheel also featured a painted finish and looked similar to its predecessor, save for slightly thicker spokes.

Optional magnesium wheels rolled over from 1999 but with a lower price tag of $2,000. Other news on the 2000 options list involved improvements to the Z51 suspension package consisting of enlarged stabilizer bars front and rear, and stiffer springs and shocks. New as well was yet another extra-cost paint scheme as the truly radiant Millennium Yellow joined Magnetic Red Metallic, both of which remained limited to coupes and convertibles.

Standard improvements for 2000 included reducing the LS1's emissions and exchanging 1999's passive keyless entry for the new active keyless entry system, which no longer included a backup key lock for the passenger-side door. The same three-body lineup reappeared that year, but demand for the lower-priced hardtop dropped by half. Meanwhile, sales of the most expensive Corvette, the convertible, went up by 21 percent. Go figure.

2001

As impressive as the C5 was from the get-go, David Hill still wasn't satisfied. Looking at the car from a pure performance perspective, he recognized that not all Corvette owners cared about making concessions to comfort and convenience—that some drivers simply wanted to be the baddest in the valley and to heck with everything else. Chevrolet's no-nonsense Corvette hardtop, introduced for 1999, represented an initial step toward appeasing uncompromising customers, but it wasn't quite bad enough, and its

CHAPTER SIX

cost-conscious nature ruled out any major muscle-maximizing mechanical improvements.

Cost-cutting measures were then thrown out the window in 2001, resulting in the Z06 hardtop, a pumped-up Corvette that was, in Chevrolet's words, "aimed directly at diehard performance enthusiasts at the upper end of the high-performance market." Now the most expensive Corvette offered, priced $500 more than the 2001 convertible, the Z06 apologized to no one for its balls-out, inconsiderate nature. It was nearly every bit as hard on the seat of the pants as it was on the wallet, but it also was, again according to Chevrolet, "simply the quickest, best handling production Corvette ever."

"We've enhanced Corvette's performance persona and broken new ground with the Z06," explained Hill with due pride. "With 0–60 [times] of four seconds flat and more than 1 g of cornering acceleration, the Z06 truly takes Corvette performance to the next level. In fact, the Corvette Team has begun referring to it as the C5.5, so marked are the improvements we've made and the optimization of the car in every dimension."

"The new Z06 will have great appeal for those who lust after something more—that indefinable thrill that comes from being able to drive competitively at 10/10ths in a car purpose-built do to exactly that," added new Corvette Brand Manager Jim Campbell.

Like its 385-horse LS6 V-8, this lean, mean coupe borrowed its name from a past legend, the one created by Zora Duntov in 1963. Duntov's original Z06 package included Chevrolet's hottest injected small block working in concert with beefed brakes and a seriously stiffened suspension. Hill's plan was similar 38 years later. Standard for the modern Z06 was Chevy's new M12 six-speed

2001

Model availability	coupe, Z06 hardtop, and convertible
Construction	fiberglass body on steel frame with central backbone frame and hydroformed perimeter rails; transmission located at rear axle
Wheelbase	104.5 inches
Length	179.7 inches
Width	73.6 inches
Height	47.7 inches (coupe), 47.7 inches (convertible), 47.8 inches (hardtop)
Curb weight	3,246 pounds (coupe), 3,248 pounds (convertible), 3,173 pounds (hardtop)
Tread (front/rear, in inches)	61.9/62; 62.4/62.6, Z06
Tires	Goodyear Eagle F1 GS Extended Mobility; P245/45ZR-17, front; P275/40ZR-18, rear
Z06 tires	Goodyear Eagle F1 SC asymmetric; P265/40ZR-17, front; P295/35ZR-18, rear
Brakes	power-assisted four-wheel discs with Bosch ABS
Brake dimensions	12.6 inches, front ; 11.8 inches, rear
Wheels	cast aluminum; 17x8.5 inches, front; 18x9.5 inches, rear
Z06 wheels	forged aluminum; 17x9.5 inches, front; 18x10.5, rear
Fuel tank	18.5 gallons
Front suspension	short- and long-arm double wishbone (forged aluminum, top; cast aluminum, bottom), transverse leaf spring, stabilizer bar, monotube shock absorbers
Rear suspension	short- and long-arm double wishbone (cast-aluminum control arms, top and bottom), transverse leaf spring, stabilizer bar, monotube shock absorbers
Steering	speed-sensitive, power-assisted rack and pinion
Engine	5.7-liter OHV V-8 (LS1) with aluminum cylinder block and heads, sequential fuel injection
Bore and stroke	3.90x3.62 inches
LS1 output	350 horsepower at 5,600 rpm; 375 ft-lbs at 4,400 rpm
Z06 engine	5.7-liter OHV V-8 (LS6) with aluminum cylinder block and heads
LS6 output	385 horsepower at 6,000 rpm; 385 ft-lbs at 4,800 rpm
Standard transmission (LS1)	four-speed 4L60-E automatic transmission
Standard transmission (LS6)	six-speed manual
Optional transmission	(behind LS1) six-speed manual

The production breakdown for 2001 was 15,681 sport coupes, 14,173 convertibles and 5,773 Z06 hardtops. The convertible base price was $47,000.

The active-handling equipment on the 2001 Corvette added a great deal of driver safety in foul weather.

manual transmission, wider wheels and tires, special brake-cooling duct work front and rear, and the exclusive FE4 suspension, which featured a larger front stabilizer bar, a stiffer leaf spring in back, and revised camber settings at both ends. Weight was cut throughout the Z06 by about 100 pounds overall compared to a typical 2001 Corvette sport coupe.

Stretched an inch wider than standard C5 rims, the 2001 Z06's wheels measured 17x9.5 inches in front, 18x10.5 inches in back. Mounted on these huge rollers were Goodyear Eagle F1 SC tires, P265/40ZR-17 in front, P295/35ZR-18 in back. C5s in 2001 featured Eagle F1 GS rubber: P245/45ZR-17 at the nose, P275/40ZR-18 at the tail.

Available only with the Z06, the LS6 V-8 took its name from the 425-horse 454 big-block V-8 offered previously beneath Corvette hoods for one year only: 1971. Chevrolet's second-edition LS6 was a muscled-up LS1 small block that looked identical externally, save for red ornamental covers (instead of black) atop each cylinder head. Boosting output to 385 horsepower resulted from more compression (10.5:1), even slicker ports, a freer-flowing intake

CHAPTER SIX

Wider wheels and tires, the exclusive FE4 suspension, and special brake-cooling duct work were all standard as well. The mighty LS6, identified by its red ornamental covers, can be seen inside. It produced 385 horsepower in 2001. Like the Z06 tag, the LS6 code also came from the archives; in 1971, it had signified the presence of the Corvette's optional 425- horsepower 454-cubic-inch big-block V-8.

> The Z06 was "simply the quickest, best handling production Corvette ever."

manifold, a higher-lift cam, and higher-volume injectors. Maximum torque for the 2001 LS6 was 385 ft-lbs at 4,800 rpm, and its redline was 6,500 rpm, compared to 6,000 revs for the LS1.

The LS1, too, received an improved intake in 2001, resulting in an increase from 345 to 350 horsepower. Torque also rose, from 350 to 360 ft-lbs in automatic applications, and to 375 ft-lbs when the optional six-speed manual was installed. Overall performance was improved further by adding second-generation active handling (formerly RPO JL4) into the standard Corvette package.

Nine different exterior colors were offered for 2001 coupes and convertibles, including the two special schemes (Millennium Yellow and Magnetic Red Metallic), now priced at $600. The Z06, on the other hand, was limited to five shades: Speedway White, Quicksilver Metallic, Torch Red, Millennium Yellow, and black.

2002

Upgrades were minor in 2002, at least as far as base coupes and convertibles were concerned. Most prominent was a switch from stainless steel to aluminum for the automatic transmission cooler case and the inclusion of an AM/FM stereo with in-dash CD player as standard equipment. One new exterior color (Electron Blue Metallic) was added, while two (Dark Bowling Green Metallic and Navy Blue Metallic) were dropped. Available Z06 shades again numbered five, but this time Electron Blue Metallic replaced Speedway White.

Z06 improvements were plentiful, beginning with an increase to 405 horsepower for its exclusive LS6 V-8. Helping make this burst possible was a new low-restriction air cleaner, an even lumpier cam (with the highest lift in small-block history), a less-restrictive mass airflow sensor, and lighter hollow-stem valves. The smaller two of the car's four catalytic converters were removed, a modification that both cut weight and reduced back pressure by 16 percent. Yet even with those missing cats, the 2002 Z06 exhaust system still met federal emissions standards, because the two remaining converters were modified to enhance their efficiency.

The Z06-specific FE4 suspension was upgraded with an enlarged front stabilizer bar, a stiffer rear spring, and recalibrated rear shock absorbers. New front brake linings featured improved durability and

> "Few vehicles have had the staying power of Corvette," said Rick Baldick in 2003.

2002

Model availability	coupe, Z06 hardtop, and convertible
Construction	fiberglass body on steel frame with central backbone frame and hydroformed perimeter rails; transmission located at rear axle
Wheelbase	104.5 inches
Length	179.7 inches
Width	73.6 inches
Height	47.7 inches (coupe), 47.7 inches (convertible), 47.8 inches (hardtop)
Curb weight	3,246 pounds (coupe), 3,248 pounds (convertible), 3,118 pounds (hardtop)
Tread (front/rear, in inches)	61.9/62; 62.4/62.6, Z06
Tires	Goodyear Eagle F1 GS Extended Mobility; P245/45ZR-17, front; P275/40ZR-18, rear
Z06 tires	Goodyear Eagle F1 SC asymmetric; P265/40ZR-17, front; P295/35ZR-18, rear
Brakes	power-assisted four-wheel discs with Bosch ABS
Brake dimensions	12.6 inches, front; 11.8 inches, rear
Wheels	cast aluminum; 17x8.5 inches, front; 18x9.5 inches, rear
Z06 wheels	forged aluminum; 17x9.5 inches, front; 18x10.5, rear
Fuel tank	18.5 gallons
Front suspension	short- and long-arm double wishbone (forged aluminum, top; cast aluminum, bottom), transverse leaf spring, stabilizer bar, monotube shock absorbers
Rear suspension	short- and long-arm double wishbone (cast-aluminum control arms, top and bottom), transverse leaf spring, stabilizer bar, monotube shock absorbers
Steering	speed-sensitive, power-assisted rack and pinion
Engine	5.7-liter OHV V-8 (LS1) with aluminum cylinder block and heads, sequential fuel injection
Bore and stroke	3.90x3.62 inches
LS1 output	350 horsepower at 5,600 rpm; 375 ft-lbs at 4,400 rpm
Z06 engine	5.7-liter OHV V-8 (LS6) with aluminum cylinder block and heads
LS6 output	405 horsepower at 6,000 rpm; 400 ft-lbs at 4,800 rpm
Standard transmission (LS1)	four-speed 4L60-E automatic transmission
Standard transmission (LS6)	six-speed manual
Optional transmission (behind LS1)	six-speed manual

fade resistance, and the Z06 aluminum wheels were cast instead of forged. The HUD system became standard Z06 equipment in 2002.

All Z06 hardtops and convertibles with HUD were fitted that year with lighter, thinner (4.8-millimeter) windshields, which shaved off 2.65 pounds per car. All coupes and topless models without HUD featured thicker (5.4-millimeter) windshields.

2003

"Few vehicles have had the staying power of Corvette," said Rick Baldick in 2003. "We believe much of that success comes from a willingness to embrace advancing technology while remaining true to Corvette's glorious history. As we celebrate our golden anniversary in 2003, we honor our past and cast a bright eye toward the future."

Predictably honoring the Corvette's golden year were 50th Anniversary badges on all 2003 models. And for $5,000, a 2003 coupe or convertible could've been honored further with the 1SC package, a cosmetic addition that wasn't offered to Z06 buyers.

> "We believe much of that success comes from a willingness to embrace advancing technology while remaining true to Corvette's glorious history." —*Rick Baldick*

Little changed in the coupe and convertible from 2001 to 2002. Fog lamps, sport seats, power assist for the passenger seat, and dual-zone air conditioning were options in 2002, and eventually became standard in 2003 models, like this 2003 50th Anniversary commemorative edition convertible.

CHAPTER SIX

2003

Model availability	coupe, Z06 hardtop, and convertible
Construction	fiberglass body on steel frame with central backbone frame and hydroformed perimeter rails; transmission located at rear axle
Wheelbase	104.5 inches
Length	179.7 inches
Width	73.6 inches
Height	47.7 inches (coupe), 47.7 inches (convertible), 47.8 inches (hardtop)
Curb weight	3,246 pounds (coupe), 3,248 pounds (convertible), 3,118 pounds (hardtop)
Tread (front/rear, in inches)	61.9/62; 62.4/62.6, Z06
Tires	Goodyear Eagle F1 GS Extended Mobility; P245/45ZR-17, front; P275/40ZR-18, rear
Z06 tires	Goodyear Eagle F1 SC asymmetric; P265/40ZR-17, front; P295/35ZR-18, rear
Brakes	power-assisted four-wheel discs with Bosch ABS
Brake dimensions	12.6 inches, front; 11.8 inches, rear
Wheels	cast aluminum; 17x8.5 inches, front; 18x9.5 inches, rear
Z06 wheels	forged aluminum; 17x9.5 inches, front; 18x10.5, rear
Fuel tank	18.0 gallons
Front suspension	short- and long-arm double wishbone (forged aluminum, top; cast aluminum, bottom), transverse leaf spring, stabilizer bar, monotube shock absorbers
Rear suspension	short- and long-arm double wishbone (cast-aluminum control arms, top and bottom), transverse leaf spring, stabilizer bar, monotube shock absorbers
Steering	speed-sensitive, power-assisted rack and pinion
Engine	5.7-liter OHV V-8 (LS1) with aluminum cylinder block and heads, sequential fuel injection
Bore and stroke	3.90x3.62 inches
LS1 output	350 horsepower at 5,600 rpm; 375 ft-lbs at 4,400 rpm
Z06 engine	5.7-liter OHV V-8 (LS6) with aluminum cylinder block and heads
LS6 output	405 horsepower at 6,000 rpm; 400 ft-lbs at 4,800 rpm
Standard transmission (LS1)	four-speed 4L60-E automatic transmission
Standard transmission (LS6)	six-speed manual
Optional transmission (behind LS1)	six-speed manual

Included in the 1SC deal were unique fender emblems, Anniversary Red "Xirallic crystal" paint, champagne-colored aluminum wheels, and a Shale-colored interior. Embroidered 50th Anniversary logos appeared on the floor mats and headrests, and convertibles were additionally treated to Shale-colored soft tops. All 50th Anniversary Corvettes were also equipped with magnetic selective ride control, yet another high-tech approach to helping keep the car's dirty side down.

A new option for base coupes and convertibles, magnetic selective ride control (RPO F55) replaced the F45 damping system offered in 2002 and represented a head-and-shoulders improvement.

> Describing the C5 as "the best 'Vette yet," David Hill claimed, "You won't find a car in Corvette's price range that provides the same level of quality, power, ride, handling, and refinement."

Actor Jim Caviezel drove the Indianapolis 500 pace car on May 26, 2002, giving customers a sneak peek at the upcoming 50th Anniversary 2003 color scheme.

C5 1997–2004

Above: What a difference 50 years can make. Standard power in the 50th commemorative model came from a 350-horsepower LS1 V-8. The heart of its forebear was a 150-horse, inline six-cylinder.

Left: The 2004 Corvette was the last of the C5 breed.
Mike Mueller

225

CHAPTER SIX

2004

Model availability	coupe, Z06 hardtop, and convertible
Construction	fiberglass body on steel frame with central backbone frame and hydroformed perimeter rails; transmission located at rear axle
Wheelbase	104.5 inches
Length	179.7 inches
Width	73.6 inches
Height	47.7 inches (coupe), 47.7 inches (convertible), 47.8 inches (hardtop)
Curb weight	3,246 pounds (coupe), 3,248 pounds (convertible), 3,118 pounds (hardtop)
Tread (front/rear, in inches)	61.9/62; 62.4/62.6, Z06
Tires	Goodyear Eagle F1 GS Extended Mobility; P245/45ZR-17, front; P275/40ZR-18, rear
Z06 tires	Goodyear Eagle F1 SC asymmetric; P265/40ZR-17, front; P295/35ZR-18, rear
Brakes	power-assisted four-wheel discs with Bosch ABS
Brake dimensions	12.6 inches, front; 11.8 inches, rear
Wheels	cast aluminum; 17x8.5 inches, front; 18x9.5 inches, rear
Z06 wheels	forged aluminum; 17x9.5 inches, front; 18x10.5, rear
Fuel tank	18.0 gallons
Front suspension	short- and long-arm double wishbone (forged aluminum, top; cast aluminum, bottom), transverse leaf spring, stabilizer bar, monotube shock absorbers
Rear suspension	short- and long-arm double wishbone (cast-aluminum control arms, top and bottom), transverse leaf spring, stabilizer bar, monotube shock absorbers
Steering	speed-sensitive, power-assisted rack and pinion
Engine	5.7-liter OHV V-8 (LS1) with aluminum cylinder block and heads, sequential fuel injection
Bore and stroke	3.90x3.62 inches
LS1 output	350 horsepower at 5,600 rpm; 375 ft-lbs at 4,400 rpm
Z06 engine	5.7-liter OHV V-8 (LS6) with aluminum cylinder block and heads
LS6 output	405 horsepower at 6,000 rpm; 400 ft-lbs at 4,800 rpm
Standard transmission (LS1)	four-speed 4L60-E automatic transmission
Standard transmission (LS6)	Tremec T56 six-speed manual
Opt. transmission (behind LS1)	Tremec T56 six-speed manual

Damping changes in the F45 system were achieved by mechanically controlling the fluid flow within each shock absorber. The F55 system, on the other hand, relied on what was called "magnetorheological" (MR) fluid, which contained iron particles that could adjust flow rates when acted upon by electrical inputs, and this electronic control could react to changing road conditions in lightning-fast fashion. Reportedly, the new magnetic damping could make as many as 1,000 adjustments per second.

Other options seen the previous year became standard on coupes and convertibles in 2003. Among these were fog lamps, sport seats, power assist for the passenger seat, and dual-zone air conditioning. A parcel net and luggage shade became standard for the coupe. The Z06 hardtop carried over essentially unchanged from 2002.

2004

Everything that made the Corvette so great during its 50th anniversary year returned for the last of the C5 models in 2004. As before, three bodies were offered that year: coupe, convertible, and Z06 hardtop. But new for 2004 was a Commemorative Edition package offered for all three models.

Created to honor the C5-R Corvette's racing success, the Commemorative Edition coupes and convertibles featured new Le Mans Blue paint with contrasting Shale interiors, special badges, and polished wheels In the convertible's case, the folding top also was done in Shale. Additional touches on the Commemorative Edition Z06 Corvettes included a special Le Mans stripe accent and a lightweight carbon-fiber hood. All Z06s, commemorative or not, also were treated to revised shock valving that, according to Corvette Product Manager Harlan Charles, made the car feel "more tied down, more glued to the road."

Eight years and out was the C5's fate. But what a great eight they were.

New colors for 2004 included Arctic White, Machine Silver, Magnetic Red, and the Commemorative Edition's Le Mans Blue.

C5 1997–2004

The 2004 Commemorative Edition hardtop also featured a new carbon-fiber hood and revised shock valving.

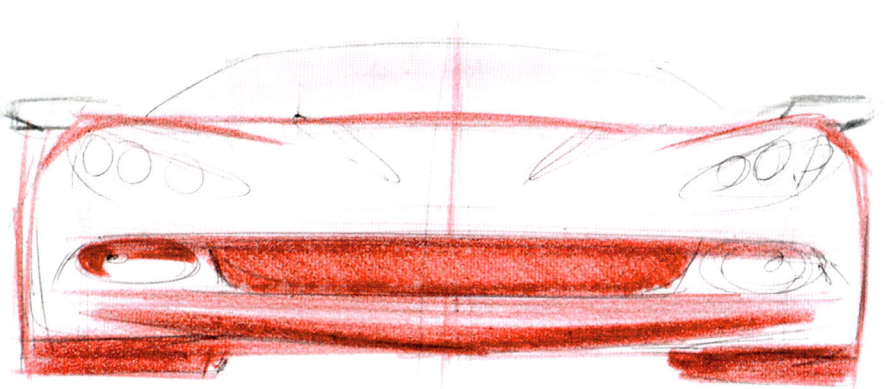

By 2004, these sixth-generation sketches had almost become a reality.

CHAPTER SEVEN

07 Thoroughly Modern

C6
2005–2013

David Hill wasn't blowing smoke in 1997 when he called Chevrolet's latest Corvette the best 'Vette yet. From nose to tail, from the roof down to the ground, the fabulous C5 was as new and improved as it got. Even so, Hill and his team were already busy working on the next best 'Vette just two years later. According to the Corvette's third chief engineer, various things about the superb C5 needed fixing, and those upgrades could only come about as part of a redesign.

ALONG WITH EXPECTED PERFORMANCE ENHANCEMENTS, C6 development involved even more refinement, this to make a more-refined customer base happy. That the C5 was arguably the world's best performance buy was only the beginning; its improved comfort and convenience features made the deal as attractive, and classy, as it had ever been during the car's half-century history. Thus, according to Corvette Marketing Director Rick Baldick, a new breed of buyer had emerged, a higher class, if you will. With these more discriminating minds in mind, engineers and designers sought to dial in further comfort, improved overall platform tightness, and extra quietness.

"The C6 represents a comprehensive upgrade to the Corvette," explained David Hill late in 2003. "Our goal is to create a Corvette that does more things well than any performance car. We've thoroughly improved performance and developed new features and capabilities in many areas, while at the same time systematically searching out and destroying every imperfection we could find." In Hill's opinion, the C5 had been "90 percent perfect." That figure, in his estimation, then became 99 percent for the C6. Furthermore, Hill claimed that as much as 70 percent of the C6's part numbers were new, a fact that may or may not have proved significant. "We frequently go through this with manufacturers," explained Daniel Pund in *Car and Driver*. "If the suspension control arms look exactly the same but carry a different parts number, are they new? We'll compromise and say the 2005 Corvette is the C5 and 11/16ths."

Initial plans called for the near-perfect, round-numbered C6 to show up in time for the Corvette's 50th birthday celebration in 2003. But then the September 11, 2001, terrorist attacks occurred. Economic aftershocks cut GM budgets to the bone, leaving officials no choice but to hold off on C6 development. When cash began to flow again, the Corvette team resorted to working nights and weekends to have the next-generation machine ready for public sale by the fall of 2004. All told, as many as 200 engineers and 1,000 workers in Bowling Green burned mucho midnight oil to guarantee a 2005 model year debut.

Working long hours under Hill and Assistant Chief Engineer Tadge Juechter were the likes of Bill Nichols, Dave Zimmerman, and Fernando Krambeck. Nichols oversaw powertrain work,

- First standard Corvette V-8 to reach the 400-horsepower level (2005)
- C6.R racer debuts at Sebring (2005)
- Chief Engineer David Hill retires, replaced by Tom Wallace (2006)
- New Z06 breaches the 500-horsepower barrier (2006)
- A Corvette convertible paces the Indianapolis 500 for the fourth straight year and ninth time overall (2007)
- Ron Fellows ALMS GT1 Champion Edition Z06 coupe introduced (2007)
- Two Corvettes serve as official Indianapolis 500 pace cars in 2008: a conventional convertible and a Z06 coupe; the Z06 is a concept vehicle powered by E85 ethanol fuel
- 6.2-liter LS3 V-8 replaces the 6.0-liter LS2 small-block (2008)
- 427 Special Edition Z06 coupe debuts midyear (2008)
- Tadge Juechter named chief engineer (2008)
- Competition Sport option offered for Z06 and LS3 coupes (2009)
- GT1 Championship Edition package introduced for LS3 coupes/convertibles and Z06 coupes (2009)
- Reborn ZR1 coupe becomes first Corvette to reach 200-mph top end (2009)
- Reborn Grand Sport debuts in coupe and convertible forms (2010)
- Carbon Limited Edition Z06 coupe introduced to mark the 50th anniversary of the Corvette's first run at Le Mans (2011)
- Centennial Edition package introduced for all Corvette models to help mark Chevrolet's 100th birthday (2012)
- A 2013 ZR1 becomes the 11th Corvette to pace the Indianapolis 500 (2012)
- LS7-powered 427 Convertible Collector Edition introduced (2013)
- Commemorative 60th Anniversary design package offered for all Corvette models (2013)

Above: Back for 2006, the Z06 was powered by a 400-horse 7.0-liter V-8. Those 7 liters translated into 427 cubic inches, a familiar figure to Corvette fans.

Opposite: The only black Corvette available in 2012 was the Centennial Edition rendition, created as expected to mark Chevrolet's 100th year in business—a momentous event that had actually arrived the year before. Featuring exclusive Carbon Flash Metallic paint, the Centennial Edition package (ZLC) was available on all models, in all body styles, including the Z06 coupe (left) and Grand Sport convertible (right).

CHAPTER SEVEN

The C6 coupe's removable roof panel was 15 percent larger than its C5 counterpart. Options included a dual-roof package and a tinted clear panel.

Zimmerman did the chassis, and Krambeck handled the interior. Though it wasn't really his job, Juechter commonly worked in the wind tunnel, where aerodynamicist Tom Froling reportedly stayed until 8 p.m. many nights to get the job done.

Tom Peters was the designer in charge of giving the C6 a truly new look, which first and foremost involved deleting the hideaway headlights that had been a Corvette tradition since 1963. Like the Sting Ray III, the new C6 featured exposed lamps housed behind cat's-eye clear covers, making more than one thoroughly entrenched purist cringe. Overall impressions, nonetheless, were hot and were further enhanced by softly rounded contours that help lessen the sharp impact made by the C5. And with a drag coefficient of 0.286, the C6 body ranked as the most aerodynamic Corvette shell yet.

Beneath that slick skin was Chevrolet's latest variation on its small-block theme: the Gen IV

Like its wheels and tires, the Z06's brakes were equally humongous, with 14-inch rotors in front, 13.4 in back.

C6 2005–2013

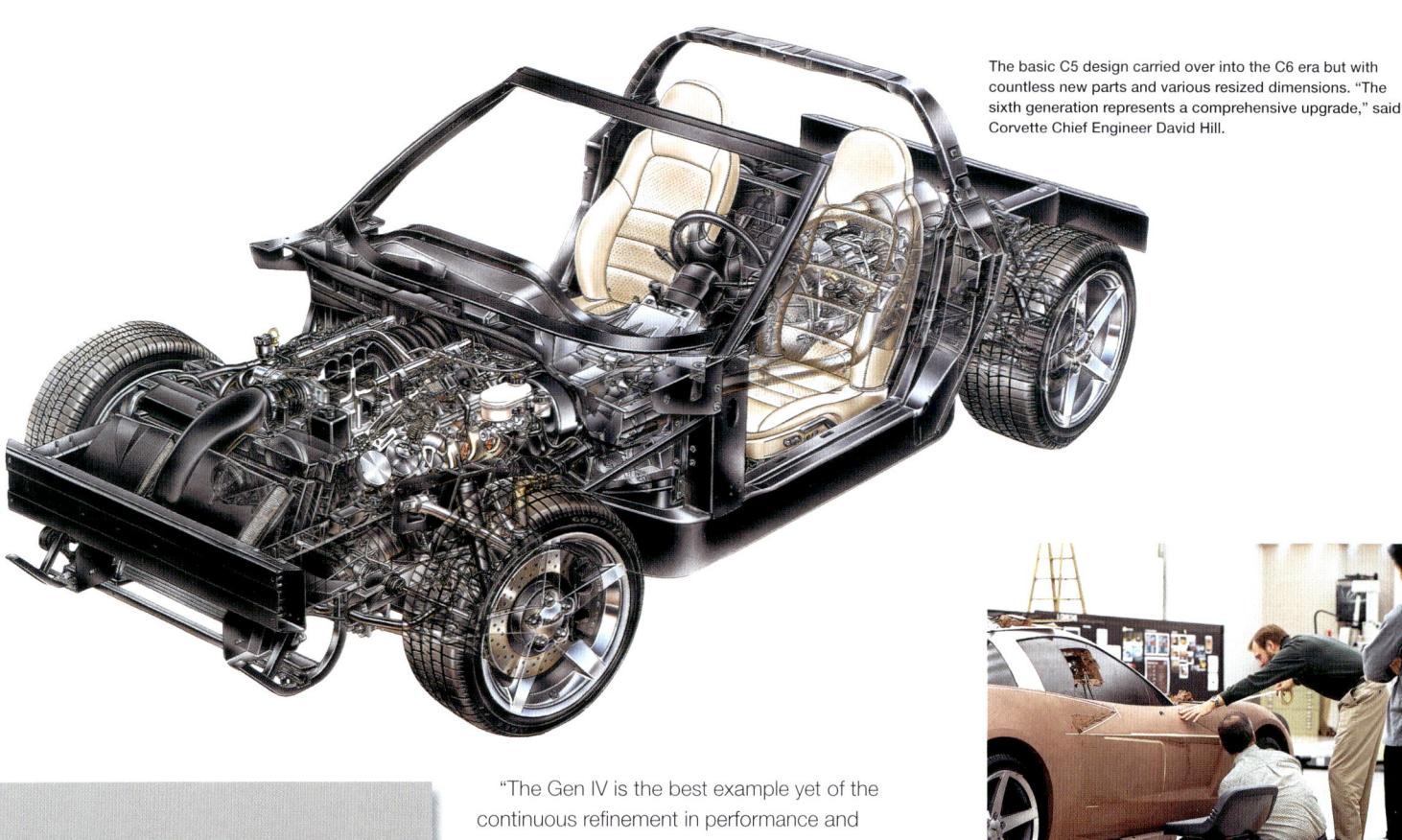

The basic C5 design carried over into the C6 era but with countless new parts and various resized dimensions. "The sixth generation represents a comprehensive upgrade," said Corvette Chief Engineer David Hill.

Though noticeably resculpted, the C6 bore considerable resemblance to its C5 forerunner.

> "The Gen IV is the best example yet of the continuous refinement in performance and efficiency that has been part of the small block's legacy since day one," said engineering chief Sam Winegarden in October 2003.

V-8. Tabbed "LS2," this new mouse motor was basically an upgraded, bored-out LS1. It displaced 6.0 liters and produced 400 horsepower, making it the strongest standard Corvette engine to date. It also was nicely fuel efficient, even more so than the 2004 C5's base 350-horse 5.7-liter LS1. Estimated combined city/highway fuel economy for the 2005 Corvette was 22.6 mile per gallon, a major plus considering today's high-flying gas prices.

Imagine that—more displacement, more horses, and more miles per gallon. Clearly Chevy engineers knew what the hell they were doing, as had most of their predecessors dating back to 1955.

"The Gen IV is the best example yet of the continuous refinement in performance and efficiency that has been part of the small block's legacy since day one," said engineering chief Sam Winegarden in October 2003. "[This] long history is one of the reasons the new generation of engines is so powerful and efficient. GM has almost 50 years of experience with its valve-in-head design, and that has provided immeasurable detail for keeping the small block a viable, relevant engine for today and the future."

More than one Corvette crazy stood disappointed in 2003 when the C6 failed to appear as the breed's 50th anniversary model. Fortunately, it was dominating magazine covers by the end of the year, and its official public unveiling then came at the Detroit auto show in January 2004.

2005

Nearly all witnesses were well aware that the new C6 wasn't a total redesign, like its C5 forerunner. That wasn't a bad thing, however, considering the long, long list of changes that added up to one thoroughly modern performance package.

"The sixth-generation Corvette blends technical sophistication with expressive style," explained promotional paperwork. "Five inches shorter than the current car, the 2005 Corvette cuts a tighter, more taut profile—with virtually no loss of usable space. More than just visual, the new dimensions make the car more agile and 'tossable,' with upgrades in handling, acceleration, and braking."

Additional dimensional changes included 1 inch less width and a wheelbase stretch of 1.2 inches. Along with a more sure-footed feel, that latter increase teamed up with the body's decreased overhangs (down 2 inches in front, 3 in back) to produce the aforementioned taut profile. More important was a reduction in weight, down to 3,179 pounds for the coupe, 3,199 for the convertible.

New for 2005 were 18-inch front wheels and 19-inch rears. Widths were 8.5 inches at the nose, 10 inches out back. Goodyear Eagle F1 GS Extended Mobility tires went onto these rims measuring P245/40ZR-18 up front, P285/35ZR-19 at the tail. Bigger, better brakes were standard, too, with the front rotors measuring 12.8 inches across, the backs 12. The C5's suspension layout carried over, but reportedly nearly all components were revised or new.

Boosting LS2 displacement to 6.0 liters (364 cubic inches) was the result of more bore (4.00 inches instead of 3.90 inches) working with the same stroke (3.62 inches) used within the LS1. Helping make those 400 horses was more compression (10.9:1), a higher-lift cam (with higher-rate valve springs), and intake and exhaust flows improved by 15 and 20 percent, respectively. Overall engine weight, meanwhile, went down by about 15 pounds, courtesy of, among other things, a smaller water pump and thinner-walled exhaust manifolds.

Behind the LS2 was either the four-speed 4L65E Hydra-matic automatic or Tremec T56 six-speed manual. The wide-ratio T56 box used by the 2004 Z06 remained on the options list but only when the Z51 performance package was ordered. Included in the Z51 deal was a stiffer suspension and even bigger (13.4-inch front rotors, 13-inch rears) brakes. Working together, these parts made for some

CHAPTER SEVEN

Chevrolet's sixth-generation Corvette debuted for 2005 in coupe and convertible forms. The new C6 was both lighter and shorter (by 5 inches) than its C5 predecessor.

seriously improved performance. According to Chevrolet officials, a Z51-equipped C6 could nearly match a 2004 Z06's lap time on the test track, a convenient occurrence considering the Z06 didn't carry over into 2005. Thankfully, it did return.

C6.R racer

Corvettes always have been in racing, even if their parent company hasn't. AMA officials could ban factory racing involvement all they wanted in 1957; that didn't stop Zora Duntov from propping Engineering's back door open for racers bent on keeping America's sports car out in front on the track. Then there was GM's 1963 edict demanding that all its divisions cease and desist competition activities. Corvettes still kept racing, and winning, with more than a little support from their makers in Warren, Michigan.

Scoring SCCA championships and endurance victories in the United States was one thing. But from the beginning, Duntov always imagined a much grander legacy for his baby. His 1957 SS was meant not just to win at Sebring, but to move on to bigger, better venues, the most important being Le Mans. AMA action that summer derailed his plan, however, and glory in France continued to elude production-based Corvettes for decades to come.

All that changed in 2001 when C5-R Corvettes finished first and second in the GTS class (8th and 14th overall) at Le Mans. Officially introduced in 1998, the C5-R factory racer first competed in France's fabled 24-hour endurance test in June 2000, placing third and fourth in the GTS field. By 2005, the Corvette Racing team had added three more firsts and a second at Le Mans, as well as three straight Sebring class wins (2001–2003) and

2005

Model availability	hatchback coupe (with removable roof panel) and convertible
Construction	composite body on hydroformed steel frame with aluminum and magnesium structural components
Wheelbase	105.7 inches
Length	174.6 inches
Width	72.6 inches
Height	49 inches
Curb weight	3,179 pounds (coupe), 3,199 pounds (convertible)
Tread (front/rear, in inches)	62/1/60.7
Tires	Goodyear Eagle F1 GS Extended Mobility (EMT); P245/40ZR-18, front; P285/35ZR-19, rear
Brakes	power-assisted four-wheel discs with ABS and electronic traction control
Brake dimensions (in inches)	12.8x1.26 front, 12x1 rear
Brake dimensions with Z51 option (in inches)	13.4x1.26, front; 13x1, rear; cross-drilled rotors
Wheels	18x8.5 front, 19x10 rear
Fuel tank	18 gallons
Suspension	short- and long-arm (SLA) double-wishbone cast-aluminum control arms, transverse composite leaf springs, and stabilizer bars front and rear; monotube shock absorbers, active handling electronics (stiffer springs and shocks, thicker stabilizer bars with Z51 option)
Steering	speed-sensitive, magnetic power-assisted rack and pinion (16.1:1 ratio)
Engine	6.0-liter (364 cubic inches) sequential fuel-injected (SFI) LS2 V-8 with overhead valves and cast-aluminum cylinder block and heads
Compression	10.9:1
Bore and stroke	4.00x3.62 inches
Output	400 horsepower at 6,000 rpm, 400 ft-lbs of torque at 4,400 rpm
Standard Transmission	six-speed manual with 2.66:1 low
Optional Transmission	Hydra-matic 4L65E four-speed automatic
Transmission with Z51 option	six-speed manual with 2.97:1 low

2005 C6.R

Body style	coupe
Construction	composite body on hydroformed steel frame with aluminum and magnesium structural components
Wheelbase	105.7 inches
Length	177.6 inches
Width	78.7 inches
Height	45.8 inches
Tread (front/rear, in inches)	62.2/63.1
Tires	Michelin racing; 290/33-18, front; 310/41-18 ,rear
Brakes	four-wheel discs with monoblock caliper and carbon-fiber rotors and pads
Fuel tank	26.4 gallons
Suspension	independent fabricated steel short- and long-arm double-wishbone, machined aluminum knuckles, coil-over multi-adjustable shocks, anti-roll bar, front and rear
Engine	7.0-liter all-aluminum LS7.R overhead-valve V-8 with electronic sequential fuel injection and dry-sump oiling
Bore and stroke	4.180x3.875 inches
Output	590 horsepower at 5,400 rpm, 640 ft-lbs torque at 4,400 rpm

five consecutive ALMS championships. All told, Corvette Racing's two cars won 45 of 66 events from 2000 to 2005, including 31 one-two finishes.

The two Velocity Yellow machines campaigned by Corvette Racing in 2005 were new C6.R models, which stood head and shoulders above their C5-R forerunners. "We developed this car for a year before it appeared in public, so it had many miles of testing before its first race at Sebring," explained Program Manager Doug Fehan. "You only have to look at the differences between the C5-R and the C6.R—a one-inch-longer wheelbase and five-inch-shorter body—to appreciate how much work had to be done. You can't just re-skin an old car and expect it to win."

"Both the chassis and aerodynamic package changed considerably," added GM Racing road racing group manager Steve Wesoloski. "The new regulations required more extensive use of the production car's chassis structure, retaining items such as the central drivetrain tunnel, the windshield frame, and the rear bumper. We also introduced new low-friction suspension attachments that made the car quicker to react."

Beneath the 2005 C6.R's nose was the 7.0-liter LS7.R V-8, a 590-horsepower, 640-ft-lbs-torque beast that had many Gen IV basics but was lighter than the regular-production LS2. "We took a lot amount of weight out of the engine, helping the balance of the car while improving performance," continued Wesoloski. "Internal components [also] were designed to reduce horsepower losses due to friction and to reduce rotating mass."

"History will remember the C5-R as one of the best sports racing cars of all time, and we've raised the bar even higher with the C6.R," said GM Racing Director Mark Kent. "Just when I think the Corvette team has done all that can be done, they surprise me. I just can't wait to see what they do next, and I'm glad that our racing efforts will be part of making the next-generation Corvette even better."

Indeed, apparently racing does improve the breed.

2006

The C6 coupe and convertible rolled over essentially unchanged from 2005, save for one major new optional feature: a six-speed "paddle shift" automatic transmission. Featuring clutch-to-clutch operation, an integrated 32-bit electronic controller, and manual-shift capabilities, this advanced gearbox switched between three operating modes: Drive, Sport, and Paddle Shift. The first two worked automatically, with the Drive mode concentrating on smoothness, while its Sport upgrade made shifts firmer for improved acceleration. Activating Paddle Shift allowed drivers to choose gear changes themselves using manual controls located on the steering wheel.

Yet, as impressive as the Paddle Shift option appeared, it didn't represent the really big news for 2006. That honor went to the new Z06, which was

The Z06 returned for 2006 after taking one year off. The C6-based Z06 featured new exclusive bodywork that measured 3 inches wider than the garden-variety C6 shell. Production was 6,272. Base price was $65,800.

CHAPTER SEVEN

2006

Model availability	hatchback coupe (with removable roof panel), convertible, and Z06 coupe (with fixed magnesium roof)
Construction	composite body on hydroformed steel frame with aluminum and magnesium structural components (unique carbon-fiber body panels and hydroformed aluminum frame for Z06)
Wheelbase	105.7 inches
Length	174.6 inches
Length (Z06)	175.6 inches
Width	72.6 inches
Height	49 inches
Curb weight	3,179 pounds (coupe), 3,199 pounds (convertible), 3,132 pounds (Z06)
Tread (front/rear, in inches)	62.1/60 (coupe and convertible); 63.5/62.5 (Z06)
Tires (coupe and convertible)	Goodyear Eagle F1 Supercar Extended Mobility (EMT); P245/40ZR-18, front; P285/35ZR-19, rear
Tires (Z06)	Goodyear Eagle F1 Supercar EMT; P275/35ZR-18, front; P325/30ZR-19, rear
Brakes	power-assisted four-wheel discs with ABS and electronic traction control
Brake dimensions (inches)	12.8x1.26, front; 12x1, rear
Brake dimensions with Z51 option (inches)	13.4x1.26, front; 13x1, rear; cross-drilled rotors
Brake dimensions, Z06 (in inches)	14x1.3, front; 13.4x1, rear; cross-drilled rotors
Wheels, coupe and cvt. (inches)	18x8.5, front; 19x10, rear
Wheels, Z06 (in inches)	18x9.5, front; 19x12, back
Fuel tank	18 gallons
Suspension	short- and long-arm (SLA) double-wishbone cast-aluminum control arms, transverse composite leaf springs, and stabilizer bars, front and rear; monotube shock absorbers, active handling electronics (stiffer springs and shocks, thicker stabilizer bars with Z51 option)
Steering	speed-sensitive, magnetic power-assisted rack and pinion (16.1 1 ratio)
Engine (LS2)	6.0-liter (364 cubic inches) sequential fuel-injected (SFI) V-8 with overhead valves and cast-aluminum cylinder block and heads
LS2 compression	10.9 1
LS2 bore and stroke	4.00x3.62 inches
LS2 output	400 horsepower at 6,000 rpm, 400 ft-lbs of torque at 4,400 rpm
Engine (LS7)	7.0-liter (427 cubic inches) SFI V-8 with overhead valves and cast-aluminum cylinder block and heads, dry-sump oiling
LS7 compression	11.0:1
LS7 bore and stroke	4.125x4.00 inches
LS7 output	505 horsepower at 6,300 rpm, 470 ft-lbs of torque at 4,800 rpm
Standard transmission	six-speed manual with 2.66:1 low
Transmission with Z51 option	six-speed manual with 2.97:1 low
Optional transmission	six-speed automatic with paddle shift

> "In many ways, [this] is a racing engine in a street car," added Muscaro.

back and better than ever after a one-year hiatus.

Included in the new Z06 deal was the most impressive list yet of unique features, including bad-to-the-bone bodywork that made the car three inches wider than typical C6s. Fashioned from carbon-fiber composites, those body panels helped trim overall weight down to a reasonably svelte 3,132 pounds. Huge disc brakes were also part of the deal (14-inch rotors in front, 13.4 in back) with six-piston calipers at the nose, four-piston units at the tail. Humongous wheels (18x9.5 inches in front, 19x12 in back) went on at the corners. Coolers for all fluids, including power steering, were included, as was a rear-mounted battery for improved weight distribution.

Last but obviously not least was the Z06's main attraction: the LS7 V-8. Along with being the most powerful small-block ever, this all-aluminum monster was also the largest. Increasing both the Gen IV's bore and stroke (to 4.125 and 4.00 inches, respectively) produced 427 cubic inches, which translated into 7.0 liters. While its 505 maximum horses arrived at 6,300 rpm, this big, bad small-block could confidently rev up to 7,100 rpm, making it one of the world's first regular-production pushrod motors able to regularly breach the seven-grand barrier and return alive. The previous Gen IV V-8 extreme was 6,600 rpm for the LS2.

"For a production engine to run at this high of an rpm blurs the lines even more between [pushrod] and overhead-cam design," said Dave Muscaro, assistant chief engineer for small-block engines. "We took a complete systems approach to achieve the high rpm. We have a tight valvetrain design along with some race-inspired materials for the reciprocating components like titanium intake valves and connecting rods."

Blurred as well by the LS7 was the barrier separating road and track. "In many ways, [this] is a racing engine in a street car," added Muscaro. "We've taken much of what we've learned over the years from the 7.0-liter C5R racing program and instilled it here. There really has been nothing else like it offered in a GM production vehicle."

LS7 heads also looked like they came right off a race track. "We consulted with our Motorsports group on numerous aspects of the cylinder head design," said valvetrain design engineer Jim Hicks. "We adopted some of the latest ideas that have been successful in the Nextel Cup and the American Le Mans series, including valve centerline positions, valve angles, valve sizes, and rocker arm ratio." "The heads are simply works of mechanical art," added Muscaro. "We left nothing on the table when it came to ensuring the best airflow through the engine."

Ports were considerably larger than those in the LS2 heads and were precisely massaged for exceptionally high flow. Valve sizes were 2.20 inches for the titanium intake unit, 1.61 for its sodium-filled running mate. Valve lift was a ridiculous 0.591 inches for intake and exhaust. Hydroformed stainless-steel headers completed things on the exhaust end.

And like its lightweight yet bulletproof reciprocating mass, the LS7's cylinder block also was unique, being specially cast to preserve strength while making room for all that bore diameter. Pressed-in cylinder sleeves and tough forged-steel cross-bolted main bearing caps also set the LS7 block apart from its LS2 little brother: the latter featured cast-in sleeves and powder-metal main caps. LS7 pistons were cast-

C6 2005–2013

Right: Basically a bored-out LS1 V-8, the C6's LS2 displaced 6.0 liters and produced 400 horsepower— the highest standard output in Corvette history. *Below:* Little changed from 2005 save for an optional six-speed "paddle-shift" automatic transmission option.

In 2005, the C6.R Corvette picked up where the C5-R left off, finishing 1-2 in class (GT1) at Le Mans that summer

235

CHAPTER SEVEN

aluminum flat-top units that brought compression to 11.0:1. On the bottom end was a balanced forged-steel crank, and lubricating all those wildly rotating parts was a competition-type dry-sump oil system, a first for street-going Corvettes.

Building the LS7 represented a job as unique as the engine itself. Each engine was hand-assembled with care by a single, specially trained technician at GM Powertrain's new 100,000-square-foot Performance Build Center, located in Wixom, Michigan. According to the facility's site manager, Timothy Schag, this intriguing process represented "a premium manufacturing technique for premium products. [It] brings a higher level of quality, because each builder is personally involved in every aspect of the assembly."

Only about 30 LS7 V-8s were produced in a day at Wixom, but quantity clearly wasn't a priority. "It was important to step away from the high-volume world we all had lived in for so long and soak in the cadence of these specialized environments," continued Schag. "We learned a lot and established a low-volume manufacturing system on par with the world's best niche builders, but we didn't lose sight of the quality practices already in place at GM."

2007

C6 changes for 2007 predictably appeared nearly nonexistent at a glance. Easily the most noticeable was a truly fresh chromatic finish, Atomic Orange, a shockingly (in some opinions) radioactive shade that cost an extra $750 due to the additional application of a special tint coat that had first shown up atop the 2000 Corvette's optional Millennium Yellow paint. Specially tinted Monterey Red and Velocity Yellow, each also priced at $750, carried over on the extra-cost palette for the third straight year.

Fashion-conscious customers additionally could've installed Z06-style two-tone upholstery inside their non-Z06 coupes and convertibles—as long as they additionally popped for either the upscale 2LT or 3LT preferred equipment group. Offered separately that year, a power soft top was included as part of the 2007 3LT convertible package.

Whatever the skin color, topless or not, an

2007

Model availability	coupe (with removable roof panel), convertible, and Z06 coupe (with fixed magnesium roof)
Construction	composite body on hydroformed steel frame with aluminum and magnesium structural components (unique carbon-fiber body panels and hydroformed aluminum frame for Z06)
Wheelbase	105.7 inches
Length	174.6 inches (175.6 inches, Z06)
Width	72.6 inches (75.9 inches, Z06)
Height	49 inches
Track (front/rear, in inches)	62.1/60.7 (coupe and convertible); 63.5/62.5 (Z06)
Curb weight	3,179 pounds (coupe), 3,199 pounds (convertible), 3,132 pounds (Z06)
Wheels (coupe and cvt)	18 x 8.5 inches, front; 19 x10 inches rear
Wheels (Z06)	18 x 9.5 inches, front; 19 x 12 inches rear
Tires (coupe and cvt)	Goodyear Eagle F1 Supercar Extended Mobility (EMT); P245/40ZR18 front, P285/35ZR19 rear
Tires (Z06)	Goodyear Eagle F1 Supercar EMT; P275/35ZR18 front, P325/30ZR19 rear
Brakes	power-assisted four-wheel discs with ABS and electronic traction control
Brake dimensions	12.8x1.26 inches, front; 12x1 inches, rear
Brake dimensions w Z51	13.4x1.26 inches, front; 13x1 inches, rear; cross-drilled rotors
Brake dimensions, Z06	14x1.3 inches, front; 13.4x1 inches, rear; cross-drilled rotors
Fuel tank	18 gallons
Suspension	short- and long-arm (SLA) double wishbone cast-aluminum control arms, transverse composite leaf springs, stabilizer bars front and rear, monotube shock absorbers, Active Handling electronics (stiffer springs and shocks, thicker stabilizer bars with Z51 option)
Steering	speed-sensitive, magnetic power-assisted rack and pinion (16.1:1 ratio)
Engine (LS2)	6.0L (364 cubic inches) sequential fuel injected (SFI) V8 with overhead valves and cast-aluminum cylinder block and heads,
Compression	10.9:1
LS2 bore and stroke	4.00x3.62 inches
LS2 output	400 horsepower at 6,000 rpm, 400 ft-lbs of torque at 4,400 rpm
Engine (LS7)	7.0L (427 cubic inches) SFI V8 with overhead valves and cast-aluminum cylinder block and heads, dry-sump oiling
Compression	11:1
LS7 bore and stroke	4.125x4.00 inches
LS7 output	505 horsepower at 6,300 rpm, 470 ft-lbs of torque at 4,800 rpm
Standard transmission	six-speed manual with 2.66:1 low
Transmission w Z51	six-speed manual with 2.97:1 low
Optional transmission	six-speed automatic with paddle shift
Axle ratio	3.42:1 (2.56:1 with paddle-shift automatic)

The Fellows Z06 featured the driver's name on the front fender.

ought-seven Vette did remain thoroughly identical to its forerunners underneath as Chevy's superb C6 mechanicals kept on keeping on. Four hundred healthy horses reappeared for LS2 coupes and convertibles, five-oh-five was again the norm for the latest Z06's LS7. The most notable technical upgrade arrived on the options list as enlarged cross-drilled brake rotors were added to the carryover F55 magnetic adjustable suspension package. A bit bigger than base brakes, a tad smaller than the Z06's serious stoppers, these drilled discs were only available previously by way of the Z51 deal, which also returned for 2007, again wearing a $1,695 price tag, while the F55 option was $1,995.

Additional sound insulation and a slightly larger glove box came standard inside a 2007 Corvette. New interior options included Bose audio enhancements and steering-wheel-mounted premium stereo controls. OnStar became available for top-shelf 2007 Z06s equipped with the 2LZ preferred equipment group; it had been limited to 3LT coupes/convertibles in 2006. Nostalgic Z06 customers also could've enhanced their cars'

Top: The production breakdown for 2007 included 21,484 base coupes and 10,418 convertibles. Base prices were $44,995 for the former, $52,910 for the latter. Some 3,790 customers forked over the extra $750 for the special Atomic Orange paint color pictured here.

Above: Two special-edition Corvettes were unveiled at the 2007 Chicago Auto Show: the Ron Fellows ALMS GT1 Champion Edition Z06 (left) and yet another Indianapolis 500 pace car replica convertible (right). May marked the Vette's ninth appearance as the 500 pace car, and for the first time since '98, 500 commemorative vehicles were produced for Walter Mitty types.

hoods with familiar "427" badges, one of many "Genuine Corvette Accessories" offered by dealers in 2007.

Two special-edition Corvettes appeared as well that year; one a plainly pale Z06 coupe, the other a radiant convertible. The former, listed as RPO Z33, was created to honor legendary Chevrolet racer Ron Fellows, hence its full name: the Ron Fellows ALMS GT1 Champion Edition Z06. All 399 Z33 coupes were done in Arctic White paint complemented with a pair of red fender stripes that mimicked the graphics on Fellows' various GT1-class champion Corvettes. Fellows' autograph graced the inside console lid, making this the first "signed" special edition in Corvette history. More would soon follow.

The special 2007 convertible, offered only in Atomic Orange, commemorated the Corvette's ninth appearance (and fourth in a row, both unmatched records) at The Brickyard as the prestigious Indianapolis 500 pace car. Not since 1998 had Chevrolet offered an Indy pace car replica to Walter Mitty types. Commemorative graphics were printed for the 2003 Corvette, but the Memorial Day pace lap tours in 2004, 2005, and 2006 were not replicated in regular production. Then in February 2007, Chevy officials announced a run of 500 Z4Z pace car convertibles, all adorned additionally with splashy gold ribbon graphics, expected Indy 500 fender badges and interior embroidery, and a Z06 rear spoiler. A manual-trans LS2 was standard; the paddle-shift automatic was the model's only option. Base price for the 2007 Indy pace car replica was $66,695. Adding the six-speed automatic bumped that number up to $68,245.

CHAPTER SEVEN

2008

A Corvette again paced the annual Indy 500 in May 2008, and Chevrolet again celebrated with a limited run of special-edition Z4Z replicas, this time offered in both coupe and convertible forms. Creating the unprecedented duo was only right considering that Chevy's overall record 19th Indy pace lap honor also represented the first time in Indianapolis Motor Speedway (IMS) history that two different pace cars appeared on race day, in this case a typical ragtop and a unique Z06 concept coupe that ran on eco-friendly E85 ethanol fuel.

> According to Chevrolet, 190 mph was no prob for an off-the-lot LS3 Corvette in 2008.

"Although not a production FlexFuel vehicle, the Corvette Z06 E85 concept car is a high-performance example of Chevrolet's gas-friendly to gas-free initiative, demonstrating viable fuel solutions," announced Chevy General Manager Ed Peper in December 2007. Announced at the same time was the pace car coupe's driver: two-time Indy 500 winner Emerson Fittipaldi, who just happens to be an ethanol refiner back home in Brazil.

"It's only fitting that the Corvette will be the first car to earn the distinction of having two models pace the Indianapolis 500 in the same year," added IMS President and CEO Joie Chitwood. "Chevrolet and Corvette are a vital part of the rich history of 'The Greatest Spectacle in Racing,' and we're honored to have a great champion of the race and of alternative fuels, Emerson Fittipaldi, as this year's pace car driver."

New standard wheels, first shown off on the 2007 Indy pace car wearing a Sterling Silver finish, appeared on the 2008 model, this time done in Sparkle Silver. A Competition Gray finish was optional. A polished forged aluminum wheel was a new option that year.

2008

Model availability	hatchback coupe (with removable roof panel), convertible, and Z06 coupe (with fixed magnesium roof)
Construction	composite body on hydroformed steel frame with aluminum and magnesium structural components (unique carbon-fiber body panels for Z06)
Wheelbase	105.7 inches
Length	174.6 inches (175.6 inches, Z06)
Width	72.6 inches (75.9 inches, Z06)
Height	49 inches
Track (front/rear, in inches)	62.1/60.7 (coupe and convertible); 63.5/62.5 (Z06)
Curb weight	3,179 pounds (coupe), 3,199 pounds (convertible), 3,132 pounds (Z06)
Wheels (coupe and convertible)	18 x 8.5 inches, front; 19 x10 inches, rear
Wheels (Z06)	18 x 9.5 inches, front; 19 x 12 inches, rear
Tires (coupe and convertible)	Goodyear Eagle F1 Supercar Extended Mobility (EMT); P245/40ZR18 front, P285/35ZR19 rear
Tires (Z06)	Goodyear Eagle F1 Supercar EMT; P275/35ZR18 front, P325/30ZR19 rear
Brakes	power-assisted four-wheel discs with ABS and electronic traction control
Brake dimensions	12.8x1.26 inches, front; 12x1 inches, rear
Brake dimensions with Z51 option	13.4x1.26 inches, front; 13x1 inches, rear; cross-drilled rotors
Brake dimensions, Z06	14x1.3 inches, front; 13.4x1 inches, rear; cross-drilled rotors
Fuel tank	18 gallons
Suspension	short- and long-arm (SLA) double wishbone cast-aluminum control arms, transverse composite leaf springs, stabilizer bars front and rear, monotube shock absorbers, Active Handling electronics (stiffer springs and shocks, thicker stabilizer bars with Z51 option))
Steering	speed-sensitive, magnetic power-assisted rack and pinion (16.1:1 ratio)
Engine (LS3)	6.2L (376 cubic inches) sequential fuel injected (SFI) V8 with overhead valves and cast-aluminum cylinder block and heads
Compression	10.7:1
LS3 bore and stroke	4.00x3.62 inches
LS3 output	430 horsepower at 5,900 rpm, 424 lb-ft of torque at 4,600 rpm (436/428 with NPP dual mode exhaust option)
Engine (LS7)	7.0L (427 cubic inches) SFI V8 with overhead valves and cast-aluminum cylinder block and heads, dry-sump oiling
Compression	11:1
LS7 bore and stroke	4.125x4.00 inches
LS7 output	505 horsepower at 6,300 rpm, 470 ft-lbs of torque at 4,800 rpm
Standard transmission	six-speed manual with 2.66:1 low
Transmission with Z51 option	six-speed manual with 2.97:1 low
Optional transmission	six-speed automatic with paddle shift
Axle ratio	3.42:1 (2.56:1 with paddle-shift automatic)

Fittipaldi even signed and numbered every 2008 pace car replica built: 266 convertibles, 234 coupes. Included this time around in the Z4Z package were new chromed five-spoke forged-aluminum wheels, silver checkered-flag graphics on black paint (further honoring the first Indy pace car Corvette from 1978), and a choice between a six-speed stick or paddle-shift automatic. A Z06 spoiler returned, as did requisite "Indy 500" fender badges. All 2008 Z4Z Corvettes also came standard with posh 3LT interior features and a new C6 powerplant, the 6.2-liter LS3 small-block.

Increasing the bore (from 4.00 inches to 4.06) was just the beginning as far as the transformation from LS2 to 376-cid LS3 was concerned. High-flow LS7-style heads worked in concert with, among other things, a lumpier cam, a revised valvetrain (with enlarged intake valves and offset intake rockers), a high-flow intake, and high-flow injectors (also copped from the Z06's LS7) to up the C6's standard output ante to 430 horsepower. According to Chevrolet, 190 mph was no prob for an off-the-lot LS3 Corvette in 2008. And with a 0–60 clocking of 4.3 seconds, a paddle-shifted

C6 2005–2013

The Corvette's small-block V-8 was bored up to 6.2 liters (376 cubic inches) for 2008, gaining 30 horsepower in the process. The name also changed from LS2 to LS3. Highflow LS7-style heads were among LS3 upgrades.

239

CHAPTER SEVEN

For the first time in the event's history, two different models paced the Indianapolis 500 race in 2008: a Corvette convertible and a Z06 coupe, the latter a concept vehicle that ran on E85 ethanol fuel.

Below: Two-time Indy 500 winner Emerson Fittipaldi poses with Chevrolet's 2008 Indianapolis 500 Pace car replica convertible. The newest Indy 500 commemorative also honored the 30th anniversary of the first Corvette to pace the annual Memorial Day motorsport spectacular.

LS3 represented the quickest automatic-equipped Corvette ever.

Yet there was more. Installing the optional dual-mode exhaust system (RPO NPP) increased LS3 output further to 436 horsepower. Another LS7-style upgrade, the new dual-mode exhausts came standard beneath the Indy pace car coupes and convertibles. But with or without those 6 extra ponies, the LS3 C6 clearly redefined "best Vette yet."

"Corvette is an uncompromising sports car that rewards its owners with impeccable performance and great comfort," bragged Ed Peper. "The changes and enhancements to the 2008 Corvette reflect continual improvements that speak to Chevrolet's unflagging commitment to building the best sports car—and with nearly 55 years of experience, the Corvette just keeps getting better and better."

One of the most eye-catching wheel designs to ever appear on a Vette were these chrome units that came standard on all 505 Z44/Z06 coupes in 2008.

Technical betterment included improved shift effort for the six-speed manual, quicker shifts for the automatic, and greatly enhanced steering. That latter feature alone was enough to have *Car and Driver*'s Larry Webster writing home to ma. "The [latest] Vette no longer has to be muscled through the turn," he scribbled. "It now glides along with far less drama. Now the Corvette has improved tactile feel to go with its outrageous performance."

The latest new C6 also wore new standard wheels, featuring a split-spoke design originally shown off (with a Sterling Silver finish) on the 2007 Indy pace car. Sparkle Silver was the standard finish in 2008; Competition Gray was optional. Polished forged-aluminum wheels were new on the 2008 options list.

Improvements inside consisted of freshened trim for the door sills and instrument panel center plate, joined by a host of new standard equipment: OnStar, XM satellite radio, an auto-dimming rearview mirror (for coupes), and an audio input jack for all radio systems save for those additionally complemented with navigation assistance. Optional interior equipment package choices, first offered in 2001, expanded in 2008 as the 2LT and 3LT groups listed (along with their 2LZ counterparts in Z06 ranks) in 2006 and 2007 were joined by an even tastier 4LT deal.

Among 2LT features were side airbags, leather upholstery, a six-way power seat for the passenger, and a wireless cell phone link. The 3LT option included all this plus head-up instrument display, memory convenience controls, heated seats, a Bose AM/FM stereo with six-disc CD changer, and a tilt-telescopic steering wheel. The new 4LT package added special interior trim and extra leather touches to all this. A 3LZ equipment group also joined the carryover 2LZ option for the 2008 Z06.

Much bigger news on the Z06 front hit the front page midyear in 2008 after production of Chevrolet's "427 Special Edition" Corvette (RPO Z44) commenced. Even though the Z06's LS7 V-8 technically displaced more like 428 cubic inches

C6 2005–2013

Above: Okay, so the Z06's LS7 V-8 technically displaced about 427.6 cubic inches. But Chevy product planners couldn't resist honoring past big-block Corvette glories with their "427 Special Edition" coupe, introduced midyear in 2008 with exclusive chrome wheels, Crystal Red tintcoat paint, unique graphics, and Wil Cooksey's signature. Cooksey retired in March 2008 after 15 years as Bowling Green plant manager.

Left: Both OnStar navigation and XM satellite radio became standard features inside the 2008 Corvette. Revised trim for the doors sills and instrument panel center plate appeared too.

(427.6), engineers just couldn't help but make the connection between this brute and the vaunted big-block bullies that had made earlier Corvettes kings of the streets from 1966 to 1969. Along with Crystal Red tintcoat paint and exclusive chrome wheels, all 505 Z44 Z06 coupes built for 2008 featured hood/fascia graphics and badges that were plainly reminiscent of the "stinger" hoods seen on the 1967 427 Vette.

"The heritage of the 427 designation with the Corvette is legendary," explained Corvette product manager Harlan Charles. "Recognizing the tie-in of the original 427 engine and the LS7's displacement has been on the Corvette team's mind since the Z06 was introduced, and we're thrilled to express it in this special model."

The Z44 price was $84,195, and for that stack of cash, a customer also received the 3LZ interior accented with a console armrest signed and numbered by Wil Cooksey, who retired in March 2008 after 15 years of shepherding Corvette production in Bowling Green. Chevy's $1,750 navigation system was the only option.

Yet another special-edition Corvette made the scene in 2008, this one via Hertz. GM supplied Hertz with 500 "ZHZ" coupes that year, all featuring a black-accented yellow finish to honor the veteran rental car firm. All also were equipped with 436-horse LS3 V-8s backed by paddle-shift automatics. Another 375 similarly adorned ZHZ convertibles were rented out by Hertz in 2009.

2009

On February 11, 2009, Chevrolet introduced its latest Stingray concept vehicle at the Chicago Auto Show. Modeled after Bill Mitchell's sensational show car of the same name, the new Stingray appeared as Sideswipe in the 2009 summer blockbuster *Transformers: Revenge of the Fallen*.

Reportedly on GM Design Center drawing boards for some six years, the 2009 Stingray also reminded some witnesses of Mitchell's regular-production 1963 Sting Ray with its split rear window layout. Ties to Italian exotics were evident as well thanks to its upward-opening, scissor-style, carbon-fiber doors.

Meanwhile, off-stage, "rollover" once again was the big word out of Bowling Green in 2009, as

CHAPTER SEVEN

nearly nothing notable marked the arrival of the base fifth-edition C6, save for the introduction of optional Bluetooth phone connectivity for those who apparently had better things to do than fully enjoy the ride in their fantastic plastic two-seaters. Introduced in the summer of 2008 as a 2009 model, the wretchedly excessive, reborn ZR1 was turning heads everywhere it went, but that story will need to wait for another page or so. Back in the real world, Chevy's 2009 LS3 Corvette looked an awful lot like its predecessor—not a bad thing considering it also still appeared to be one mighty bang for the buck. Same for its 505-horse Z06 running mate, the former cock of the walk now almost lost in the shadow of Chevy's latest biggest, baddest Vette yet.

With further apologies to the new ZR1, about all that remained to report in 2009 was the debut of two more limited-edition models: the Competition Sport Package and GT1 Championship Edition. Specifically created for "enthusiasts who attend driving schools and track events," the Competition Sport option was offered for the Z06 1LZ and 436-horse 1LT coupe, with prices listed at $77,500 and $55,655, respectively. In both cases, the package included special striping, wheels, and headlamp covers, as well as various Corvette Racing logos inside and out. LS3 versions were painted either Arctic White or Blade Silver; their Z06 running mates came in black or Blade Silver. Added as well to LS3 coupes (along with the dual-mode exhausts) was the Z51 package, a differential cooler, head-up display, red-painted brake calipers, and the Z06 spoiler. Production was 52 for the LS3 Competition Sport, 20 for its 505-horse cousin.

Available in black or Velocity Yellow, the GT1 Championship Edition served as a commemorative exclamation point to 10 years of domination in American Le Mans Series racing, where Corvettes piled up more than 70 wins and scored eight GT1 Manufacturers Championships. Another five GT1-class wins also came in France at the real Le Mans endurance event during that time.

The GT1 package was listed for 4LT coupes/convertibles and 3LZ Z06 coupes, with LS3 versions incorporating the Z51 performance package, dual-mode exhausts, and Z06 spoiler. In all cases, Corvette Racing graphics graced the

2009

Model availability	hatchback coupe (with removable roof panel), convertible, and Z06 hatchback coupe (with fixed magnesium roof)
Construction	composite body on hydroformed steel frame with aluminum and magnesium structural components (unique carbon-fiber body panels and hydroformed aluminum frame for Z06)
Wheelbase	105.7 inches
Length	174.6 inches (175.6 inches, Z06)
Width	72.6 inches (75.9 inches, Z06)
Height	49 inches
Track (front/rear, in inches)	62.1/60.7 (coupe and convertible); 63.5/62.5 (Z06)
Curb weight	3,217 pounds (coupe), 3,246 pounds (convertible), 3,210 pounds (Z06)
Wheels (coupe and cvt)	18x8.5 inches front, 19 x10 inches rear
Wheels (Z06)	18x9.5 inches front, 19x12 inches rear
Tires (coupe and cvt)	Goodyear Eagle F1 Supercar Extended Mobility (EMT); P245/40ZR18 front, P285/35ZR19 rear
Tires (Z06)	Goodyear Eagle F1 Supercar EMT; P275/35ZR18 front, P325/30ZR19 rear
Brakes	power-assisted four-wheel discs with ABS and electronic traction control
Brake dimensions	12.8x1.26 inches front, 12x1 inches rear
Brake dimensions (w Z51)	13.4x1.26 inches front, 13x1 inches rear, cross-drilled rotors
Brake dimensions (Z06)	14x1.3 inches front, 13.4x1 inches rear, cross-drilled rotors
Fuel tank	18 gallons
Suspension	short- and long-arm (SLA) double wishbone cast-aluminum control arms, transverse composite leaf springs, and stabilizer bars front and rear; monotube shock absorbers, Active Handling electronics (stiffer springs and shocks, thicker stabilizer bars with Z51 option)
Steering	speed-sensitive, magnetic power-assisted rack and pinion (16.1:1 ratio)
Engine (LS3)	6.2-liter (376-ci) sequential fuel injected (SFI) V-8 with overhead valves and cast-aluminum cylinder block and heads
Compression	10.7:1
LS3 bore and stroke	4.06x3.62 inches
LS3 output	430 horsepower at 5,900 rpm, 424 lb-ft of torque at 4,600 rpm (436/428 with NPP dual mode exhaust option)
Engine (LS7)	7.0-liter (427-ci) SFI V-8 with overhead valves and cast-aluminum cylinder block and heads, 11:1 compression, dry-sump oiling
LS7 bore and stroke	4.125x4.00 inches
LS7 output	505 horsepower at 6,300 rpm, 470 lb-ft of torque at 4,800 rpm
Standard transmission	six-speed manual with 2.66:1 low
Transmission with Z51 option	six-speed manual with 2.97:1 low
Optional transmission	six-speed automatic with paddle shift
Axle ratio	3.42: 1 (2.56:1 with automatic)

Left: The GT1 edition's seats, dash, and instrument panel received special "GT1" embroidery as well as the menacing Corvette Racing logo . . .

Right: . . . while the Corvette's total domination of the ALMS GT1 class was displayed in the form of irrefutable numerical evidence.

Little changed from the 2008 model for the 2009 base Corvette, but two new limited-edition models were available: the Competition Sport Package, and the GT1 Championship Edition.

Yet another special model, the GT1 Championship Edition, emerged in 2009 to honor 10 years of domination in American Le Mans Series racing. This package was available for top-shelf 4LT LS3 coupes/convertibles and 3LZ Z06 coupes.

exterior, chrome wheels went on at the corners, and inside was ebony leather with exclusive yellow accent stitching. "GT1" embroidery complemented the seats, instrument panel, and center armrest, and the engine was topped by a special "carbon pattern" cover sporting yellow "Corvette" lettering.

Pricing was $65,310 for the GT1 LS3 convertible, $71,815 for its topless counterpart, and $86,385 for the Championship Edition Z06. Production was 53, 17, and 55, respectively.

2009 ZR1

In 2006, *AutoWeek* called the Z06 "the best supercar buy of all time." But those words apparently still weren't good enough for General Motors CEO Rick Wagoner. If Chevy could offer so much Corvette for 60 grand, he thought, what could David Hill and right-hand man Tadge Juechter do for 100 Gs? Hill retired in January 2006, leaving his assistant to answer Wagoner's question with the aforementioned world-class ZR1.

Base priced at $106,000, Chevy's latest ZR1 far and away ranked as the most expensive GM product ever. But so what? At an unworldly 638 horsepower, it was also the most powerful, and in turn it didn't take a rocket scientist to determine that all those ponies made this major investment The General's fastest offering of all time.

"The ZR1 is the first Corvette to step up into the rarefied air of truly unearthly performance," claimed *Motor Trend*'s Arthur St. Antoine, who additionally

243

CHAPTER SEVEN

2009 ZR1

Model availability	hatchback coupe with fixed magnesium roof
Construction	composite body with carbon-fiber panels on hydroformed aluminum frame with aluminum and magnesium structural components
Wheelbase	105.7 inches
Length	176.2 inches
Width	75.9 inches
Height	49 inches
Track (front/rear, in inches)	63.5/62.5
Curb weight	3,350 pounds
Wheels	19x10 inches front, 20x12 inches rear
Tires	Michelin Pilot Sport PS2 ZP; P285/30ZR19 front, P335/25ZR20 rear
Brakes	power-assisted four-wheel drilled ceramic discs with ABS and electronic traction control; six-piston calipers front, four-piston calipers rear
Brake dimensions	15.5 inches front, 15.0 inches rear
Fuel tank	18 gallons
Suspension	short- and long-arm (SLA) double wishbone cast-aluminum control arms, transverse composite leaf springs, and stabilizer bars front and rear; monotube shock absorbers, Magnetic Selective Ride Control
Steering	speed-sensitive, magnetic power-assisted rack and pinion (16.1:1 ratio)
Engine (LS9)	6.2-liter (376-ci) supercharged SFI V-8 with overhead valves and cast-aluminum cylinder block and heads, dry-sump oiling
Bore and stroke	4.06x4.00 inches
Compression	9.1:1
Output	638 horsepower at 6,500 rpm, 604 lb-ft of torque at 3,800 rpm
Standard transmission	close-ratio six-speed manual with 2.29:1 low
Axle ratio	3.42: 1

reported a 0–60 run of only 3.3 seconds, a quarter-mile pass at a scant 11.2 ticks (at 130.5 miles per hour), and a real top end of 200.4 miles per hour. It went without saying that the last stat also represented a GM first, and it probably wasn't the actual tip-top.

"There's something ridiculous about a street car that can do 66 miles per hour in 1st gear and has a tight-ratio 6-speed that allows for a top speed in excess of 200 miles per hour," added *Road & Track*'s Shaun Bailey. "Who can afford the gasoline or the speeding tickets? However, it must be noted that for 'social responsibility,' said chief engineer Tadge Juechter, there is an electronic limiter set for 210 miles per hour. Yet if one disabled the limiter and managed to run the boost beyond 10.5 psi, the ZR1 runs out of gear at 215 mph. It's only a matter of time before a tuned ZR1 breaks that barrier."

Hand-assembled with care along with the Z06's LS7 V-8 at GM's Performance Build Center in Wixom, Michigan, the ZR1's 6.2-liter LS9 V-8 was topped by an Eaton R2300 twin-rotor supercharger specially designed to fit snugly beneath the Corvette's hood. Even with its "dual brick" intercooler system straddling it on both sides, the sixth-generation Eaton blower still stood only slightly taller than the LS3's electronic injection hardware.

On top of that, the ZR1's carbon-fiber hood featured a transparent polycarbonate panel that showed off the LS9's beautiful blue engine cover and aluminium intercooler. The highly visible cover's sparkling color commemorated the "Blue Devil" reference used internally during ZR1 development, itself a nickname reportedly chosen in honor of Wagoner's alma mater, North Carolina's Duke University.

The Corvette lineup remained impressive as ever in 2009 with the 505-horsepower Z06 coupe (pictured) still in action, along with the base 430-horse LS3 model, again available with or without a top. The new-for-2009 ZR1 (introduced the year before) represented an outrageous exclamation point to the latest chapter in Chevy's long-running Corvette tale.

C6 2005–2013

Simply doing the Z06's existing LS7 V-8 one better wasn't enough, so GM label-makers opted to assign the LS9 tag to the ZR1's awesome 638-horsepower 6.2-liter V-8. An intercooled Eaton supercharger hid beneath that blue "beauty cover."

The polycarbonate window in the ZR1's carbon-fiber hood allowed curious gawkers a look at the supercharged 638-horsepower beast beneath. Its widened fenders were also made of carbon fiber.

With a top speed surpassing 200 miles per hour, the 2009 ZR1 stood taller than tall as the fastest General Motors' regular-production vehicle ever. Car and Driver called it "a 638-horsepower flip-off to those in the mainstream media who think GM doesn't build any good cars." At $106,000, Chevrolet's 2009 ZR1 represented the most expensive General Motors product ever. Production was 1,415.

Weight-saving carbon fiber was used throughout the body to help compensate for the nearly 200 pounds of extra weight added beneath that peek-a-boo hood to help make those 638 horses. This race-bred material also made up the widened fenders, roof panel and bows, rocker moldings, and the wind-tunnel-tested splitter added to the Z06-style front fascia. A special clearcoat also was developed to allow body designers to leave the weave patterns inherent to carbon-fiber construction exposed—to further enhance the ZR1's "visual identity"—without any fears of the stuff yellowing or dulling like it normally does upon exposure to ultraviolet light.

Additional highlights include a specially tuned suspension capable of more than 1.05g on the skid pad, huge 20-spoke wheels (19-inchers in front, 20s at the rear), massive Michelin Pilot Sport 2 tires, a high-capacity dual-disc clutch, and enormous carbon-ceramic Brembo discs measuring 15.5 inches across at the nose, 15 outback. Like the Z06, the ZR1 came only with a six-speed-manual.

Summed up, the ZR1 not only predictably outperformed its Z06 running mate, it also offered a kinder, gentler nature. "The ZR1 is a car anyone can drive confidently and comfortably," claimed Corvette vehicle line executive Tom Wallace. "From the very beginning, refinement, balance, and compliance were targets that were as important as the car's maximum performance." According to chief engineer Juechter, "It's a supercar that doesn't sacrifice ride quality for performance."

Can you say "best buy"?

CHAPTER SEVEN

2010

Model availability	hatchback coupe (with removable roof panel) and convertible; Grand Sport hatchback coupe (with removable roof panel) and convertible; Z06 and ZR1 hatchback coupes (with fixed magnesium roof)
Construction	composite body on hydroformed steel (aluminum for the Z06 and ZR1) frame with aluminum and magnesium structural components (unique carbon-fiber body panels for Z06/ZR1/Grand Sport)
Wheelbase	105.7 inches
Length	174.6 inches (175.6 inches, Z06/Grand Sport; 176.2 inches, ZR1)
Width	72.6 inches (75.9 inches, Z06/ZR1/Grand Sport)
Height	49 inches (48.7 inches, Z06/ZR1/Grand Sport)
Track (front/rear, in inches)	62.1/60.7 (coupe and convertible); 63.5/62.5 (Z06/ZR1/Grand Sport)
Curb weight	3,217 pounds (coupe), 3,246 pounds (convertible), 3,210 pounds (Z06), 3,350 pounds (ZR1), 3,311 pounds (Grand Sport coupe), 3,289 pounds (Grand Sport convertible)
Wheels (coupe and cvy)	18x8.5 inches front, 19 x10 inches rear
Wheels (Z06 and Grand Sport)	18x9.5 inches front, 19x12 inches rear
Wheels (ZR1)	19x10 inches front, 20x12 inches rear
Tires (coupe and cvt)	Goodyear Eagle F1 Supercar Extended Mobility (EMT); P245/40ZR18 front, P285/35ZR19 rear
Tires (Z06 and Grand Sport)	Goodyear Eagle F1 Supercar EMT; P275/35ZR18 front, P325/30ZR19 rear
Tires (ZR1)	Michelin Pilot Sport PS2 ZP; P285/30ZR19 front, P335/25ZR20 rear
Brakes	power-assisted four-wheel discs with ABS and electronic traction control
Brakes (ZR1)	power-assisted four-wheel drilled ceramic discs with ABS and electronic traction control; six-piston calipers front, four-piston calipers rear
Brake dimensions	12.8x1.26 inches front, 12x1 inches rear
Brake dimensions (Z06 and Grand Sport)	14x1.3 inches front, 13.4x1 inches rear, cross-drilled rotors
Brake dimensions (ZR1)	15.5 inches front, 15.0 inches rear
Fuel tank	18 gallons
Suspension	short- and long-arm (SLA) double wishbone cast-aluminum control arms, transverse composite leaf springs, and stabilizer bars front and rear; monotube shock absorbers, Active Handling electronics (Magnetic Selective Ride control with ZR1)
Steering	speed-sensitive, magnetic power-assisted rack and pinion (16.1:1 ratio)
Engine (LS3)	6.2-liter (376-ci) sequential fuel injected (SFI) V-8 with overhead valves and cast-aluminum cylinder block and heads
Compression	10.7:1
LS3 bore and stroke	4.06x3.62 inches
LS3 output	430 horsepower at 5,900 rpm, 424 lb-ft of torque at 4,600 rpm (436/428 with NPP dual mode exhaust option)
Engine (LS7)	7.0-liter (427-ci) SFI V-8 with overhead valves and cast-aluminum cylinder block and heads, 11:1 compression, dry-sump oiling
LS7 bore and stroke	4.125x4.00 inches
LS7 output	505 horsepower at 6,300 rpm, 470 lb-ft of torque at 4,800 rpm
Engine (LS9)	6.2-liter (376-ci) supercharged SFI V-8 with overhead valves and cast-aluminum cylinder block and heads, dry-sump oiling
Compression	9.1:1
LS9 bore and stroke	4.06x4.00 inches
LS9 output	638 horsepower at 6,500 rpm, 604 lb-ft of torque at 3,800 rpm
Standard transmission	six-speed manual with 2.66:1 low
Optional transmission	six-speed automatic with paddle shift
Axle ratio	3.42: 1 (2.56:1 with automatic)

2010

Visible updates for 2010 were once again minimal, and included an always popular paint choice that returned after a six-year hiatus. Last seen in 2004 (technically speaking, it did show up on one 2005 model), Torch Red again graced the Corvette palette, ending up on 2,249 cars that year, making it 2010's second favorite choice behind black (2,929).

Beneath the skin, side airbags were made standard on all 2010 Corvettes, and the six-speed paddle-shift box was again revised, this time to allow a less fussy return to automatic mode. Easily 2010's top technical story involved the introduction of Launch Control, a system that modulated engine torque up to 100 times per second to help maximize traction during full-throttle take-offs. Launch Control came standard on all manual-trans models. And new for the 2010 ZR1 was Performance Traction Management, which integrated the existing Traction Control, Active Handling, and Selective Ride Control systems to further maximize off-the-line abilities.

Last, but certainly not least, on the new-for-2010 list was yet another revived Corvette legend, this one last seen in 1996. First used in 1963 for Zora Duntov's five all-out competition Corvettes, the "Grand Sport" badge returned 33 years later to both honor those legendary lightweights and help send the retiring C4 breed out with a high-profile bang. A little commemoration was also a part of the plan in 2010, but the main goal involved combining much of the Z06's roadworthiness with as many of the base Corvette's amenities as possible.

For starters, the custom-bodied 2010 Grand Sport was offered as either a removable-roof coupe or convertible; the purposeful Z06 came only as a fixed-roof coupe. Additionally, a Grand Sport customer could've picked six-speed gearbox, manual, or paddle-shift auto; all Z06 drivers' souls were stirred with a stick. Manual-trans Grand Sports featured a little more standard purpose than their automatic counterparts: they were equipped with dry-sump oiling, a rear-mounted battery, and differential cooler, items best suited, of course, for the track.

All Grand Sports were powered by the LS3, with or without optional dual-mode exhausts, and all C6 paint choices carried over as well. Setting the GS apart was a unique body, widened at the corners, with "Grand Sport" badges integrated into the front fenders and Z06-style touches (front splitter, rear spoiler) at both ends. Functional brake cooling ducts also were standard, as were special painted wheels. Chromed wheels were optional. Standard tires were 275/35ZR18 in front, 325/30ZR19 out back. Z06-size brakes and a track-ready suspension completed the package. A Grand Sport buyer also could've added the Heritage option, which featured two-tone seats and competition-style front fender hash marks.

C6 2005–2013

Left: Chevrolet built a Corvette Stingray Concept for the summer 2009 movie *Transformers: Revenge of the Fallen*. Inspired by the original Corvette Stingray race car, the Autobot concept was dubbed Sideswipe.

Below: "Sideswipe represents an exercise in exploration for the Corvette," said Ed Welburn, vice president of GM Global Design. "By giving my creative team the freedom to design no-holds-barred vision concepts, it helps them push boundaries and look at projects from different perspectives."

CHAPTER SEVEN

Right: New for 2010 was the commemorative Grand Sport, which combined much of the Z06's might with as many base Corvette conveniences and amenities as possible. Grand Sports were available in both removable-roof coupe and convertible forms. Production was 3,707 for the former, 2,335 for the latter.

Below: Among the Grand Sport's unique visual cues were special side vents, Grand Sport badging, and optional fender hash mark stripes. Wheels were a Grand Sport–only style and available in standard silver (shown here) or optional competition gray or chrome.

> "For track use, the Z06 Carbon is the best balanced Corvette yet," explained Tadge Juechter.

Another specially packaged C6, the Carbon Limited Edition Z06 coupe, emerged in 2011 to commemorate the 50th anniversary of the Corvette's first appearance at Le Mans. Various ZR1 parts were standard.

As the Grand Sport arrived, the Z51 performance handling package departed. Grand Sport pricing was $55,720 for the coupe, $59,530 for the convertible. Production was 3,707 coupes, 2,335 convertibles.

2011

The 2010 lineup rolled on unchanged into 2011, with the 430-horse LS3 coupe and convertible again forging the way, followed a second time around by the Grand Sport coupe and convertible. The two manual-trans-only fixed-roof coupes, Z06 and ZR1, carried on too, still with 505 and 638 horsepower, respectively.

Mixing and matching model features in 2010 to produce the Grand Sport proved so much fun for product planners that they tried the trick again. And again. RPO Z07, the Ultimate Performance package, made the ZR1's adaptive suspension, carbon-ceramic brakes, and wheel/tire combo available to Z06 customers. Available as well for the 505-horse Corvette was a new Carbon Fiber package (CFZ) that added most of the ZR1's lightweight body parts. Grand Sport buyers, in turn, could've opted for Magnetic Ride Control.

Further in-breeding resulted in Chevrolet's special C6, the Z06 Carbon Limited Edition, a hybrid that not only served to commemorate the 50th anniversary of the Corvette's first appearance at Le Mans but also could've perhaps made competition history itself. "For track use, the Z06 Carbon is the best balanced Corvette yet," explained Tadge Juechter. "It combines the lightweight and naturally-aspirated Z06 engine with the road-holding and braking of the ZR1."

2011

Model availability	hatchback coupe (with removable roof panel) and convertible, Grand Sport hatchback coupe (with removable roof panel) and convertible, Z06 and ZR1 hatchback coupes (with fixed magnesium roof)
Construction	composite body on hydroformed steel frame (aluminum for the Z06 and ZR1) with aluminum and magnesium structural components (unique carbon-fiber body panels for Z06/ZR1/Grand Sport)
Wheelbase	105.7 inches
Length	174.6 inches (175.6 inches, Z06/Grand Sport; 176.2 inches, ZR1)
Width	72.6 inches (75.9 inches, Z06/ZR1/Grand Sport)
Height	49 inches (48.7 inches, Z06/ZR1/Grand Sport)
Track (front/rear, in inches)	62.1/60.7 (coupe and convertible); 63.5/62.5 (Z06/ZR1/Grand Sport)
Curb weight	3,208 pounds (coupe), 3,221 pounds (convertible), 3,175 pounds (Z06), 3,333 pounds (ZR1), 3,311 pounds (Grand Sport coupe), 3,289 pounds (Grand Sport convertible)
Wheels (coupe and cvt)	18x8.5 inches front, 19x10 inches rear
Wheels (Z06 and Grand Sport)	18x9.5 inches ront, 19x12 inches rear
Wheels (ZR1)	19x10 inches front, 20x12 inches rear
Tires (coupe and cvt)	Goodyear Eagle F1 Supercar Extended Mobility (EMT); P245/40ZR18 front, P285/35ZR19 rear
Tires (Z06 and Grand Sport)	Goodyear Eagle F1 Supercar EMT; P275/35ZR18 front, P325/30ZR19 rear
Tires (ZR1)	Michelin Pilot Sport PS2 ZP; P285/30ZR19 front, P335/25ZR20 rear
Brakes	power-assisted four-wheel discs with ABS and electronic traction control
Brakes (ZR1)	power-assisted four-wheel drilled ceramic discs with ABS and electronic traction control; six-piston calipers front, four-piston calipers rear
Brake dimensions	12.8x1.26 inches front, 12x1 inches rear
Brake dimensions (Z06 and Grand Sport)	14x1.3 inches front, 13.4x1 inches rear, cross-drilled rotors (ceramic rotors for Z06 equipped with Z07 option)
Brake dimensions (ZR1)	15.5 inches front, 15.0 inches rear
Fuel tank	18 gallons
Suspension	short- and long-arm (SLA) double wishbone cast-aluminum control arms, transverse composite leaf springs, and stabilizer bars front and rear; monotube shock absorbers, Active Handling electronics (Magnetic Selective Ride control with ZR1)
Steering	speed-sensitive, magnetic power-assisted rack and pinion (16.1:1 ratio)
Engine (LS3)	6.2-liter (376-ci) sequential fuel injected (SFI) V-8 with overhead valves and cast-aluminum cylinder block and heads
Compression	10.7:1
LS3 bore and stroke	4.06x3.62 inches
LS3 output	430 horsepower at 5,900 rpm, 424 lb-ft of torque at 4,600 rpm (436/428 with NPP dual mode exhaust option)
Engine (LS7)	7.0-liter (427-ci) SFI V-8 with overhead valves and cast-aluminum cylinder block and heads, 11:1 compression, dry-sump oiling
LS7 bore and stroke	4.125x4.00 inches
LS7 output	505 horsepower at 6,300 rpm, 470 lb-ft of torque at 4,800 rpm
Engine (LS9)	6.2-liter (376-ci) supercharged SFI V-8 with overhead valves and cast-aluminum cylinder block and heads, dry-sump oiling
Compression	9.1:1
LS9 bore and stroke	4.06x4.00 inches
LS9 output	638 horsepower at 6,500 rpm, 604 lb-ft of torque at 3,800 rpm
Standard transmission	six-speed manual with 2.66:1 low
Optional transmission	six-speed automatic with paddle shift
Axle ratio	3.42: 1 (2.56:1 with automatic)

Above: On November 29, 2011, Chevrolet invited various engineering veterans, execs and VIPs to GM's Performance Build Center in Wixom, Michigan, to help bolt together Chevrolet's 100 millionth small-block V-8, a 638-howsepower LS9. Present that day was automotive scribe and longtime Corvette whisperer Don Sherman, who was responsible for installing the number 6 piston/rod assembly.

Indeed, the ZR1's Magnetic Selective Ride Control and big Brembo carbon-ceramic brakes (fitted with specific dark gray metallic calipers) were standard for the Carbon Limited Edition, as were black 20-spoke wheels measuring 19 inches wide in front, 20 in back. Included too were the CFZ body parts, a special leather/suede interior, various specific logs, and black-accented headlamps. Only two exterior finishes were offered: Inferno Orange and Supersonic Blue. Production ended at 252.

2012

Incorporated on November 3, 1911, Chevrolet wasn't about to turn 100 without commemorating the centennial anniversary. In April 2011, company officials announced the upcoming arrival of the Centennial Edition Corvette, a 2012 offering that helped mark a century of business in "sinister-looking" fashion (Chevy's own words). Exclusive to the Centennial Edition package (RPO ZLC) was Carbon Flash Metallic paint, the only darker-than-dark finish available in 2012. The ever-popular basic black (code 41U) was eliminated from the Corvette palette that year because the plant could not paint straight black and carbon flash. Satin-black exterior graphics darkened the image further, as did unique aluminum wheels, also done in satin-black.

CHAPTER SEVEN

Dressed in Carbon Flash Metallic finish, the 2012 Centennial Edition Corvette Z06 managed to be both "sinister" and sexy at the same time.

With the ZLC package offered on all 2012 models, those extra-light "Cup-style" rims varied slightly with the application: smaller ones (18-inch fronts, 19 rears) went on the base and Grand Sport versions, larger (19 fronts, 20 rears) on the Z06/ZR1. A thin red accent stripe ran around the Cup wheels' perimeters in all cases save for the Grand Sport's (yet another curious omission). Red-painted brake calipers, already standard in the Z06's case, also complemented things, and Magnetic Selective Ride Control was included in the deal to help maximize the Centennial Edition driving experience.

All ZLC Corvettes featured ebony interiors, and a black top was also required in convertible applications. Leather-wrapped dash and door panels were accented with red stitching, as were the seats, console, shifter, and steering wheel. Headrests were embossed with special centennial logos, and Louis Chevrolet's image appeared in graphics added to the steering wheel hub, wheel center caps, and B-pillars.

Chevrolet built 5,056 Grand Sport coupes for 2012, making it that year's most popular model for the third straight year. The Grand Sport convertible (2,268 built) ranked third on the 2012 list.

Right: Chevrolet's familiar trim level pecking order (standard 1LT, 2LT, 3LT, and 4LT) rolled over into 2012 for base models and Grand Sports (coupe and convertible in both cases) and each of the three options again added extra preferred equipment and features into the mix. The production count for the 2012 Grand Sport convertible (shown here) was 2,268.

250

2012

Model availability	hatchback coupe (with removable roof panel) and convertible, Grand Sport hatchback coupe (with removable roof panel) and convertible, Z06 and ZR1 hatchback coupes (with fixed magnesium roof)
Construction	composite body on hydroformed steel frame (aluminum for the Z06 and ZR1) with aluminum and magnesium structural components (unique carbon-fiber body panels for Z06/ZR1/Grand Sport)
Wheelbase	105.7 inches
Length	174.6 inches (175.6 inches, Z06/Grand Sport; 176.2 inches, ZR1)
Width	72.6 inches (75.9 inches, Z06/ZR1/Grand Sport)
Height	49.1 inches (48.7 inches, Z06/ZR1/Grand Sport)
Track (front/rear, in inches)	62.1/60.7 (coupe and convertible); 63.5/62.5 (Z06/ZR1/Grand Sport)
Curb weight	3,208 pounds (coupe), 3,221 pounds (convertible), 3,199 pounds (Z06), 3,353 pounds (ZR1), 3,311 pounds (Grand Sport coupe), 3,289 pounds (Grand Sport convertible)
Wheels (coupe and cvt)	18x8.5 inches front, 19 x10 inches rear
Wheels (Z06 and Grand Sport)	18x9.5 inches front, 19x12 inches rear
Wheels (ZR1)	19x10 inches front, 20x12 inches rear
Tires (coupe and cvt)	Goodyear Eagle F1 Supercar Extended Mobility (EMT); P245/40ZR18 front, P285/35ZR19 rear
Tires (Z06 and Grand Sport)	Goodyear Eagle F1 Supercar EMT; P275/35ZR18 front, P325/30ZR19 rear
Tires (ZR1)	Michelin Pilot Sport PS2 ZP; P285/30ZR19 front, P335/25ZR20 rear
Brakes	power-assisted four-wheel discs with ABS and electronic traction control
Brakes (ZR1)	power-assisted four-wheel drilled ceramic discs with ABS and electronic traction control; six-piston calipers front, four-piston calipers rear
Brake dimensions	12.8x1.26 inches front, 12x1 inches rear
Brake dimensions (Z06 and Grand Sport)	14x1.3 inches front, 13.4x1 inches rear, cross-drilled rotors (ceramic rotors for Z06 equipped with Z07 option)
Brake dimensions (ZR1)	15.5 inches front, 15.0 inches rear
Fuel tank	18 gallons
Suspension	short- and long-arm (SLA) double wishbone cast-aluminum control arms, transverse composite leaf springs, and stabilizer bars front and rear, monotube shock absorbers, Active Handling electronics (Magnetic Selective Ride control w/ZR1)
Steering	speed-sensitive, magnetic power-assisted rack and pinion (16.1:1 ratio)
Engine (LS3)	6.2L (376 cubic inches) sequential fuel injected (SFI) V8 with overhead valves and cast-aluminum cylinder block and heads
Compression	10.7:1
LS3 bore & stroke	4.06 x 3.62 inches
LS3 output	430 horsepower at 5,900 rpm, 424 lb-ft of torque at 4,600 rpm (436/428 with NPP dual mode exhaust option)
Engine (LS7)	7.0L (427 cubic inches) SFI V8 with overhead valves and cast-aluminum cylinder block and heads, dry-sump oiling
Compression	11:1
LS7 bore & stroke	4.125 x 4.00 inches
LS7 output	505 horsepower at 6,300 rpm, 470 lb-ft of torque at 4,800 rpm
Engine (LS9)	6.2L (376 cubic inches) supercharged SFI V8 with overhead valves and cast-aluminum cylinder block and heads, dry-sump oiling
Compression	9.1:1
LS9 Bore & stroke	4.06 x 4.00 inches
LS9 Output	638 horsepower at 6,500 rpm, 604 lb-ft of torque at 3,800 rpm
Standard transmission	six-speed manual with 2.66:1 low
Optional transmission	six-speed automatic w/Paddle Shift
Axle ratio	3.42: 1 (2.56:1 with automatic)

"Louis Chevrolet was a fearless racing pioneer who also designed our first car," explained marketing vice president Rick Scheidt while ntroducing the ZLC option. "Corvette is a natural fit to honor that legacy while creating a compelling new package for sports car drivers. The Centennial Edition not only celebrates our 100 years, it once again pushes Corvette forward for a new generation of fans."

While nowhere near as noticeable as the ZLC package, various other upgrades also helped entice that new generation, per Chevrolet's ongoing plan. "We constantly strive to make the Corvette a better car on the road and the track," said chief engineer Tadge Juechter of that year's Corvette. "For 2012, the lineup achieves its highest performance level ever, while at the same time being easier to drive and enjoy thanks to several changes and new features inside the car."

Interior updates included improved seats with larger bolsters, the better to keep occupants more firmly planted during hot-blooded heel-toe action. Optional microfiber suede seat inserts also appeared to help heighten tactile seat-of-the-pants responses. New too was a revised steering wheel (featuring wrapped spokes, streamlined switch trim, and model-specific badges) and a padded center console/armrest for all models. On the outside, Carlisle Blue Metallic paint replaced 2011's Jetstream Blue Metallic Tintcoat among 2012's exterior color choices, and buyers also could personalize their wheels with colored brake calipers done in red, yellow, silver, or gray.

The LS3 line's familiar trim-level pecking order (1LT, 2LT, 3LT, 4LT) carried over from 2011 but was enhanced with a new technology package for the second-tier option. The 2012 2LT package included a navigation system, USB port, head-up display, Bluetooth phone connectivity, and upgraded Bose audio equipment with nine speakers, two more than the previous year. As in 2011, the LS7-powered Z06 again was offered in 1LZ, 2LZ, and 3LZ trim levels, while the supercharged ZR1 stepped up only once, from 1ZR to 3ZR. The Centennial Edition option was limited to the top-shelf 3LZ and 3ZR for the two hottest 2012 Corvettes, while base model and Grand Sport customers could combine the commemorative ZLC package with either 3LT or 4LT builds.

Upgrades for the 505-horsepower Z06 in 2012 included an optional carbon-fiber hood and an improved Ultimate Performance Package (Z07) that now included a full-width racing-style rear spoiler, new Michelin Pilot Sport (PS) Cup tires on black Cup wheels, and the Performance Traction Management (PTM) system previously limited to the ZR1. Essentially street-legal racing tires optimized for warm, dry conditions, those Michelins worked in concert with the advanced automatic torque-metering PTM hardware to raise Corvette road-worthiness to all-new heights. Chevrolet engineers

Above: In 2012, the bodacious 638-horse ZR1 was back and meaner than ever, especially when equipped with the new PDE High-Performance Package. Quarter-mile performance for this beast was 11.5 seconds at 129 mph, according to *Road & Track*.

The Corvette lineup carried over unchanged into 2012: base coupe and convertible, Grand Sport coupe and convertible, and the two "Z-car" coupes, Z06 and ZR1. Total 2012 production was 11,647, including 2,820 LS3 coupes.

New super-sticky Michelin Pilot Sport Cup tires appeared along with a revised Ultimate Performance Package (Z07) for the 2014 Z06, and this race-ready rubber was mounted on equally new, super-light Cup-style wheels. The ZR1's all-new High-Performance Package (PED) also included this cutting-edge wheel and tire combination, and the 2012 Centennial Edition option featured various sizes of Cup wheels.

boldly promised both an 8-percent increase (to more than 1.1g) in maximum lateral acceleration and notably shorter braking distances. *Motor Trend* road testers proved them right on, managing a superb 1.13g on their skid pad while braking (after a 0–60 run) in an astonishing 94 feet, 8 feet less than a comparable 2011 Z06 needed to complete the same test.

"Capable tires are one of the simplest, most cost-effective performance enhancements that can be made to a car," wrote *Motor Trend*'s Rory Jurnecka in August 2011 after living the new Z07 experience. "With those wide, gooey Michelins warmed up, this thing gripped our test track with the tenacity of a wad of chewing gum on a warm summer's day, enveloping the asphalt in damn-near-race-spec, barely DOT-legal, pray-it-doesn't-rain-type rubber. . . . The downside of so much grip is that even with heavier bolstering for 2012, the 'Vette's seats are wholly inadequate for performance driving, leaving the pilot looking for support when he should be looking for apexes." Sizes for the track-ready Michelins were P285/20ZR19 in front, P335/25ZR20 in back.

A similar maximum-effort option—the High-Performance Package (PDE)—was introduced in 2012 for the LS9-powered ZR1 and included the same spoiler, lighter-than-light wheels (each weighing about 5 pounds less than a standard ZR1 rim), and Michelin PS rubber offered in the Z07 deal, plus a special six-speed manual transmission featuring even closer gear ratios than the 205-mph Corvette's standard close-ratio six-speed. Road test results for the Z07/Z06's big, bad, blown brother were predictably eye-popping. In June 2011, a PDE-equipped ZR1 lapped Germany's legendary 12.9-mile Nürburgring road course in 7 minutes, 19.63 seconds, beating the 638-horse breed's previous best time there (set in 2008) by more than 6 seconds. That's 6 seconds, not 0.6.

As for numbers closer to home, according to *Road & Track*, a 2012 PDE ZR1 required a scant 3.5-second to run 0–60 and finished the quarter mile in 11.5 seconds, topping out at 129 mph. In a *Motor Trend* test, a Z07/Z06 ran from rest to 60 mph in 3.8 seconds and posted a quarter-mile time slip of 11.9/122.5. How out-of-this-world fast you wanted to go hinged simply on how much cash you had stashed. *Road & Track*'s seriously option-loaded ZR1 carried a $129,949 price tag, while a fully loaded Z06 went for at least a hundred grand in 2012. Base prices for the two in relative bare-bones form were $75,600 for the Z06, $112,600 for the ZR1.

Eye-popping indeed.

2013

Honorary pace cars have led the field around "the Brickyard" prior to the start of the Indianapolis 500 ever since the race's inaugural running in 1911, but it wasn't until 1948 that the first Chevrolet, a Fleetmaster convertible, did the trick. More

2013

Model availability	hatchback coupe (with removable roof panel) and convertible, Grand Sport hatchback coupe (with removable roof panel) and convertible, 427 Convertible Collector Edition, Z06 and ZR1 hatchback coupes (with fixed magnesium roof)
Construction	composite body on hydroformed steel frame (aluminum for the Z06 and ZR1) with aluminum and magnesium structural components (unique carbon-fiber body panels for Z06/ZR1/Grand Sport/427 Convertible Collector Edition)
Wheelbase	105.7 inches
Length	174.6 inches (175.6 inches, Z06/Grand Sport/427 Convertible Collector Edition; 176.2 inches, ZR1)
Width	72.6 inches (75.9 inches, Z06/ZR1/Grand Sport/427 Convertible Collector Edition)
Height	49.1 inches (48.7 inches, Z06/ZR1/Grand Sport/427 Convertible Collector Edition)
Track (front/rear, in inches)	62.1/60.7 (coupe and convertible), 63.5/62.5 (Z06/ZR1/Grand Sport/427 Convertible Collector Edition)
Curb weight	3,208 pounds (coupe), 3,239 pounds (convertible), 3,175 pounds (Z06), 3,333 pounds (ZR1), 3,311 pounds (Grand Sport coupe), 3,289 pounds (Grand Sport convertible), 3,355 pounds (427 Convertible Collector Edition)
Wheels (coupe and cvt)	18x8.5 inches front, 19 x10 inches rear
Wheels (Z06 and Grand Sport)	18x9.5 inches front, 19x12 inches rear
Wheels (ZR1 and 427 Convertible Collector Edition)	19x10 inches front, 20x12 inches rear
Tires (coupe and cvt)	Goodyear Eagle F1 Supercar Extended Mobility (EMT); P245/40ZR18 front, P285/35ZR19 rear
Tires (Z06 and Grand Sport)	Goodyear Eagle F1 Supercar EMT; P275/35ZR18 front, P325/30ZR19 rear
Tires (ZR1 and 427 Convertible Collector Edition)	Michelin Pilot Sport PS2 ZP; P285/30ZR19 front, P335/25ZR20 rear
Tires (Grand Sport)	P275/35ZR18 front, P325/30ZR19 rear
Brakes	power-assisted 4-wheel discs with ABS and electronic traction control
Brakes (ZR1)	power-assisted four-wheel drilled ceramic discs with ABS and electronic traction control; 6-piston calipers front, 4-piston calipers rear
Brake dimensions	12.8x1.26 inches front, 12x1 inches rear
Brake dimensions (Z06/Grand Sport/427 Convertible Collector Edition)	14x1.3 inches front, 13.4x1 inches rear, cross-drilled rotors (ceramic rotors for Z06 equipped with Z07 option)
Brake dimensions (ZR1)	15.5 inches front, 15.0 inches rear
Fuel tank	18 gallons
Suspension	short- and long-arm (SLA) double wishbone cast-aluminum control arms, transverse composite leaf springs, and stabilizer bars front and rear; monotube shock absorbers, Active Handling electronics (Magnetic Selective Ride control with ZR1)
Steering	speed-sensitive, magnetic power-assisted rack and pinion (16.1:1 ratio)
Engine (LS3)	6.2-liter (376-ci) sequential fuel injected (SFI) V-8 with overhead valves and cast-aluminum cylinder block and heads
Compression	10.7:1
LS3 bore and stroke	4.06x3.62 inches
LS3 output	430 horsepower at 5,900 rpm, 424 lb-ft of torque at 4,600 rpm (436/428 with NPP dual mode exhaust option)
Engine (LS7)	7.0-liter (427-ci) SFI V-8 with overhead valves and cast-aluminum cylinder block and heads, dry-sump oiling
Compression	11:1
LS7 bore and stroke	4.125x4.00 inches
LS7 output	505 horsepower at 6,300 rpm, 470 lb-ft of torque at 4,800 rpm
Engine (LS9)	6.2-liter (376-ci) supercharged SFI V-8 with overhead valves and cast-aluminum cylinder block and heads, dry-sump oiling
Compression	9.1:1
LS9 bore and stroke	4.06x4.00 inches
LS9 output	638 horsepower at 6,500 rpm, 604 lb-ft of torque at 3,800 rpm
Standard transmission	six-speed manual with 2.66:1 low
Optional transmission	six-speed automatic with paddle shift
Axle ratio	3.42: 1 (2.56:1 with automatic)

CHAPTER SEVEN

curiously, a Corvette didn't pace the race until 1978. Go figure.

Fortunately, Chevrolet wasted little time catching up. By May 2011, 22 Chevys had served as Indy 500 pacer cars, including 10 in a row beginning in 2002—both runaway records. And 10 of those 22 pacers were Corvettes, including 5 straight from 2004 to 2008, records too for one single model. Can you say "dynasty"?

A ZR1 coupe made it 11 in a row for the bowtie brand in May 2012, and for the second time an Indy pace car Corvette also predicted the future. In 2002, a preproduction 2003 C5 coupe had hit the bricks on race day to demonstrate the upcoming 50th Anniversary appearance package. Ten years later, Chevrolet did it again: 2012's Indy pacer was a 2013 ZR1 adorned in 60th Anniversary graphics. In both cases, some customers ended up feeling a bit jilted, considering hopes had been high that each of those birthday celebrations also would've marked the arrival of the latest next-generation Corvette. Oh well.

Company officials, on the other hand, were once again pleased as punch to see a C6 make the pace car grade in 2012. "I can't think of a better way to mark the 60th anniversary of Corvette than having it lead the starting field of the Indianapolis 500," beamed Jim Campbell, General Motors' vice president of Performance Vehicles and Motorsports. "Corvette embodies pure performance, so the ZR1 is a perfect car to pace the most prestigious auto race in the world," added Indianapolis Motor Speedway Corporation President and CEO Jeff Belskus.

The 638-horsepower Corvette easily represented the most powerful (not to mention most expensive) production-based Indy 500 pace car to date, and street-going copies surely would've made prized collectibles. Chevrolet product planners, however, opted against offering a replica package, but that wasn't necessarily a bad thing considering what did end up on customers' plates in 2013. Joining the aforementioned commemorative package that year was a new model, the 427 Convertible Collector Edition, a mean machine meant to honor an end of an era.

"The 2013 model year will be historic for Corvette, marking its 60th anniversary and the final year for the current 'C6' generation," explained Chevrolet marketing vice president Chris Perry in January 2012 while announcing the breed's third Collector Edition rendition (following predecessors in 1982 and 1996). "We couldn't think of a more fitting way to celebrate these milestones than bringing back one of the most-coveted combinations in the brand's history—the Corvette convertible and a 427-cubic-inch engine."

Incorporating various Z06 and ZR1 components, the 427 Convertible represented the most potent topless Corvette yet. It was based on the LS3 line's steel foundation, but it relied on a collection of

A 2013 ZR1 coupe became the 11th Corvette to pace the Indianapolis 500 in May 2012, and this appearance also marked the 11th straight year a Chevrolet product served as the prestigious Indy pace car. ZR1 production for 2013 was 482.

carbon-fiber body panels (hood, fenders, and door panels) to help keep weight way down, making it easier for its 505-horsepower LS7 engine to propel it where few ragtops had gone before. Its excellent power-to-weight ratio, 1 horsepower for every 6.64 pounds of mass, bettered some of the world's most coveted convertibles, including the Porsche 911 Turbo S Cabriolet (6.90:1), Audi R8 RSI Spyder (7.58:1), Aston Martin DBS Volante (7.82:1), and Ferrari California (8.31:1). Performance estimates from Chevrolet included a 190-mph top end, 0–60 in 3.8 seconds, an 11.8-second quarter-mile, and 1.04g lateral acceleration.

Additional standard equipment included the LS7's requisite six-speed manual transmission, Magnetic Selective Ride Control, a rear-mounted battery, and Michelin PS2 tires mounted on the big lightweight Cup wheels that had debuted the year before as part of the Z07 and PDE options. Standard Cup wheels in the 427 Convertible's case featured machined faces with gray-painted pockets. Black Cup wheels and chromed ZR1 rims were optional, as was the Carbon Fiber Package (CFZ) introduced in 2011 as standard equipment for the Carbon Limited Edition Z06. The CFZ option included lightweight rockers, a front splitter, and a full-width rear spoiler, all done in carbon-fiber.

Optional too, of course, was the $1,425 60th Anniversary Design Package (RPO Z25), which was available on all 2013 models, with top or without. Included in the Z25 deal was Arctic White paint, a Blue Diamond leather interior accented in suede, a ZR1-style spoiler, expected anniversary badges, gray-painted brake calipers, and "60th" logos on the wheel centers, steering wheel, and headrests. A blue top was added to Z25 convertibles, and a pair of Pearl Silver Blue racing stripes (Z30) also could go on at extra cost. On a Z25 coupe, the Z30 stripes ran the full length of the car; Z30-equipped 60th Anniversary convertibles featured "tonal stitch" accents on their soft tops.

All 2013 Corvettes, Z25-equipped or not, received special 60th Anniversary identification inside and out, while most other general features and options predictably carried over unchanged from 2012. Exceptions included that year's Carbon Flash Metallic (exclusive to the one-hit-wonder Centennial Edition Corvette) and Carlisle Blue paints, which both faded into the archives. Night Race Blue Metallic was new for 2013, and good ol' black (the Corvette's most popular shade from 2004 to 2011) was back after a one-year hiatus. Also new were two options—black-painted wheels and the ZR1-style rear spoiler—for base LS3 Corvettes and Grand Sports alike.

With the close of the 2013 model run came the end of the sixth chapter in an epic book six decades in the making. Once again, it was time to turn the page.

Like the ZR1, the Z06 coupe also made one final appearance in 2013 to help send off the Corvette's sixth generation with a major bang. The Z06 production count that final year was 471.

CHAPTER SEVEN

Above: New for 2013 was the 427 Collector Edition convertible, the hottest topless Corvette yet thanks to the inclusion of various lightweight body panels and the Z06's 505-hp LS7 engine. The production count for this 190-mph droptop was 2,552.

Right: Available on all 2013 models, with or without tops, was the Z25 60th Anniversary Design Package, which among other things added Arctic White Paint. Optional Pearl Silver Blue racing stripes also could've accented the Z25 package. All 2013 Corvettes, Z25-equipped or not, received commemorative 60th anniversary badges.

Also rolling over for one final fling in 2013 were the two Grand Sport Corvettes. Chevrolet rolled out 1,756 Grand Sport convertibles that year. The tally for the last C6 Grand Sport coupe was 4,908.

CHAPTER EIGHT

08 World Beater

C7 2014–2019

Most recognize that history often repeats itself, but how many know it also sometimes stutters? Consider the Corvette legacy, now in its seventies. Chevrolet has always had trouble issuing its eagerly awaited next-generation models, with delays ranging from one year for the C3 to about five for the radically redesigned C5. Not to tinker with tradition, the C6 obligingly followed suit, once more to the dismay of some customers, who'd this time hoped to kick off the breed's second half-century with a bang. As in 1993, Chevy again served up leftovers during the Corvette's latest big birthday bash in 2003.

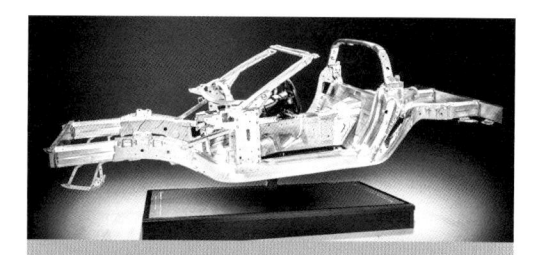

- Stingray concept vehicle debuts at Chicago auto show (February 2009)
- Last seen in 1976, the fabled "Stingray" nameplate returns as C7 debuts at Detroit North American International Auto Show (January 2013)
- Stingray convertible debuts at Geneva Motor Show (March 2013)
- Preproduction Corvette Stingray coupe becomes 12th Corvette to pace the Indianapolis 500 (May 2013) — C7s also serve as Indy 500 pace cars in 2015, 2017, 2018, and 2019
- Manufactured in Tonawanda, New York, the C7's new gen 5 V-8 becomes the third Corvette small-block to wear the familiar "LT1" label (2014)
- The C7 coupe features rear quarter windows, last seen on the solid-axle Corvette's optional removable hardtop. (2014)
- Z06 model returns, this time in coupe and convertible forms with 650 horsepower (2015)
- Grand Sport coupe and convertible return (2017)
- ZR1 is reborn (also in both coupe and convertible forms) with 755-hp LT4 small-block (2019)

FOR SOME, THE DISAPPOINTMENT didn't end there. When sixth-gen models finally did show in 2005, more than one journalist tepidly described it as a warmed-over C5, albeit reheated to record highs. But, while technically correct to some degree, those hard-to-please critics couldn't deny that the 400-horse C6 nonetheless clearly represented the best Vette yet, and their mild complaints soon were completely drowned out by rave reviews for the reborn Z06 in 2006, including *AutoWeek*'s conclusion that it was "the best supercar buy of all time." The reincarnated ZR1 then bumped the mercury even higher in 2009, leaving some witnesses asking just how much more the thermometer could stand.

By then the rumor mill already had been churning overtime in response to such queries. As early as 2007, press sources had been reporting yet another bit of déjà vu: the next great Corvette finally would feature the midengine platform long imagined by Zora Duntov. Indeed, GM Vice Chairman Bob Lutz reportedly was very much in favor (though not at first — see chapter 9) of taking the Corvette "global," of moving its engine amidships to help it match up more seriously with the likes of Ferrari's supercars. And the initial word had this unquestionably new C7 arriving in time to help mark Chevrolet's 100th anniversary in 2011.

What happened?

Yes, a midengine C7 (perhaps fitted with a Saab-sourced dual-clutch transaxle) was in initial planning stages before GM's finances came crashing down in 2008, leading to the mega-corporation's Chapter 11 bankruptcy filing in June 2009. No way, at that time, was the funding available for such an expensive redesign, and some curbside kibitzers were even heard wondering (again, yet again) whether the Corvette was finally about to roll off into the sunset, Washington's bailout notwithstanding.

Doomsayers also were inspired by a new Corporate Average Fuel Economy (CAFE)

Opposite: A preproduction Stingray coupe became the 12th Corvette (and 24th Chevrolet vehicle) to pace the Indianapolis 500 in May 2013. No plans for street-going replicas were announced.

CHAPTER EIGHT

Previously, Chevrolet had introduced its upcoming C6 coupe in January 2004 at the Detroit Auto Show and unveiled its convertible running mate in Geneva, Switzerland, two months later. The same introductory plan was used in the 2014 Stingray's case, and the new convertible again went on sale late in the year a few months after its coupe counterpart.

standard—signed into law by President George W. Bush in December 2007 as part of his Energy Independence and Security Act—which was scheduled to reach 35 miles per gallon by 2020. Low-production models like the Corvette aren't inhibited by CAFE ratings (originated by Congress in 1975) that much, but implications still weighed heavily on future development, and not just of Chevy's two-seater. As Chevrolet engineer Tadge Juechter told the *Kansas City Star* late in 2007, the entire sports car breed could become an endangered species due to rising CAFE demands. "High-performance vehicles may actually be legislated out of existence," said the man with a finger most firmly planted on the Corvette's pulse.

Though it has yet to morph into the death knell Juechter spoke of, the heightened 2020 CAFE minimum did have a predictably immediate impact on Corvette development. Coincidence or not, the C7's scheduled intro was pushed back to 2012, right about the time word of the upcoming 35-miles-per-gallon standard first made news. Additional rumors of a switch from eight cylinders to six also started making the rounds soon after, to the utter dismay of nearly every Corvette crazy who caught wind of such outrageousness.

Of course, GM's developing recession-era cash crunch obviously did the most to influence early postponement decisions, as did the plain fact that the C6 was selling better than ever in 2007. Production dropped slightly (from 37,372 in 2005) to 34,021 in 2006, but it soared the following year to 40,561 and remained healthy (35,310) in 2008 before plummeting demand late that year inspired GM to temporarily shutter the Bowling Green plant more than once. The depressed count for 2009 was 16,956; followed by 12,194 in 2010; 13,598 in 2011; and 11,647 in 2012.

Alarming news late in 2008 also included an unexpected changing of the guard as four key players retired on November 1, including top man Tom Wallace, after only a couple years leading the engineering team. Retiring along with Wallace were marketing man Gary Claudio, Ron Meegan (lead engineer for the LS3, LS7, and LS9 V-8s) and longtime GM performance guru John Heinricy, while Tadge Juechter stepped up to become the Corvette's fifth chief engineer.

GM officials made no comment concerning Wallace's untimely abdication, but some industry-watchers speculated that his exit was the result of a difference of opinion concerning C7 plans, which had been on indefinite hold since 2007. "Will the Corvette live on?" asked *Motor Trend*'s Matt Stone in November 2008 while blogging about Wallace's departure. "Yes, but there's no question that its development will stutter, and likely be pushed back several years beyond the C7's projected 2012 launch date." Wallace made it clear that Chevy's niche-market machine would roll on regardless, telling *AutoWeek*'s Mac Morrison in October 2008 that "we're still bullish on Corvettes." Maybe so, but "sheepish" sure looked then to be a fair description for the breed's redesign plans, at least as far as the impatient automotive press was concerned.

"No one has suggested that the model is in danger of becoming extinct," added Morrison. "But with the next car's characteristics not even

C7 2014–2019

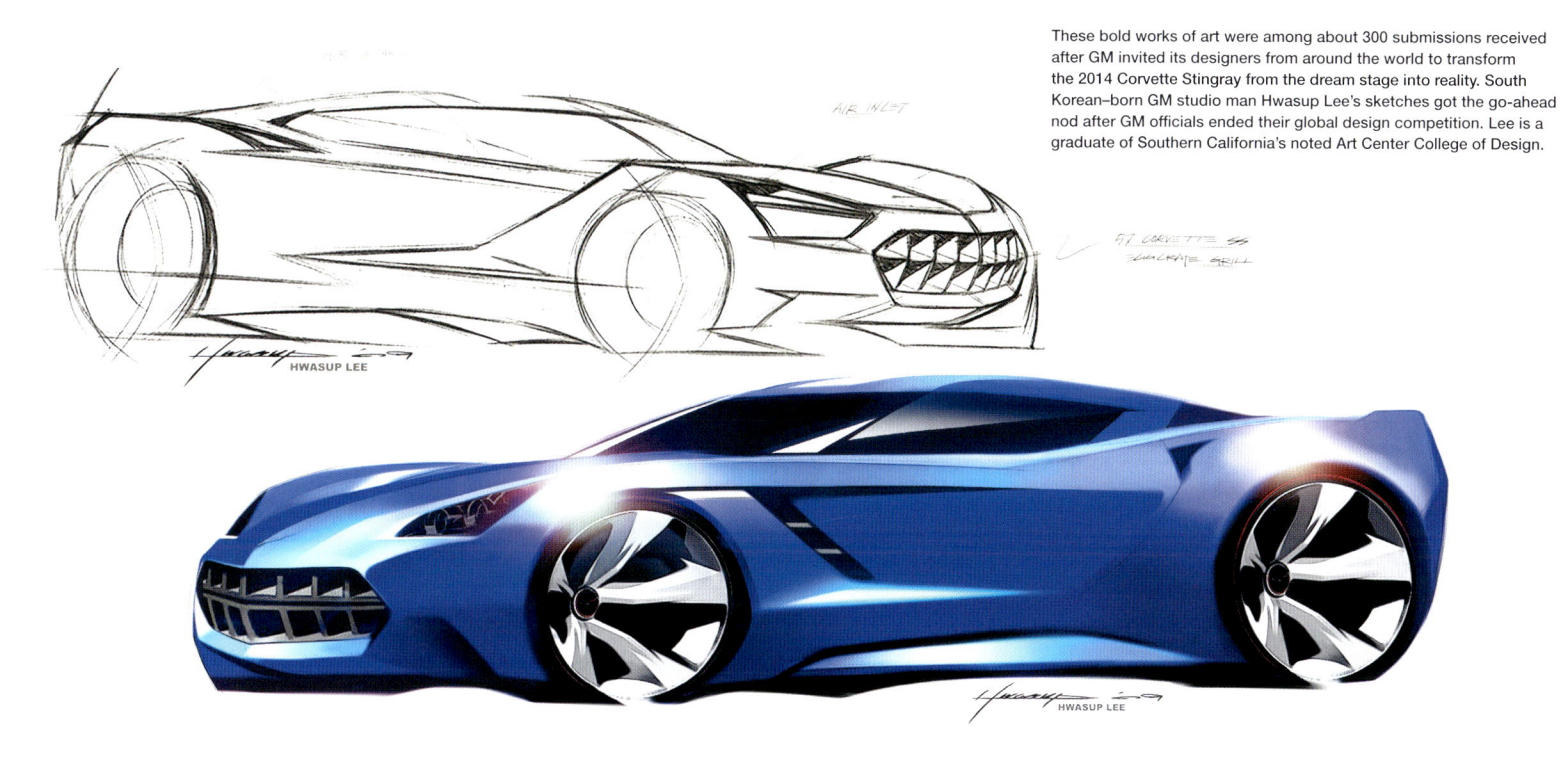

These bold works of art were among about 300 submissions received after GM invited its designers from around the world to transform the 2014 Corvette Stingray from the dream stage into reality. South Korean–born GM studio man Hwasup Lee's sketches got the go-ahead nod after GM officials ended their global design competition. Lee is a graduate of Southern California's noted Art Center College of Design.

GM officials again searched the globe for ideas when it was time to develop a new interior, with the goal being to create an out-of-this-world cockpit capable of putting the Stingray on par with this planet's highest-flying supercars. Aviation themes clearly dominated among the countless submissions.

decided and GM in financial disarray, we don't anticipate the next incarnation of America's sports car to bow before 2014—and we won't be surprised if that estimate turns out to be optimistic."

GM CEO Fritz Henderson reiterated Wallace's promise in May 2009, again to *AutoWeek*, explaining further that "Corvette pays its rent." C7 development was "on track," he said, but an intro date remained unavailable, as did technical specifics, leaving imaginative journalists to continue filling in the blanks. Talk of a midengine chassis—if not for the C7, then the C8 to follow—was still being bandied about into 2010, as was speculation about a turbocharged V-6.

Juechter took special offense in the latter case after it appeared to him that *Automobile* magazine had erroneously credited him with a claim that the upcoming C7 definitely would be armed with a six-shooter. "Don't believe any of what you read, most of it'll be wrong," he explained in May 2010 before a gathering at the National Corvette Museum in Kentucky. "It can even be attributed to me and be totally wrong." In defense, *Automobile* deputy editor Joe DeMatio pointed out that it was author Don Sherman who made the prognostication (a speculative one at that) in print, not the Corvette's latest chief engineer. "At no point did Don quote Mr. Juechter as definitely stating that a V-6 is in the works for C7," wrote DeMatio. Great theater, huh?

Further thickening the plot was GM's Stingray show car, introduced with its gas/electric hybrid powertrain in February 2009. Lutz was quick to explain that this was a "pure concept" not necessarily meant to predict eventual production plans, and designer Tom Peters echoed his claim, saying simply that this was "not the C7." But that didn't stop those rascals at *Car and Driver* from publishing renderings early in 2011 of so called C7 "prototypes"—which clearly resembled 2009's Stingray.

As for a hybrid Corvette, Karl-Friedrich Stracke, GM's global vehicle engineering vice president, told *Automotive News* in August 2010 that this logical advancement represented "an interesting idea" but wouldn't say when such interest might arise. He was, on the other hand, more than willing to speak up about the long-rumored midengined platform, dual-clutch transmission, and V-6, all of which, according to him, were plain and simply not included in future plans.

Additional news eager customers truly could use arrived in May 2011 with the announcement that GM would invest $131 million to refurbish the Bowling Green plant in preparation for next-generation production. When that process would start remained unsaid, clearly or otherwise, for more than a year.

On October 18, 2012, GM officials finally let the C7 out of the bag: it would debut as a 2014 model on Sunday evening, January 13, 2013, at Detroit's North American International Auto Show. Another high-profile public showing would follow in New York the next week, coming exactly 60 years to the day after Chevy's original Corvette was unveiled there on January 17, 1953. Car-lovers from coast to coast were then given a chance to ogle the C7 on May 26, 2013, when it became the 12th Corvette to pace the Indianapolis 500.

CHAPTER EIGHT

Next-generation Corvette testing always has involved both comparisons to previous models and all-weather extremes, taking the cars from the heat of Death Valley to the cold of the great white north. Here three Stingray prototypes join two C6 Corvettes for a little fun in the snow at a GM facility in Michigan's Upper Peninsula.

In typical fashion, the finalized Stingray design was sculpted into a full-size clay model, which then was measured by computer in order to create a mathematical "buck" for production-ready body panel fabrication.

Integration Vehicle Engineering Release (IVER) prototypes typically were disguised to keep prying public eyes a-boggle whenever GM took the Stingray into the wild for testing. The effect shown here was known in-house among engineering and design teams as "Cartoon Network" camouflage.

First official mention of C7 features came along with the Detroit debut announcement as GM global design vice president Ed Welburn introduced a new crossed-flag logo, a Corvette trademark from the get-go. More than 100 variations on this venerable theme had been considered before a suitable rendition won out. "The all-new, seventh generation Corvette deserved an all-new emblem," explained Welburn. "The [new] flags are much more modern, more technical, and more detailed than before—underscoring the comprehensive redesign of the entire car."

Apparently the C7 also deserved a new name: Stingray. Well, maybe not really new, but certainly momentous. "Stingray is one of the hallowed names in automotive history," continued Welburn in January 2013. "We knew we couldn't use [it] unless the new car truly lived up to the legacy."

Before gracing 2009's concept car, the revered Stingray label had last shown up on a regular-production Corvette in 1976. Thirty-seven years later it was back on the street to enhance the identity of a newer-than-new sports car, a warmly welcomed milestone still powered by a he-man V-8 located up front, where most witnesses believed it belonged. All appeared well as another long-anticipated next-gen model once more sure looked well worth the wait.

In the end, the C7's development delay amounted to a little more than two years. But the lag sure seemed longer, perhaps due to all the reading required to keep up with the seemingly endless speculative reporting. Furthermore, the stuttering this time around ultimately proved to be a blessing. As Juechter told *USA Today*'s ace automotive reporter James Healey in January 2013, "some of the technology might not have been there" had original C7 deadlines been met. Most importantly, those extra years allowed engineers to refine the lightweight material applications that helped the 2014 Corvette Stingray rise sky high above its predecessors. Just as planned.

2014

Worries that the C7 would end up being, in *AutoWeek*'s words, a "modest revamp of [the] C6" grew especially whiny after GM ran dangerously low on cash in 2008. More fears followed as the resulting federal bailout kept the lights on but left some wondering if government meddling would inhibit development of a car deemed "frivolous" more than once over the years by callous industry-watchers. Such woeful rumination, however, proved foolish, as Tadge Juechter told *Road & Track* in February 2013: "It was in the interest of U.S. taxpayers to continue the [car]." Not to mention build it better.

Of course, soaring profits, which quickly worked Washington out of the picture, didn't hurt things in the least. Resurging cash flow predictably made it more than possible to kick-start the developmental roller coaster, pushing the C7 to somewhat surprising new heights. Perhaps *Automobile*'s Ezra Dyer said it best in 2013 after his first look at the reincarnated Stingray: "This is not a C6 with 25 more horsepower and LED strips draped along the headlights."

"Since 1953, through good times and bad for this company, there was always Corvette, demonstrating what it means to win," added GM North America president Mark Reuss during the 2014 Stingray's introduction in Detroit in January 2013. "And now, here comes the best Corvette ever. [It] is all new from the ground up [and] absolutely the best performance car we know how to build."

Bragging? Not hardly. When the company line read "all-new," it meant it. For starters, Juechter was quick to point out that only two parts carried over from the C6: the interior compartment's air filter and the rear latch for the C7 coupe's removable roof. Greeting the eye first and foremost was a fully fresh facade that, according to Ken Parkinson, executive director of global design, "breaks new ground yet remains true to the fundamental elements that make a Corvette a Corvette."

"For the new Corvette to be called Stingray, it had to deliver an incredible, purposeful visual impact—just as the original did in 1963," added exterior design director Tom Peters. Form truly followed function as each aesthetically pleasing crease and scoop played its part to make this the slickest, most slippery Vette yet. "Every square inch of the 2014 Corvette's exterior is designed to enhance high-performance driving," claimed exterior design manager Kirk Bennion. "The team delivered a great balance of low drag for efficiency and performance elements for improved stability and track capability."

Up front, the hood vent allowed at least a third of the airflow coming off the radiator (now slanted forward instead of back) to whip over the car, not underneath, thus reducing aero lift at speed. Anything that needed cooling (brakes, transmission, etc.) received appropriate ductwork, and the coupe's removable roof panel incorporated a channel to draw more air pressure down onto the body towards the rear spoiler. Almost lost in all

2014–2019 Stingray

Model availability	hatchback coupe (with removable roof panel) and convertible; aero package included with Z51 models
Construction	composite/carbon-fiber body panels, hydroformed aluminum frame with aluminum and magnesium structural/chassis components
Wheelbase	106.7 inches
Length	176.9 inches
Width (without mirrors)	73.9 inches
Height	48.8 inches (coupe), 48.9 inches (convertible)
Track (front/rear, in inches)	63.0/61.7
Curb weight	3,298 pounds (coupe), 3,362 pounds (convertible) **NOTE:** new composite body panels in 2016 reportedly deleted about 20 pounds, but published totals did not change during C7 run
Wheels	**2014–2015:** silver-painted aluminum 5-spokes; 18x8.5 inches front; 19x10 inches rear **2016-2017:** silver-painted aluminum split 5-spokes (Z51-style); 18x8.5 inches front, 19x10 inches rear **2018–2019:** silver-painted aluminum split 5-spokes (Z51-style); 19x8.5 inches front, 20x10 inches rear
Wheels (Z51)	silver-painted aluminum split 5-spokes; 19x8.5 inches front; 20x10 inches rear
Tires	**2014–2017:** Michelin Pilot Super Sport ZP summer-only; P245/40ZR18 front, P285/35ZR19 rear **2018–2019:** Michelin Pilot Super Sport ZP summer-only; P245/35ZR19 front, P285/30ZR20 rear
Tires (Z51)	Michelin Pilot Super Sport ZP summer-only; P245/35ZR19 front, P285/30ZR20 rear
Brakes	power-assisted four-wheel discs with four-piston calipers and ABS (slotted rotors with Z51)
Brake rotor diameter	12.6 inches front, 13.3 inches rear **Z51:** 13.6 inches front, 13.3 inches rear
Fuel tank	18.5 gallons
Suspension	short/long-arm (SLA) double wishbone cast-aluminum control arms, transverse composite leaf springs, stabilizer bars front and rear, monotube shock absorbers, and StabiliTrak electronic stability control **Z51:** added Magnetic Selective Ride control; specific spring, shocks, and stabilizers; and electronic limited-slip differential
Steering	power-assisted, speed-sensitive, variable-ratio rack and pinion
Engine (LT1)	6.2-liter (376-ci) direct-injection overhead-valve V-8 with cast-aluminum cylinder block and heads **Z51:** added performance exhausts (beginning in 2015) and dry-sump oiling **Compression:** 11.5:1
LT1 bore and stroke	4.06x3.62 inches
LT1 output	455 horsepower at 6,500 rpm, 460 lb-ft of torque at 4,500 rpm (460 horsepower and 465 lb-ft torque with optional NPP performance exhausts, available alone or with Z51 beginning in 2015)
Standard transmission	seven-speed manual with 2.66:1 low (2.97:1 low with Z51) and Active Rev Matching
Optional transmission	**2014:** 6L80 six-speed paddle-shift automatic with 4.02:1 low (RPO MYC) **2015–2019:** 8L90 eight-speed paddle-shift automatic with 4.56:1 low (RPO M5U)
Axle ratio	3.42: 1 (2.56:1 with six-speed automatic*, 2.73:1 with Z51 automatic) *2.41:1 with eight-speed automatic beginning in 2015

CHAPTER EIGHT

Every aspect of the 2014 Stingray was designed for aesthetic beauty and superior high-speed performance. Carrying over from the sixth-generation was a similar trim-level pecking order (1LT, 2LT, and 3LT—this time, there was no "4") that again added more standard goodies into the mix at each step up. Among other things, the 2LT deal included a Bose 10-speaker stereo and heads-up display, while 3LT customers were treated to a navigation system and leather interior touches. Repeating as well in seventh-gen terms was the 1LZ/2LZ/3LZ coding that appeared in 2015 for the reborn Z06.

the technical upgrades was the Stingray's new rear-quarter glass, last seen on the C1's optional removable hardtop in 1962.

The wind-tunnel-proven body's makeup enhanced functionality further by keeping pounds down. Both the hood and the coupe roof were formed from weight-saving carbon fiber, and underbody sections consisted of a composite carbon-nano material that simultaneously lightened the load and increased strength. Sheet-molded compounds again made up the fenders, doors, rear-quarter panels, and rear hatch but were less dense compared to the C6 stuff.

These advanced materials not only limited overall mass, but they also helped the C7 achieve (according to Chevrolet) an ideal 50/50 front/rear weight balance. *Car and Driver*'s apparently more precise calculations, meanwhile, read 49.4 percent in front, 50.6 out back, making this the first Corvette to claim a rearward bias, something of which Duntov could only dream.

Beneath the better-balanced Stingray was an aluminum frame, a feature previously reserved for the C6's Z06/ZR1 renditions. And this time the frame wasn't shipped in. Reportedly $52 million of the $130-million-plus spent on refurbishing the C7's ol' Kentucky home went toward a new shop to house construction of this reconfigured foundation, made up of hydroformed, cast-and-extruded aluminum sections that were computer tailored to maximize strength while minimizing weight. Overall, the C7 frame was 99 pounds lighter than the C6's standard steel structure, yet it remained 57 percent stiffer, and all that extra torsional toughness meant that no excess reinforcement was required to transform a coupe into a convertible.

By keeping an eye on the Stingray's figure, designers guaranteed yet another all-time best claim as its power-to-weight ratio surpassed the standard C6's, as well as those of many imported rivals. Powertrain engineers did their part, too, increasing output to 455 horsepower, the most standard Corvette muscle ever. Producing those ponies was the truly new Gen 5 V-8, the latest variation on Chevrolet's undying small-block theme. Also new for the Gen 5 was another familiar label: "LT1."

Like its LS3 forerunner, the Gen 5 V-8 featured all-aluminum construction and displaced 6.2 liters. Helping set the reborn LT1 apart from the LS3 were its new direct injection (to ensure greater combustion efficiency); continuously variable valve timing (to simultaneously optimize performance, efficiency, and emissions); and Active Fuel Management system, which imperceptibly shut down four of those eight cylinders in light-load situations to conserve fuel. Suffice it to say that "all new" again applied in spades. As did "world class."

"We feel we have [delivered] a true technological masterpiece that seamlessly integrates a suite of advanced technologies that can only be found on a handful of engines [on Earth]," added global product development senior v.p. Mary Barra. "What makes this engine truly special is the advanced combustion system that extracts the full potential of these technologies. The art and science behind that system makes the LT1 one of the most advanced V-8 engines in the world."

More than 10 million hours of computational analysis time went into the LT1, including 6 million alone dedicated to those superbly refined combustion dynamics, which helped allow compression to climb to long-forgotten levels (11.5:1), rekindling fond memories of quarter-a-gallon gas and Cragar S/S mags. Fuel efficiency, at the same time, also went up, while emissions decreased, enhancing the wonderment.

"By leveraging technology, we are able to get more out of every drop of gasoline," claimed Juechter in October 2012. "And because of that we expect the [C7] will be the most fuel-efficient 450-horsepower car on the market." Indeed, its EPA-estimated ratings of 17 mpg city, 29 highway simply couldn't be beat by any vehicle of equal or greater power from any company in any country. As for the opposite end of the spectrum, initial Chevrolet estimates claimed 0–60 in less than 4 seconds and 1g of lateral acceleration, making the Stingray "the most capable standard Corvette model ever." Additional Corvette firsts included a seven-speed manual transmission (still found in back) fitted with Active Rev Matching technology, which electronically "blipped" the throttle in advance of a gear change, up or down, to help make shifts seamless. Like the optional six-speed automatic, the manual box's shift-assist feature

C7 2014–2019

Upgrading the cockpit was a prime goal, and to this end the 2014 Stingray was treated to better seats (that held occupants in place more firmly) and more exciting, plusher surroundings. The Stingray's standard stitched-and-grained vinyl dash wrap was, according to *Car and Driver*, "good enough to fool a cowhide inspector."

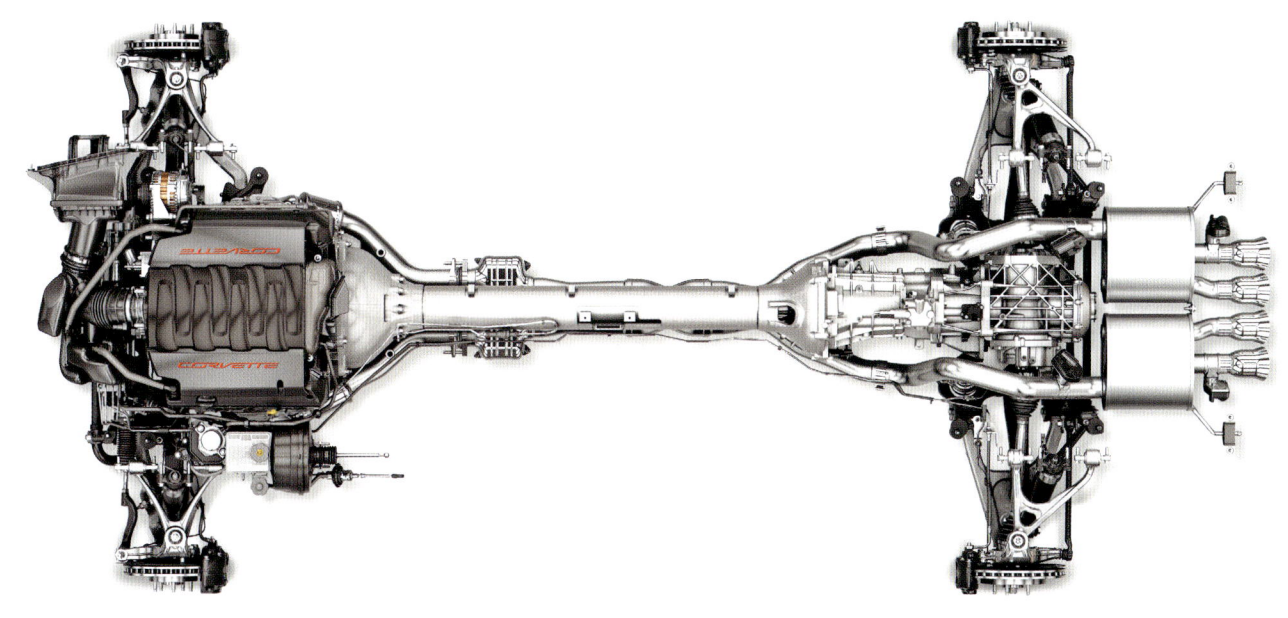

Like its C5 and C6 forerunners, the seventh-generation model again featured a rear-mounted transmission. New for the Stingray was a standard seven-speed manual gearbox, a Tremec TR60670 unit fitted with Active Rev Matching technology to help drivers row their way through the gears in seamless fashion.

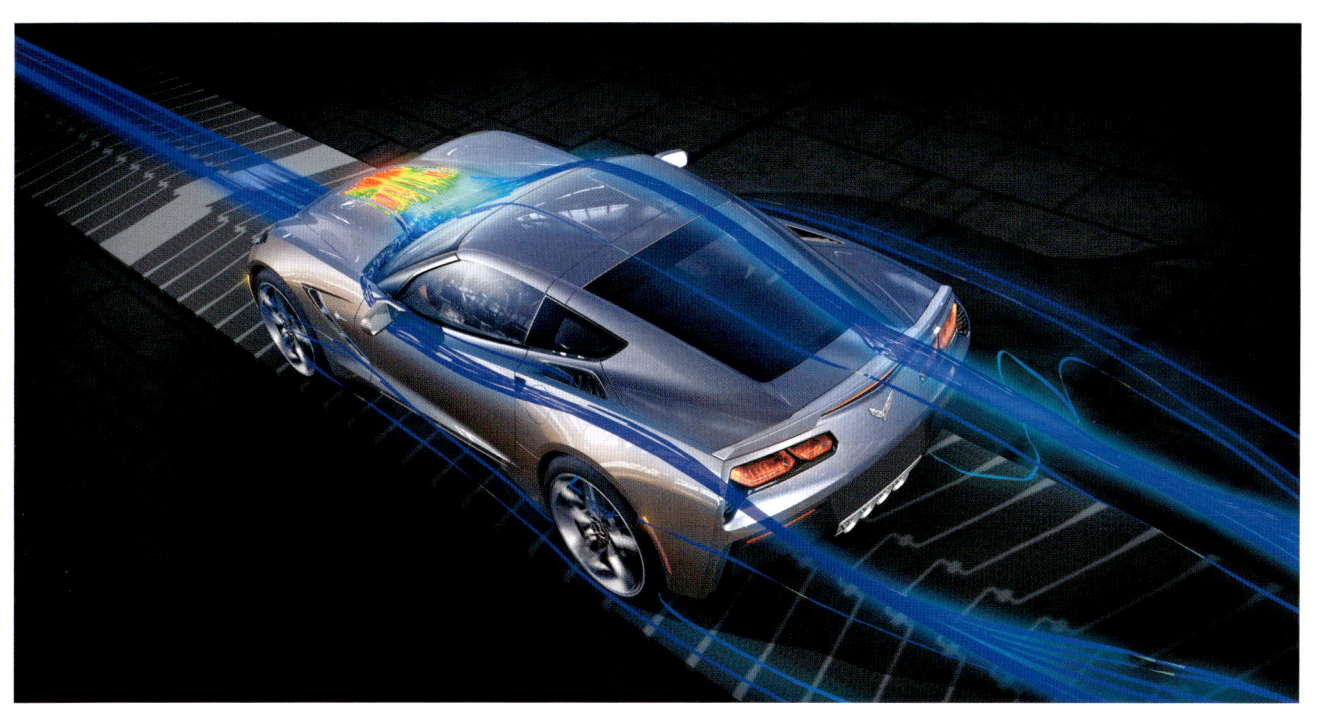

Serious wind-tunnel testing yet again played a major role in creating the latest and best 'Vette yet. The front hood vent helped reduce front-end lift, and the roof's shape induced airflow down tight towards the rear spoiler for extra downforce on the tail. Quarter-panel intakes also helped vent cooling air to the rear-mounted transmission and (in optional applications) differential. This flow then exited the body through vents near the taillights.

265

CHAPTER EIGHT

Enlarged Brembo brakes and 35-millimeter Bilstein shocks were standard for the 2014 Stingray. Hollow lower control arms deleted about 9 pounds from the model's curb weight, and a new aluminum toe-link in back shaved off another 2.4 pounds.

Used in 1970–1972, and again from 1992 to 1996, Chevrolet's famed "LT1" label returned in 2014 for the new Stingray's 455-horsepower, 6.2-liter V-8—an all-aluminum small-block featuring new direct injection, Active Fuel Management (which allowed operation on only four cylinders for optimum fuel economy), and an enlarged (2.75-inch-diameter) active exhaust system. A dual-mode active exhaust system was optional.

C7 2014–2019

The base Stingray coupe/convertible duo (back) was joined in 2015 by the latest, greatest Z06, also available this time with or without a top—a first for the breed, save reportedly for one ancient ancestor released way back in 1963. A removable roof panel also was passed over from standard Stingray to Z-car coupe, another first. Notice the widened wheelhouses and "high hoods" (needed to clear the supercharger hiding below) on the Z06 pair in front.

could be controlled (engaged or disengaged) by paddles on the steering wheel.

All-time-high levels of driver control also were made possible by the new Driver Mode Selector system, activated by a rotary knob located near the shifter. Offering five modes—Tour, Weather, Eco, Sport, and Track—these electronics automatically adjusted up to 12 parameters to (in product manager Harlan Charles's words) "give drivers an easy way to tailor virtually every aspect of the car to fit their driving environment."

New seats with lighter, stronger magnesium frames were fitted to drivers' rears far better than ever before, created in response to complaints that previous designs simply didn't inspire enough occupant confidence whenever "sport" or "track" environments were encountered. GT buckets came standard while Competition Sport seats (featuring larger, "more aggressive" side bolsters) were optional, as were full leather appointments, carbon-fiber accents and microsuede trim touches, all also inspired by critiques claiming the C6 cockpit paled in comparison to classier interiors found in most imported supercars.

Back on the Corvette options list in 2014 was the Z51 Performance Package, last seen five years earlier. Included in the C7's Z51 deal were special brakes (larger front discs, slotted rotors, black-painted calipers), a beefed suspension, bigger wheels and tires (19- and 20-inchers front/rear, as opposed to the standard 18/19 combo), dry-sump oiling for the LT1, and an electronic limited-slip differential (eLSD) with cooler. The eLSD system featured a hydraulically actuated clutch that went from open (one rear wheel gripping) to fully locked (both biting) in "tenths of a second." Optional Magnetic Selective Ride Control also carried over for the C7 but only in Z51 applications.

Again, according to Chevrolet's tests, 2014's Z51 Stingray could do 0–60 in 3.8 seconds, scorch the ¼ mile in 12 ticks (at 119 mph), and pull 1.03g on a skid pad. Braking from 60 mph back to rest required only 107 feet. Z51 equipment was available for both coupe and convertible, which, like its topless C6 forerunner in 2005, debuted later in March 2013 at the Geneva Motor Show and then went on sale a few months after its full-roofed counterpart.

2015

Save for the expected return of the Z06, 2015's headlines were few and primarily of meager point sizes. Horsepower hounds probably noticed that the NPP performance exhaust option that had boosted LT1 output to 460 horsepower in 2014 was now included in the Z51 package. And they might've additionally seen that NPP plumbing was standard Z06 fare. Once they read all about it, that is.

Word of an optional transmission trade-out featured more prominently in the mainstream. Gone was 2014's MYC six-speed automatic, superseded by the GM-designed/built (in Toledo, Ohio) 8L90 eight-speed auto, also able to function manually via steering-wheel-mounted paddles. Lighter and more efficient than its MYC forerunner, the 8L90 (RPO M5U) was also constructed much more like a brick outhouse, meaning it didn't need to be limited to Chevy's standard Stingrays. Seventh-gen Z06 buyers (trust us, more on this real soon) could check off the M5U option, too, and worry not one wit about ending up in line at their local service department hoping GM's warranty still applied

267

CHAPTER EIGHT

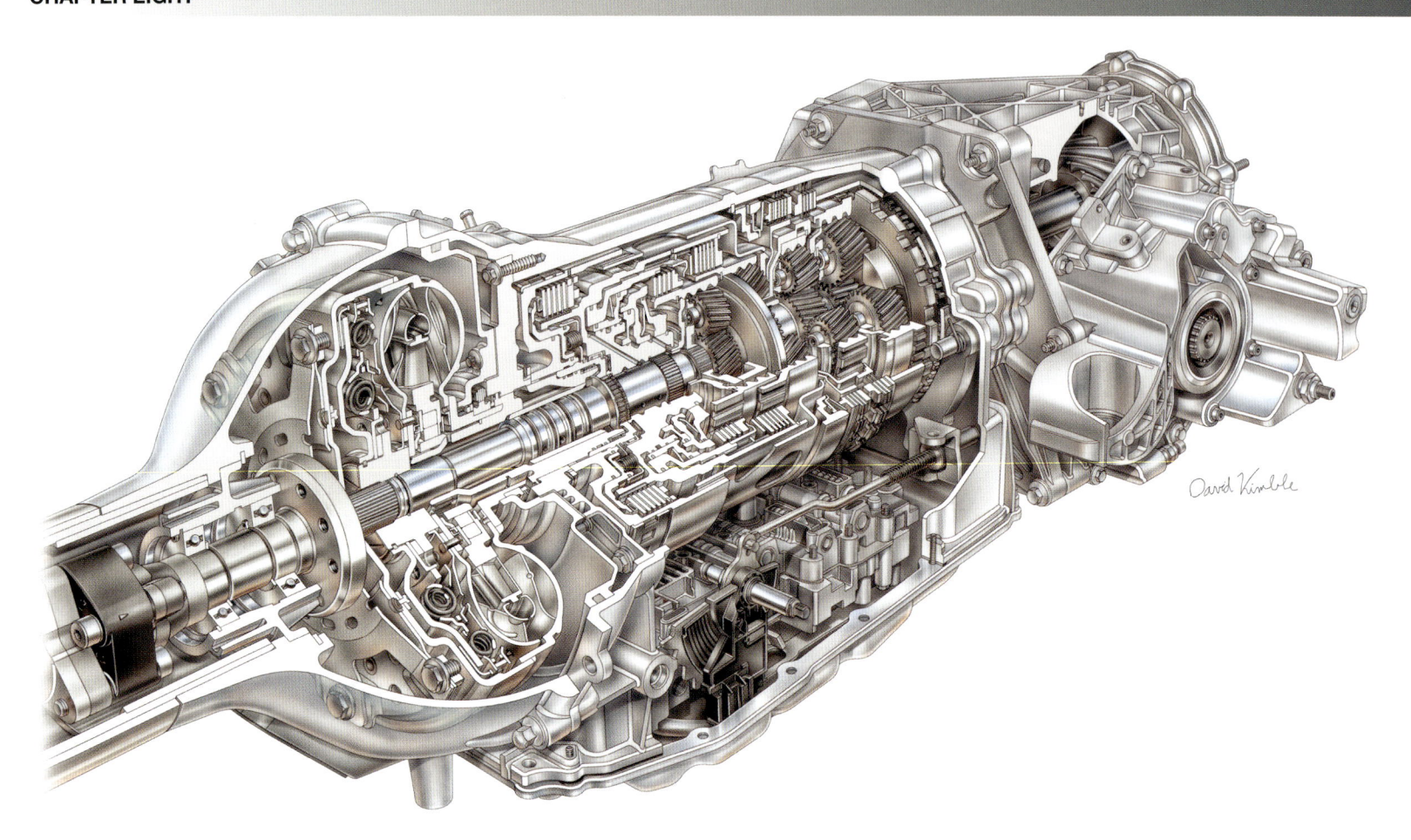

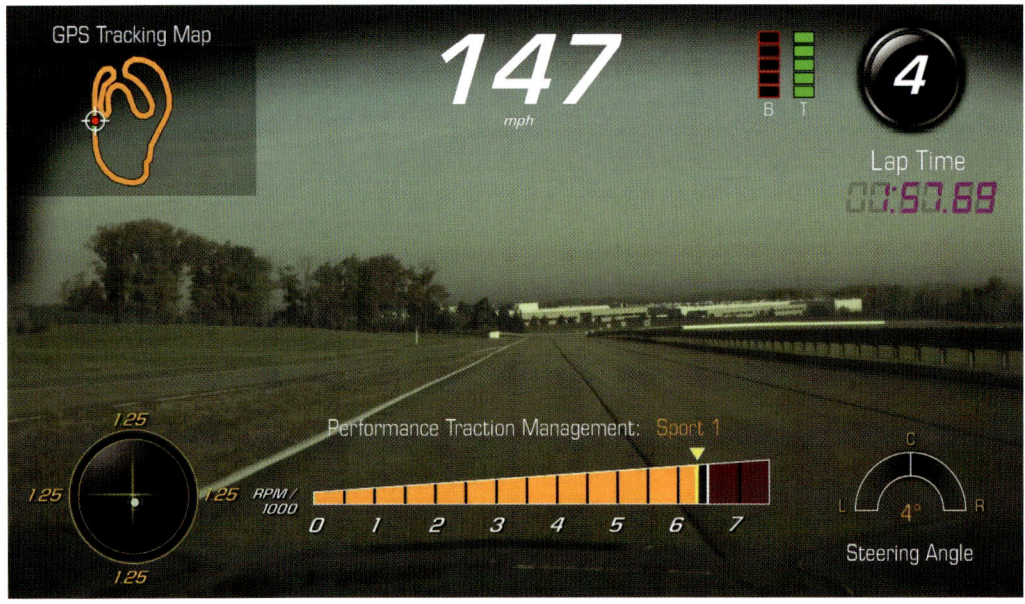

Top: Extensive use of aluminum and magnesium made GM's new-for-2015 eight-speed, paddle-shift 8L90 automatic transmission about eight pounds lighter than the six-speed auto found in 2014's Stingray. It also was both more efficient and quicker on the shifts, making it faster and, perhaps, easier on gas. That latter result depended, of course, on how often and hard a driver produced the former.

Left: New for 2015 was a Performance Data Recorder that was included with top-shelf trim choices (3LT or 3LZ) or added (RPO UQT) to other Stingrays in exchange for 1,795 extra bucks. With this toy you could record your on-track adventures, complete with all pertinent readouts, and play 'em back at a later date—hopefully when you were parked back home in a recliner next to a bowl of popcorn.

to a grenaded trannie overstressed by too many wild horses. According to assistant chief engineer Bill Goodrich, the 8L90 qualified as "the highest-capacity automatic transmission ever offered in a Chevrolet car."

Able to compare and/or contrast with lightning like no GM trans before, the M5U's electronic shifts (analyzed, commanded, and executed as many as 160 times a second) were as quick, if not quicker, than some rival dual-clutch manuals. And those two extra gears understandably translated into "shorter steps" between ratio changes, enhancing said efficiency while also exploiting the engine's power band more precisely than ever.

Fully documenting just how well a Vette-ster put this electro-automatic through its paces became possible in 2015 via another new extra-cost addition, the UQT Performance Data Recorder. Incorporating 2014's optional navigation hardware, the UQT package consisted of a camera (mounted in the windshield header), a cockpit voice recorder, another to record telemetry, and a secure digital (SD) card slot in the glovebox. Walter Mitty–types could play back high-definition video—overlaid with pertinent data readouts—of their on-track (or not) adventures on their C7's central display screen. When parked, of course. Alongside the road directly ahead of a constabulary cruiser perhaps?

Those who prefer home viewing—accompanied by some of Orville Redenbacher's finest, maybe?—needed only take advantage of UQT's SD-card transfer-ability. Steve McQueen didn't even have it this good. (Google the movie *Bullitt* if you're hearing crickets here.) As for Corvette customers who play race drivers only on TV, this equipment also doubled as surveillance gear whenever someone else was at those paddles (or stick), thanks to a "valet mode" that captured their every move, along with all comments. Like "how 'bout we take 'er out on the super-slab to see what she'll do?" People's exhibit A, right?

Giving technocrats even more to crow about was a new-for-2015 standard feature: high-speed 4G LTE (long-term evolution) OnStar connectivity. This rolling WiFi hotspot capability was, in GM's words, "intended [only] for passenger use when vehicle is in operation."

Two specially focused Design Packages appeared midyear in 2015: the Atlantic luxury convertible and the Pacific performance coupe, both offered exclusively for Z51-equipped Stingrays upgraded with either 2LT or 3LT trim. As the descriptions above imply, the Atlantic Design Package concentrated more on attracting attention in polite society. Its Pacific Design counterpart was intended to look way-too-cool-for-school parked, say, alongside a racetrack. To those ends, Atlantic convertibles featured (among other things) a Z06-style front splitter, chrome wheels and custom luggage. Most prominent on Pacific coupes, meanwhile, were racing stripes, specially adorned Z51 wheels, red brake calipers, and the Carbon Flash treatment.

2015 Z06

This year's really big news surprised no one, save perhaps those living in their parents' basements. Most everyone else residing above ground during C7 development understood that a Z06 rebirth had to be a part of the plan. Eventually. And, as promised, reporting of that rumor-turned-reality starts now—too bad there's simply not enough column inches available here to do this planet-sized tale justice.

Where to begin? Its state-of-the-pushrod-art supercharged LT4 small-block alone could fill a small book with its feature presentations, not the least of which involved its output, 650 horsepower—yet another all-time high for the revered engine legacy then marking six decades in action. But not so fast.

2015–2019 Z06

Model availability	hatchback coupe (with removable roof panel) and convertible
Construction	composite/carbon-fiber body panels, hydroformed aluminum frame with aluminum and magnesium structural/chassis components
Body modifications	Fender width increased 2.2 inches in front, 3.15 inches in back to house wider wheel/tire combo; rear fascia widened, moving taillights* about 3 inches farther apart * smoked in Z06 application
Body additions	**standard:** front splitter, front wheel opening spats, unique carbon-fiber "high hood" with enlarged vent, unique grille with brake-cooling intakes, and Z51 rear spoiler **optional:** CFZ aero package* added front splitter (with aviation-style winglets), rocker panels (done in carbon-fiber), and large rear spoiler with fixed "wickerbill," a small, vertical tab that notably increased downforce * available in either black or visible carbon-fiber finish **Z07:** required CFZ package and added "tall" front-splitter winglets and an adjustable see-through center section to the rear spoiler
Wheelbase	106.7 inches
Length	176.9 inches (177.9 with Z07)
Width (without mirrors)	77.4 inches
Height	48.6 inches (coupe), 48.7 inches (convertible)
Track (front/rear, in inches)	63.5/62.5
Curb weight	3,524 pounds (coupe), 3,582 pounds (convertible)
Wheels	exclusive pearl-nickel aluminum split ten-spokes; 19x10 inches front, 20x12 inches rear
Tires	Michelin Pilot Super Sport ZP summer-only; P285/30ZR19 front, P335/25ZR20 rear
Tires (Z07)	Michelin Pilot Super Sport Cup 2 summer-only; P285/30ZR19 front, P335/25ZR20 rear
Brakes	power-assisted four-wheel slotted Brembo discs with aluminum calipers (six-piston front with monobloc fixed calipers, four-piston rear) and ABS **Z07:** added carbon-ceramic Brembo cross-drilled rotors with monobloc fixed aluminum calipers front/rear—reportedly these rotors saved 23 pounds compared to their steel counterparts
Brake rotor diameter	14.6 inches front, 14.4 inches rear **Z07:** 15.5 inches front, 15.3 inches rear
Fuel tank	18.5 gallons
Suspension	short/long-arm (SLA) double wishbone cast-aluminum control arms, transverse composite leaf springs, stabilizer bars front and rear, monotube shock absorbers, and Magnetic Selective Ride control
Steering	power-assisted, speed-sensitive, variable-ratio rack and pinion
Engine (LT4)	6.2-liter (376-ci) direct-injection overhead-valve V-8 with cast-aluminum cylinder block and heads, intercooled Eaton R1740 supercharger, performance exhausts, and dry-sump oiling **Compression:** 10.1:1
LT4 bore and stroke	4.06x3.62 inches
LT4 output	650 horsepower at 6,400 rpm and 650 lb-ft of torque at 3,600 rpm
Standard transmission	seven-speed manual with 2.29:1 low and Active Rev Matching
Optional transmission	8L90 eight-speed paddle-shift automatic with 4.56:1 low (RPO M5U)
Axle ratio	3.42:1 with electronic limited-slip differential (2.41:1 with eight-speed automatic)

Base C7 brakes were steel rotors at the corners, each fitted with four-piston calipers. Z51 and Z06 models featured slotted rotors (shown here) in larger diameters, and six-piston front calipers were added in the latter's case. Standard rotor diameters were 12.6 inches, front; 13.3, rear. Z51s kept those rears but got 13.6-inch fronts, and Z06s were fitted with 14.6-inchers at the nose, 14.4 out back. Dark gray metallic calipers were standard for base Stingrays, while red calipers were the norm for Z06s. Z51 calipers were black if not superseded by an optional color, either red or yellow. The base gray shade was not available with the Z51 stuff but could be added to a Z06 via the Z07 Performance Package. Non-Z07s could be fitted with black calipers, and red/yellow were available in all applications.

CHAPTER EIGHT

Record-setting pony production represented only half of the attraction. Arguably much more historic was the 6.2-liter LT4's Earth-rotating torque, the true measure of how rapidly any supercar trades rest for supersonic speeds. At 650 lb-ft, LT4 maximum torque not only represented another Corvette milestone, it notably overshadowed (at far less cost) many of Europe's highest flyers.

Consider Ferrari's F12 Berlinetta, priced at nearly $320K. This wallet-immolating Italian exotic offered about 12 percent more horsepower than Chevy's latest Z06, but its 6.3-liter V-12 generated 28 percent less torque and not until those dozen pint-sized slugs were dizzily reciprocating at more than 6,000 rpm. Priced at *only* $79K, 2015's Z06 coupe was already spitting out 457 lb-ft right off idle and hit 625 pounds at a mere 2,800 revolutions. Furthermore, 90 percent of the force-fed LT4's kick-in-the-pants capabilities remained on hand from 2,500 up to 5,400 revs.

What did this mean? "The LT4's abundance of [torque] at every rpm in [its] speed range helps the 2015 Z06 accelerate quicker and respond nearly instantaneously," explained small-block chief engineer Jordan Lee in June 2014. "It's the very definition of power on demand." As for unquestionably unbiased opinions, the 650-horsepower Z06 was, in the words of *Road & Track*'s Sam Smith, "equal parts high-revving sociopath and low-rpm sweetheart."

Demanding lunatic results from a Z06 served only to piss off its LT4 mouse motor like no other previous small-block. When angered, according to Chevrolet tests, 2015's Z06 established new performance standards unimaginable from an American muscle machine not all that many years prior. Think 0–60 in 2.95 seconds, the quarter-mile in 10.95 at 127 mph. Talk about time-trippin'. As alt-rocker Frank Zappa once put it, "great googlie-mooglie!"

Know, too, that breaking into the superhuman sub-three realm was accomplished with an optional 8L90 eight-speed, the first automatic offered to Z06 customers during any generation. Clearly a new epoch had arrived, an era every bit as unthinkable not long before where an auto-trans was, as mentioned earlier, able to easily withstand mucho horsepower *and* simultaneously make surreal track-worthy performance possible. Curiously, few eyebrows arched when Chevrolet announced lesser (yet still unsettling) timings for 2015's seven-speed-stick Z06: 0–60 in 3.2 seconds and an 11.2-click quarter-mile. Double googlie-mooglie!

Above: Allowing customers to "tailor their car to their personalities" (Corvette exterior design manager Kirk Bennion's words) were two design packages following in the tracks of 2014's Premiere Edition, both announced in April 2015. The Pacific coupe (captured here) featured satin-black full-length racing stripes, satin-black Z51 wheels adorned with red stripes, Carbon Flash treatment (rear spoiler, outside mirrors, and badges), and red brake calipers. Its Atlantic convertible cousin, among other things, incorporated a Z06-style front splitter, chrome Torque wheels, and Shark Gray exterior accents, but that wasn't all: Like the Premiere Edition, the Atlantic Design Package also included custom luggage.

Opposite Top: Like the LS9 small-block introduced along with the C6 generation's ZR1 in 2009, the 650-hp LT4 V-8 barely contained beneath the Z06's hood in 2015 was force-fed by an intercooled Eaton supercharger, a 1.7-liter unit capable of winding upwards of 20,000 rpm, 5-grand more than its LS9 forerunner. The LT4 parts list additionally included Rotocast aluminum heads, titanium intake valves, forged-aluminum pistons, dry-sump oiling, and rather-high (for a blown motor) 10.0:1 compression, made possible by the additional use of direct-injection.

Opposite Bottom Left: They didn't call 'em "summer-only" tires for nothing. Suited better for the track, a dry one hopefully, C7 Michelin tires were predictably worthless in snow—a plain reality Chevrolet paperwork announced to all comers before they went home with their new Corvettes in states not named Florida. Included in the balls-out Z07 package, Michelin's Pilot Sport Cup 2 rubber (shown here) were especially disclaimer-worthy when it came to use in anything other than the warmest, driest conditions. Guess jingle all the way was out of the question.

Opposite Bottom Right: Initially listed only for the Z06, the Z07 Performance Package among other things included these huge-gantic Brembo carbon-ceramic matrix brakes. Z07 rotor diameters were 15.6 inches, front; 15.3, rear. Gray calipers were Z07 standard fare.

C7 2014–2019

271

Epochal as well was 2015's Z06 convertible, a sexy beast that easily qualified as Chevy's hottest topless model ever. Reportedly one of the 199 original Z06 Sting Rays built for 1963 went public sans roof, but every other rendition to follow during later generations appeared more modestly in full-bodied fashion. Until now. Blame the C7's ultra-strong aluminum foundation for this shameless development. And all that digitally mastered underlying rigidity additionally allowed the seventh-gen Z06 coupe to wear a removable roof panel, also unavailable for its C5/C6 predecessors.

Much of the latest Z06's makeup followed in similar C6 traditions: bigger wheels and tires, more brutal brakes, and a widened body, all the better to house a shipload of extra Michelin rubber. Newer, however, once again meant better, at least in application. While some critics poked fun at what they called "cartoonish" looks, none could deny that 2015's Z06 shell was as pragmatically purposeful as anything ever let loose by a U.S. automaker. Cheating the wind—not to mention bullying breezes into allies—was a prime promise, made and kept, by Chevy's newest supreme Corvette.

"The [Z06's] aerodynamics produce the most downforce of any production car GM has ever tested," went Tadge Juechter's tout during the model's introduction in January 2014. "And we are closing in on the aero performance of a dedicated race car."

That superbly functional form, working in concert with the Vette's best chassis yet, helped produce additional performance records, most notably the

Top: Winner of NASCAR's inaugural Brickyard 400 race, Jeff Gordon was introduced as the latest honored driver to pace the field of the Indianapolis 500 in April 2015. He then piloted Chevrolet's all-new Z06 coupe across the Bricks on May 24 prior to the 99th running of the fabled Memorial Day motorsport spectacular. (*IMS photo, for Chevy Racing*)

Bottom: "Until recently, it was not possible to create a lightweight, open-roof structure strong enough to cope with the braking, cornering and acceleration of Corvette's top performance models," said the breed's chief engineer, Tadge Juechter, in April 2014. But no longer, not with an aluminum foundation that now stood 20 percent stiffer (with a top or otherwise) than the sixth-gen Z06's skeleton—and it incorporated a fair dose of overhead bracing. Creating a topless Z06 for 2015 was a piece o' cake. No extra reinforcement was needed to stand the strain, and this in turn meant that the 650-hp convertible weighed nearly the same as its fully enclosed C7 running mate. *Mike Finkelstein photo, for General Motors*

fastest lap ever around GM's 2.9-mile Milford road course, besting the C6 ZR1 by a *full* second. Who needed high-tech, ultraprecise digital timing when road-testers could determine how much faster/quicker the Z06 was relative to its formidable forerunner by simply counting "1-Mississippi, 2". . . ?

New standards also were established for lateral acceleration (1.2g) and stopping power—60 mph to zero in only 99.6 feet. Next question: who was giggling now?

Chevrolet folk were more than willing to admit that their Z06 test results were produced by a 650-horsepower Corvette further enhanced with the Z07 Performance Package, not offered for Stingrays. Along with Brembo carbon-ceramic brakes and ultra-sticky Michelin Pilot Sport Cup tires, the Z07 option pushed aerodynamic capabilities even higher with larger winglets for the front splitter and an adjustable, see-through section in the center of the rear spoiler. Carbon-fiber ground effects (RPOs CFV and CFZ) were mandated along with the Z07 deal, an addition that further bolstered Juechter's claims.

More prestigious lap time followed on May 24, 2015, when Chevy's latest Z06 served as the pace car for the 99th running of the Indianapolis 500. NASCAR legend Jeff Gordon was at the wheel of the 13th Corvette (and 26th Chevrolet) to lead the Indy field around the fabled Brickyard. More records—who'da thunk it?

C7.R racer

Introduced early in 2014 alongside the C7 Z06 was a second supercar, this one a competition-dedicated alter ego. Or not. Even blind witnesses recognized by then that lines between road and track in the Corvette realm had blurred almost beyond recognition, with 2014's Stingray serving as rapidly rolling proof. Earlier lessons learned while building and racing the C6.R helped morph street-going seventh-gen models into the closest things American mainstream machines had ever come to trading stoplights for checkered flags with minimal fuss/muss. Then came the latest Z06, which basically didn't even need to stop at a phone booth to reemerge faster than a speeding bullet. (Try googling "Superman"—and perhaps "phone booth?") When chief engineer Juechter called it "the most track-capable production Corvette ever," he was shooting straighter than straight.

So anyone caught asking "is that a bird, or a plane, or a Z06 in tights and cape?" during that January 2014 introduction wasn't too far off target. At least in the latter analogical case. Chevrolet's newest race-ready Corvette, the predictably labeled C7.R, was "co-developed" (Chevy's term) along with 2015's Z06 and hence shared various aspects more closely than the C5.R and C6.R did with their regular-production alternates.

C7.R

Model availability	coupe (produced in collaboration between Bowling Green Assembly and Pratt & Miller)
Construction	composite/carbon-fiber body panels, hydroformed aluminum frame with aluminum and magnesium structural/chassis components
Wheelbase	106.6 inches
Width	80.7 inches
Height	45.3 inches
Track (front/rear, in inches)	63.5/62.5
Wheels	18x12.5 inches front, 18x13 inches rear
Tires	Michelin racing summer-only; 30/68-18 in front, 31/71-18 in back
Brakes	steel rotors with six-piston monobloc front calipers, six-piston rear calipers
Brake rotor diameter	14.0 inches front, 14.0 inches rear
Suspension	short/long-arm (SLA) double wishbone fabricated steel upper/lower control arms and coil-over adjustable shock absorbers
Engine	5.5-liter (336 ci) naturally aspirated direct-injected pushrod V-8
Output	500 horsepower (approximate)
Bore and stroke	4.104x3.150 inches
Transmission	six-speed manual

Which came first, chicken or egg? As Chevrolet explained it, the street-going Z06 (bottom left) and its C7.R competition-ready sibling (above) were "co-developed," sharing an unprecedented level of parts, characteristics, and engineering expertise. The two were even introduced to the public together on the same auto show stage in Detroit in January 2014.

CHAPTER EIGHT

Above: Among the things shared by C7.R and Z06 was its aluminum frame, which now was being built in-house at Bowling Green. Reportedly the C7.R chassis was 40 percent more rigid than its C6.R predecessor.

Left: First seen inside 2015's Z06, this flat-bottom steering wheel became a standard Stingray feature for 2016. Also, now in the base-model norm were Z51-style split-spoke wheels, still measuring 18 and 19 inches tall (front/back) per existing C7 specs. Z51 wheel diameters were 19 inches in front, 20 in back. *Mike Mueller*

Below Left: Continental Structural Plastics began supplying non-carbon-fiber Corvette body panels in 2016, resulting in a weight-savings of about 20 pounds per car. This 2016 Stingray is one of 10,415 Z51 coupes built that year. Base price for this combination was $61,395. *Mike Mueller*

Below Right: Easily handled by most, a lightweight removable carbon-fiber roof was standard for C7 coupes from the get-go. But which roof? The basic color-coordinated lid could have been superseded, at extra cost, by a transparent single panel, an exposed-carbon-fiber single panel (shown here), an exposed-fiber dual roof, and another dualie done in body color. In the latter two cases, a transparent center section was framed by body-color edges or a combination fiber/body-color border. *Mike Mueller*

At Right, Top: **Three more prestigious design packages debuted for 2016, each based on exclusive interior treatments done in Twilight Blue (RPO ZLD), Jet Black Suede (ZLG), or Spice Red (ZLE), with the first appearing here. Special accents included unique Spice Red brake calipers for the ZLE package. All three were available for 3LT/3LZ Stingrays and Z06s, top or not. Production was 371 for the ZLD option, 454 for the ZLD, and 535 for the ZLE.**

At Right, Bottom: **Chevrolet commemorated the Corvette's proud racing heritage in 2016 with the C7.R Edition, a dress-up package for Z06 coupes and convertibles. Yellow-striped black wheels, yellow brake calipers, various bits of visible carbon-fiber, leather interior with yellow striping, and competition seats were included. And, oh yeah, so was the Z07 equipment. Color choices were yellow or black.**

Most notable commonalities involved aerodynamics and the new breed's superstrong aluminum frame. That latter family tie was nothing new in the .R world, but thanks to all the computer-aided work that went into 2014's water-formed production foundation, the C7.R racing chassis ranked 40 percent stronger than its sixth-gen predecessor. On the flipside, the R-car's shell remained familiar to civilian Stingray customers while predictably incorporating a few no-brainer mods, beginning with that track-proven uber-wing in back. Already ample aero-abilities were additionally improved by dropping the profile 3.3 inches, and component-cooling airflow was invited inside via revised ducting.

Functional appearance alterations also included widened wheelhouses: by another 3.3 inches side-to-side compared to the also-flared Z06, up a healthy 6.8 clicks relative to a garden-variety Stingray. As expected, shrouded beneath those ballooned quarters were wider wheels, while coil-over adjustable shocks replaced stock-spec transverse springs at both ends. Brakes were altered with, among other things, six-piston (instead of four-) calipers added in back.

Most notably setting C7.R and Z06 apart was the former's 5.5-liter small-block, a direct carryover from the C6.R. GT-class racing rules since 2010 had limited displacement for qualifying cars, as well as prohibited forced induction, so the Z-car's blown V-8 needed not apply. But new (make that "back") for the seventh-gen GT-legal Stingray was warmly welcomed direct-injection fuel delivery, last seen fanning the C6.R's flames in 2009.

2016

Some drivers surely appreciated this year's new-across-the-board standard feature: the flat-bottom steering wheel that debuted exclusively inside 2015's Z06. Go ahead, have another doughnut, Chevy has got your back. Err, front.

Meanwhile, new standard wheels at the Stingray's corners were now of the split-spoke style previously reserved for the RPO list and Z06. Added as well into 2016's basic mix was Apple CarPlay (iPhone 5 or newer) touchscreen projection

CHAPTER EIGHT

via MyLink, along with lower-tech power-cinch systems for both the coupe hatch and convertible trunk latches.

Debuting among this year's extra-cost items was a forward-aimed curb-view parking camera and three more "personally tailored" Design Packages, each adding, among other things, exclusive Twilight Blue, Spice Red, or Jet Black Suede interiors to 3LT-equipped Stingrays and 3LZ Z06s, coupes or convertibles.

Higher-tech-minded customers with a little extra moola to burn could, beginning this year, add Magnetic Selective Ride Control to all Stingrays, not just Z51-equipped models. Standard for the Z06 and listed as RPO FE4 in the Z51 C7's case, this second electronic damping option (FE2) cost twice as much ($3,495) as its FE4 cousin because it brought along with it the Z51's wheel/tire combo and rear spoiler.

Z06 fans lousy with disposable income could make their supercharged C7s appear even more ready to race by adding RPO ZCR, announced in April 2015 to pay homage to the Corvette's storied competition career. Aptly tabbed "C7.R Edition," this package cost $23,055 when added to 2016's

Z06 coupe or $24,150 in convertible applications. Z07 goodies were included, 3LZ trim was mandated, and the lengthy list of remaining ZCR additions was made up of appearance upgrades and exclusive interior touches. On the outside went Corvette Racing Yellow tintcoat paint, a new-for-2016 choice for all Stingrays that echoed the finish well-familiar by then to race fans around the world. Basic black was available, too.

2017

Yet another ol' friend made yet another return to the Corvette lineup in 2017, again in coupe and convertible forms: Grand Sport. Part Stingray and part Z06, the seventh-gen's GS was predictably priced between the two, thanks in part to the fact that it stuck with Chevy's naturally aspirated LT1. Or perhaps that tag should've read "LT2.0" to acknowledge its "standard" extras: performance exhausts (that once more boosted output to 460 horses) and the Z51 package's dry-sump oiling.

Copped as well from the Z06 were its big Brembo brakes and electronic limited-slip differential. The Grand Sport's grille, front splitter, and widened rear quarters were distinctly "Z06-esque," too. GS wheels, meanwhile, were of exclusive design, and they were wrapped in Michelin Pilot Super Sports summer-only tires. Magnetic Ride Control also was included.

Formerly limited to Z06 applications, the Z07 option was available to GS buyers, as were Carbon-Fiber ground effects. And GS grandeur could be enhanced further by checking off

2017–2019 Grand Sport

Model availability	hatchback coupe (with removable roof panel) and convertible
Construction	composite/carbon-fiber body panels, hydroformed aluminum frame with aluminum and magnesium structural/chassis components
Body modifications	included Z06-style grille, front splitter, rockers, wickerbill spoiler, and widened rear fenders/fascia*
	Z07: available for Grand Sport but did not include tall splitter end caps and clear center section in rear spoiler
	* GS taillights were red, not smoked like Z06
Wheelbase	106.7 inches
Length	176.9 inches
Width (without mirrors)	77.4 inches
Height	48.6 inches (coupe), 48.7 inches (convertible)
Track (front/rear, in inches)	63.5/62.5
Curb weight	3,428 pounds (coupe), 3,487 pounds (convertible)
Wheels	exclusive pearl-nickel aluminum Cup-Style 10-spokes; 19x10 inches front, 20x12 inches rear
Tires	Michelin Pilot Super Sport ZP summer-only; P285/30ZR19 front, P335/25ZR20 rear
Tires (Z07)	Michelin Pilot Super Sport Cup 2 summer-only; P285/30ZR19 front, P335/25ZR20 rear
Brakes	power-assisted four-wheel slotted Brembo discs with aluminum calipers (six-piston front with monobloc fixed calipers, four-piston rear) and ABS
	Z07: added carbon-ceramic Brembo cross-drilled rotors with monobloc fixed aluminum calipers front/rear
Brake rotor diameter	14.6 inches front, 14.4 inches rear
	Z07: 15.5 inches front, 15.3 inches rear
Fuel tank	18.5 gallons
Suspension	short/long-arm (SLA) double wishbone cast-aluminum control arms, transverse composite leaf springs, stabilizer bars front and rear, monotube shock absorbers, and Magnetic Selective Ride control
Steering	power-assisted, speed-sensitive, variable-ratio rack and pinion
Engine (LT1)	6.2-liter (376-ci) direct-injection overhead-valve V-8 with cast-aluminum cylinder block and heads, performance exhausts, and dry-sump oiling
	Compression: 11.5:1
LT1 bore and stroke	4.06x3.62 inches
LT1 output	460 horsepower at 6,000 rpm and 465 lb-ft of torque at 4,600 rpm
Standard transmission	seven-speed manual with 2.97:1 low and Active Rev Matching
Optional transmission	8L90 eight-speed paddle-shift automatic with 4.56:1 low (RPO M5U)
Axle ratio	3.42:1 with electronic limited-slip differential (2.73:1 with eight-speed automatic)

Opposite Top: Another triumphant return occurred in 2017, this one involving the Grand Sport, appearing again in coupe and convertible forms. A Grand Sport paced the Indianapolis 500 on May 28, 2017. This was the 14th Corvette dating back to 1978 to do pace car duties at Indy and the 28th Chevrolet dating back to 1948.

Opposite Bottom: Fitted with various Z06 and Z51 parts, the latest GS shared its "low hood" with base Stingrays—because there was no need for the Z-car's bulging lid. Power came from the naturally aspirated LT1 fitted with dry-sump oiling and NPP performance exhausts, meaning output crept up to 460 horsepower.

Left: Also new for 2017 was the Grand Sport Collector Edition, intended to more specifically commemorate the five original Grand Sports built by Zora Duntov in 1963. Special embossing on each seat's headrest helped remind occupants of those rare racers.

CHAPTER EIGHT

Opposite Top: One of the five original Grand Sports takes a breather next to 2017's limited-run Grand Sport Collector Edition. Along with its Watkins Glen Gray Metallic finish, this auction-star-in-waiting was treated to Tension Blue fender hash marks, satin-black full-length racing stripes, black Cup-style wheels, and unique interior appointments.

Opposite Bottom: All the familiar Grand Sport striping came into play via RPO Z15, the Heritage Package. Center stripe colors numbered five: white, blue, gray, red, and Carbon Flash. Fender hash marks came in six shades: Torch Red, Carbon Flash, Hyper Green, Shark Grey, Volcano Orange, and yellow. Z15 availability carried over from 2017 into 2018 and 2019.

Above: A limited production run resulted in fewer than 10,000 Corvettes hitting the streets for 2018, the first time that plateau hadn't been reached annually since 1959. Individual rarities included Z51 Stingray convertibles (only 198 built), Z06 convertibles (449), Grand Sport convertibles (512), and base Stingray convertibles (537).

either RPO Z15 or Z25, the former consisting of a reverent Heritage Package that added the bloodline's familiar fender hash tags and centerline striping.

Much more majestic, the Z25 Collector Edition paid homage even further in history, to Zora Duntov's GS racers of 1963. Its Watkins Glen gray metallic finish gave a nod to those lightweight legends by way of its Tension Blue accents, which, in Chevrolet's words, represented "a bold, modern take on the hue historically associated with the Grand Sport." Duntov's originals were additionally honored inside on special seat embossments and an instrument panel plaque that incorporated a unique build-sequence number.

Of course, Chevrolet's reborn Grand Sport was back on the Bricks in May 2017, making it 16 straight Indy 500 pace-lap appearances for Chevrolet products. And 14 overall for Corvette—a runaway record that'll probably outlive Joltin' Joe's. Once again, Google's right there at your thumb-tips. . .

2018

A trickle-down approach helped differentiate 2018 Stingrays from their forerunners as base models were treated to larger, Z51-derived wheels, up an inch in diameter (19 in front, 20 in back) compared to 2017's standard rollers. Five new optional wheels also made the scene this year, as did standard HD digital radio. And in other RPO news, the price for FE2 Magnetic Ride Control was slashed in half by offering it without Z51 accompaniment. Ceramic brakes also became a stand-alone (in this case, sold separately without Z07 equipment) option for the latest Grand Sport.

While the Grand Sport's traditional Heritage Package carried over from 2017, the pricier Collector Edition didn't, in keeping with Chevy's hope that, true to its name, this one-off commemorative would someday represent a true prize among enthusiasts. But in the Z25's place came another low-production future gem

CHAPTER EIGHT

Chevrolet marked 65 years of Corvette history in 2018 with RPO Z30, the Carbon 65 Edition. Only 650 were produced, based on both Z06s and Grand Sports.

(also tagged with a specially-dedicated vehicle identification number), this one marking the Corvette's 65th birthday.

Limited to 650 examples delivered globally, the Carbon 65 Edition option (RPO Z30) was offered for top-shelf (3LT and 3LZ, respectively) Grand Sports and Z06s with or without tops. Another new color, Ceramic Matrix gray, was the sole paint choice this time and was complemented by black wheels, blue brake calipers, appropriate graphics (on the fenders and doors), and various visible carbon-fiber touches. Inside were Jet Black suede appointments accented with additional carbon-fiber effects.

Total production for 2018 itself ended up highly limited, first due to a shutdown (August through October 2017) at the Bowling Green plant in order to upgrade its paint shop. The truncated 2018 run then took another hit, abruptly ending the following January, resulting in a final count of only 9,686 Corvettes—the lowest total since 1959. More on this in the next chapter.

Base price for a Stingray convertible in 2018 was $60,490; $65,490 if the Z51 package was tossed in for the ride. The most popular model this year was the Grand Sport coupe (2,569 built), priced at $66,490. Larger standard wheels (19-inchers in front, 20 in back) superseded the 18/19 combo installed beneath earlier C7s.

280

2019

As announced well ahead of time, 2018 production halted on January 22 that year. Then, just as reported widely months before, 2019's run commenced a week later, taking full advantage of those new paint facilities, refurbished with an eye toward—among other things—finishing carbon-fiber panels better than ever. Improved overall finish quality for all Corvettes resulted, too, making those available tintcoats introduced along with the C7 even more tantalizing.

And that was just about that concerning the latest news. Almost. Once the Bowling Green plant restarted late in 2017, Chevrolet officials opted to again make future plans (no, not those—you'll again have to wait another chapter) well known.

On November 12, 2017, came word of the "return of the king," the ZR1, previously the super-est Corvette to date before the seventh-gen Z06 rose to power. Save the best for last, huh?

"I've never driven a Corvette like this before," beamed Global Product Development executive vice president Mark Reuss while unveiling the 2019 ZR1. "And nobody else has either, because there's never *been* one like this before."

Indeed, according to *AutoWeek*, "the new king of Vettes breaks all the laws—of physics." Even though the Z06 apparently had already pushed performance parameters as tightly up against the wall as humanly possible, the reborn ZR1 managed a bit of renovation—as in 0–60 in 2.85 eye-popping (figuratively and perhaps literally) seconds.

There couldn't have been much more, if any, left in the tank, right? Wait, don't answer. Cuz this barely grounded supercar also reached 212 mph at the test track, yet another Corvette standard in a long line of recent history-making moments that seemingly wouldn't end.

How did Chevrolet manage such extremes? Not with volume, mind you, but in part with another muscled-up small-block, the LT5, rated at—you guessed it—a *record* 755 horsepower. Helping make the leap from LT4 to 5 was a larger, more efficient intercooled supercharger and another GM first: dual fuel-injection, which supplemented

Right: Horsepower hounds in the heartland got their first real look at Chevrolet's latest best 'Vette yet, the new ZR1, at where else? Yes, this 212-mph supercar paced the Indianapolis 500, this time on May 27, 2018. If that high wing in back makes you think race car, you got it.

2019 ZR1

Model availability	hatchback coupe (with removable roof panel) and convertible
Construction	composite/carbon-fiber body panels, hydroformed aluminum frame with aluminum and magnesium structural/chassis components
Body modifications	unique fascia with extra cooling inlets and carbon-fiber splitter, carbon-fiber "halo" hood and exposed engine cover, rear brake cooling ducts, widened wheelhouses and rear fascia, and low rear wing; high rear wing* available with ZTK Track Performance Package, which also added carbon-fiber end caps to the front splitter, Michelin Pilot Sport Cup 2 tires and specific chassis tuning. *ZTK high wing added 950 pounds of downforce at speed
Wheelbase	106.7 inches
Length	179.8 inches
Width (without mirrors)	77.4 inches
Height	48.5 inches (coupe), 48.7 inches (convertible)
Track (front/rear, in inches)	63.5/62.5
Curb weight	3,560 pounds (coupe), 3,618 pounds (convertible)
Wheels	exclusive pearl-nickel aluminum ten-spokes; 19x10 inches front, 20x12 inches rear
Tires	Michelin Pilot Super Sport ZP summer-only; P285/30ZR19 front, P335/25ZR20 rear
Tires (ZTK)	Michelin Pilot Super Sport Cup 2 summer-only; P285/30ZR19 front, P335/25ZR20 rear
Brakes	power-assisted four-wheel cross-drilled carbon-ceramic Brembo rotors w/aluminum calipers* (six-piston front with monobloc fixed calipers, four-piston rear) and ABS *blue calipers standard, orange calipers included w/Sebring Orange Design Package, RPO ZLZ
Brake rotor diameter	15.5 inches front, 15.3 inches rear
Fuel tank	18.5 gallons
Suspension	short/long-arm (SLA) double wishbone cast-aluminum control arms, transverse composite leaf springs, stabilizer bars front and rear, monotube shock absorbers, and Magnetic Selective Ride control
Steering	power-assisted, speed-sensitive, variable-ratio rack and pinion
Engine (LT5)	6.2-liter (376-ci) overhead-valve V-8 with cast-aluminum cylinder block and heads, intercooled Eaton R2650 supercharger, dual fuel delivery system (primary direct injection and supplemental port injection), performance exhausts, and dry-sump oiling **Compression:** 10.1:1
LT5 bore and stroke	4.06x3.62 inches
LT5 output	755 horsepower at 6,300 rpm and 715 lb-ft of torque at 4,400 rpm
Standard transmission	seven-speed manual with 2.29:1 low and Active Rev Matching
Optional transmission	8L90 eight-speed paddle-shift automatic with 4.56:1 low (RPO M5U)
Axle ratio	3.42:1 with electronic limited-slip differential (2.41:1 with eight-speed automatic)

CHAPTER EIGHT

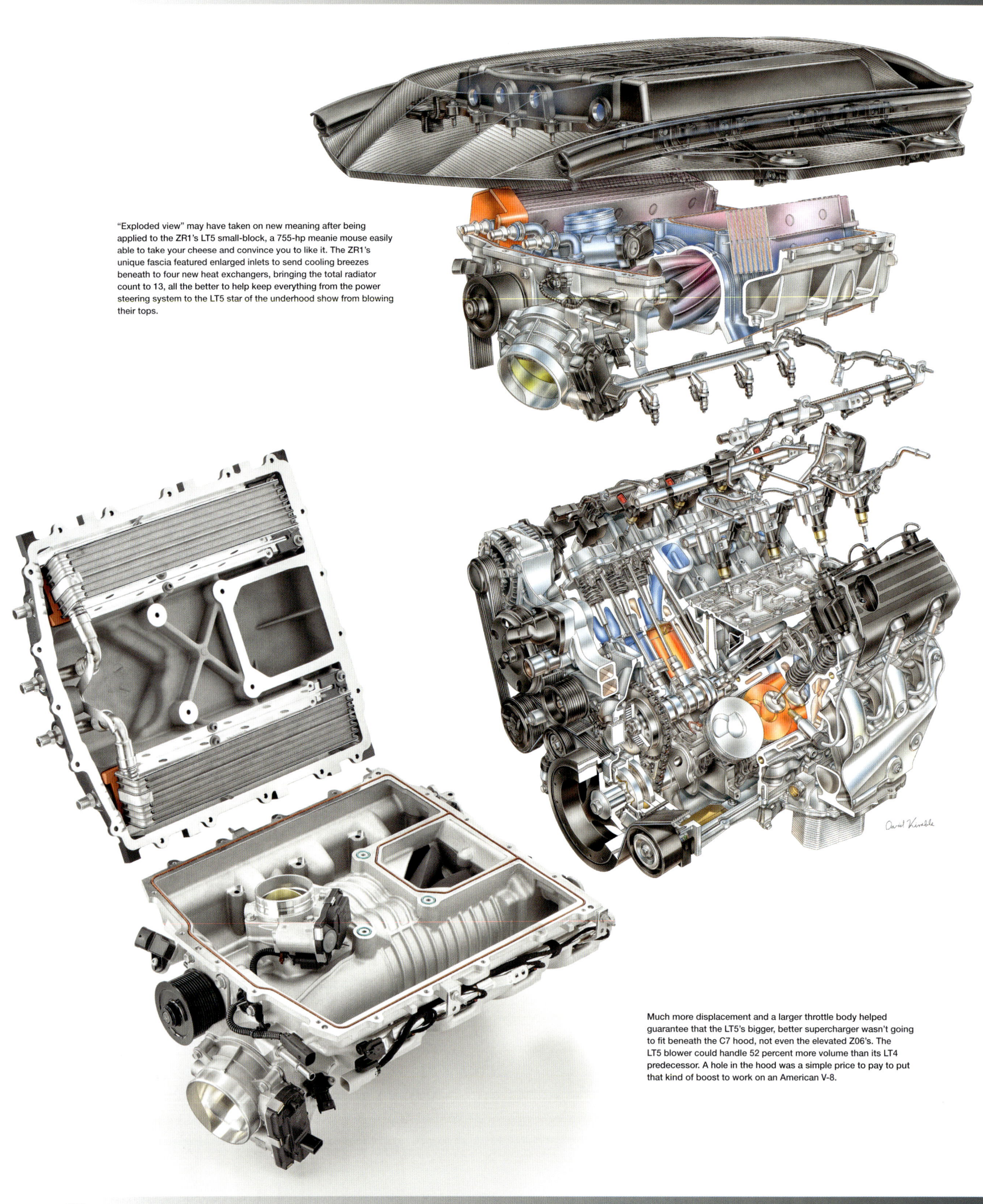

"Exploded view" may have taken on new meaning after being applied to the ZR1's LT5 small-block, a 755-hp meanie mouse easily able to take your cheese and convince you to like it. The ZR1's unique fascia featured enlarged inlets to send cooling breezes beneath to four new heat exchangers, bringing the total radiator count to 13, all the better to help keep everything from the power steering system to the LT5 star of the underhood show from blowing their tops.

Much more displacement and a larger throttle body helped guarantee that the LT5's bigger, better supercharger wasn't going to fit beneath the C7 hood, not even the elevated Z06's. The LT5 blower could handle 52 percent more volume than its LT4 predecessor. A hole in the hood was a simple price to pay to put that kind of boost to work on an American V-8.

primary direct injection with additional injectors at each port. Unprecedented, too, was the available 8L90 8-speed, the ZR1 breed's first auto trans.

But again, as demonstrated by the Z06, rounding up more ponies alone doesn't necessarily put the super in supercar. Picking up where the slippery 650-horsepower Corvette left off, the ZR1 took road-hugging to all-new heights, thanks to even more downforce induced by a front "underwing" and a fixed low wing in back.

Available, too, was the Track Performance Package (RPO ZTK), which added an adjustable high wing at the tail and carbon-fiber end caps for the underwing up front. Michelin Pilot Sport Cup 2 summer-only tires also were part of the ZTK deal, along with specific chassis calibrations and Magnetic Ride Control.

GM officials introduced the 2019 ZR1 in Los Angeles on November 28, 2017, promising that sales—of both coupe and ground-breaking convertible renditions—would start the next spring. Just in time for open-wheel fans to meet Chevy's latest, undeniably greatest Vette yet. Once more, you guessed it: the new ZR1 paced the 102nd Indianapolis 500 in May 2018.

Another C7, a 2019 Grand Sport this time, toured the Brickyard on race day in May 2019. A suitable send-off for the shortest Corvette run ever. Short time-wise, sure. But taller than tall as far as facing off against the world's best supercars was concerned.

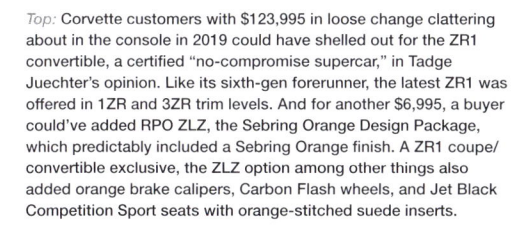

Top: Corvette customers with $123,995 in loose change clattering about in the console in 2019 could have shelled out for the ZR1 convertible, a certified "no-compromise supercar," in Tadge Juechter's opinion. Like its sixth-gen forerunner, the latest ZR1 was offered in 1ZR and 3ZR trim levels. And for another $6,995, a buyer could've added RPO ZLZ, the Sebring Orange Design Package, which predictably included a Sebring Orange finish. A ZR1 coupe/convertible exclusive, the ZLZ option among other things also added orange brake calipers, Carbon Flash wheels, and Jet Black Competition Sport seats with orange-stitched suede inserts.

Middle: In January 2019, Chevrolet announced availability of yet another group of special-edition design packages, these created with the help of the Corvette Racing team. The four drivers captured here (left to right: Oliver Gavin, Jan Magnussen, Tommy Milner, and Antonio Garcia) each teamed up with a Chevy designer to put together a special collection of interior/exterior graphics for the last Grand Sport coupe, at least in C7 terms. Production was tiny: 35 for Gavin's, 25 for Magnussen's, 21 for Milner's, and 14 for Garcia's.

Bottom: A 2019 Grand Sport helped send the C7 legacy off in proud fashion by once again pacing the Indianapolis 500. The driver on May 26, 2019, was NASCAR legend Dale Earnhardt, Jr.

CHAPTER NINE

09 Makin' the Ol' Man Proud

C8 2020–2025

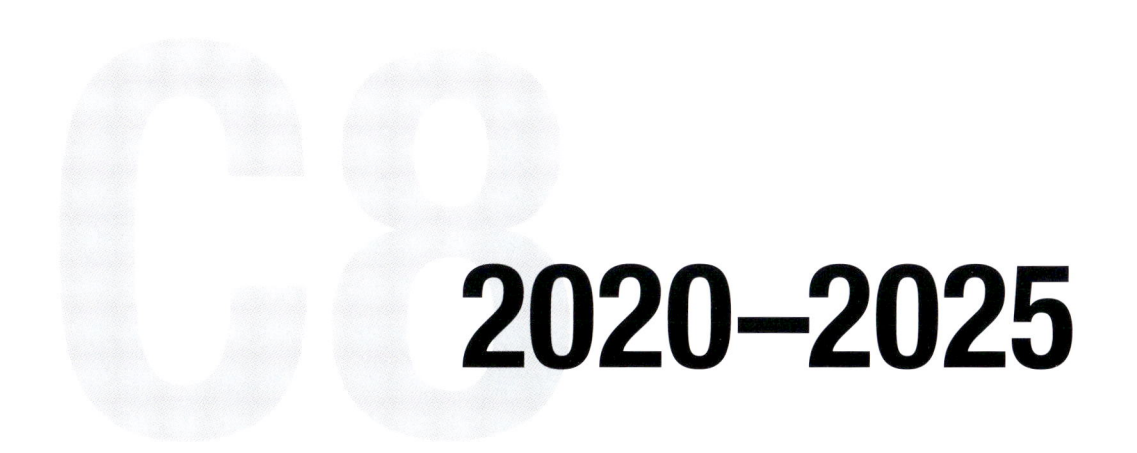

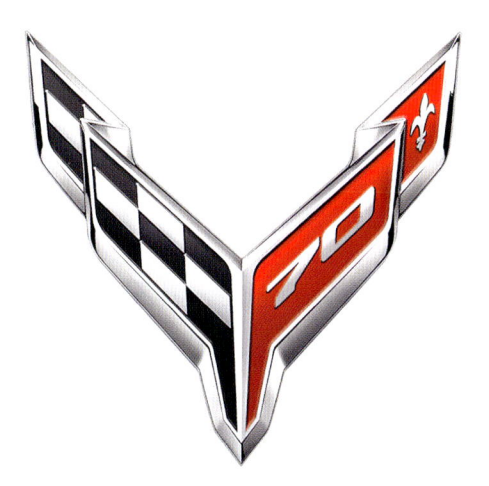

Cleanup efforts following the C7's various coming-out parties were still underway and prognosticating press people, per longstanding tradition, already were confidently announcing what surely would trigger the next confetti dump. And wouldn't you know it? Leading the way was Tadge Juechter's favorite auto writer, Don Sherman, who kick-started his career at *Car and Driver* in the summer of 1971—just in time to catch a ride on a wave that would wash up repeatedly in the Corvette reporting world over the decades since.

- Long rumored and even longer fantasized about, a midengine Corvette is finally unveiled (July 18, 2019)
- A United Auto Workers strike, GM's longest in a half-century, delays both final production of C7 Corvettes and assembly startup of C8s (September–October 2019)
- C7 production ends (November 14, 2019)
- Midengine C8 becomes third Corvette (joining 1984 and 1998 models) to garner *Motor Trend*'s Car of the Year trophy (November 2019)
- C8 dealer deliveries commence (March 2020) but are temporarily interrupted by the Covid-19 pandemic
- Bowling Green Assembly builds Chevrolet's first right-hand-drive Corvette (January 21, 2021) for export to Japan
- Eighth-gen 670-hp Z06 introduced (October 26, 2021); goes on sale later as 2023 model
- First all-wheel-drive Corvette, the 655-hp E-Ray, introduced (January 17, 2023) with electric power up front; goes on sale later as 2024 model
- Most powerful Corvette ever at 1,064 hp, the eighth-gen ZR1 debuts via worldwide web video (July 25, 2024); goes on sale later as 2025 model

STILL KEEPING HIS FINGER on the pulse of Chevrolet's two-seater five generations later, Sherman not only still stands among the most savvy, well-versed reporters in this vein, he also surely leads the league in making *the* prediction that again and again just wouldn't come true. When asked in February 2020 how many times he had put down prose foretelling the upcoming arrival of a midengine Vette during his long tenure, his lyrical reply was as short as it was quick: "Way-y-y too many."

But even a blind journalist makes it back to the tree with a nut now and again. As for Sherman, he finally cracked this case in October 2014, assuring *Car and Driver* readers that, "after 61 years of evolution, the C8 will be revolutionary." Well, at least he correctly broke the biggest news.

"The new Corvette will be the midengined American Dream Machine that Chevy couldn't, until now, muster the courage to build," he continued before forging ahead bravely himself. Once more with a decent dose of specific speculations reportedly curried from reliable inside sources. Among other things:

a) The company's indefatigable pushrod small-block would carry on. . .
b) . . .backed by only one transaxle choice; a seven-speed dual-clutch unit supplied, perhaps, by supercar transmission superstar Oerlikon Graziano.
c) The midengine C8 would appear first in top-gun "Zora ZR1" guise for 2017. . .
d) . . .then "peacefully coexist in showrooms with C7 models for a few years."
e) Lastly, by 2020, all Corvettes would morph into midshipmen.

History's verdicts?
a) Right on, brother! Old-school remained too cool for school.
b) Correcto mundo on the single transaxle and its dual-clutch design; not so much as far as its speed count ("missed it by *that* much. . .") and builder were concerned.
c) That debut year came and went, and the jury remained out concerning the Duntov honor. But Sherman simply couldn't lose by forecasting a ZR1 return—*eventual*, that is.

Opposite: Chevrolet designers dating back to C5 development had repeatedly stressed the need to preserve the Corvette's identity while powering it upward and onward into each new generation. Having succeeded in this task for 5th- through 7th-gen models, they were especially challenged to radically reshape the latest Stingray into a world-class midengine supercar. No worries. Most critics agreed that 2020's C8 surely ranked as their crowning achievement.

Above: Not even Harley Earl could've imagined that his baby would still be standing tall seven decades down the road. In 2023, Chevrolet celebrated the Corvette's 70th birthday with commemorative cross-flags on all exteriors, plus anniversary identification inside between the seats. And, of course, a special-edition 70th Anniversary package was offered too.

CHAPTER NINE

Above: Though surely dreamed of by Duntov and others dating almost back to the Corvette's birth, a midengine platform didn't truly start taking shape until a half-century before the C8's debut. First came the XP-880 Astro II (top) in 1968, joined that year by two XP-882 show cars. One of these then morphed into 1972's XP-895, a heavyweight middie that in turn was rebodied into the lighter Reynolds Aluminum Corvette (second from top) in 1973. That same year, the trendy Wankel-powered Four-Rotor Corvette appeared, only to be repackaged (with the XP-895's V-8 powertrain) as 1977's Aerovette (middle). Fitted with a 600-horsepower twin-turbo V-8, the midengine Corvette Indy hit the show circuit in 1986. C8 sketches (bottom) then began appearing even before the C7 Stingray was unveiled in January 2013.

Left: The main mind behind the longstanding midengine ideal, chief engineer Zora Arkus-Duntov poses here at 1970's New York auto show with GM's XP-882 concept car, one of various middie experiments considered to be prototypes for future next-gen Corvettes over the years since.

The designer mostly responsible for 1986's Corvette Indy, Tom Peters was leading the team by the time the C8 started forming full-size in clay. Hundreds, if not thousands, of sketches led to this model, which Peters felt still represented the key, even in our digital age, to determining a final form. Not that he didn't value the latest computer-aided design tools; he simply recognized that a hands-on approach, true sculpting rather than keypunching, was just as important. Peters retired in 2019 and was soon afterward inducted into the National Corvette Museum's Hall of Fame.

If not for GM's financial misfortunes of 2007–2008, a midengine Corvette would've appeared one generation earlier. Code-named GMX721, Chevrolet's midengine C7 program began in earnest in 2007 but only reached the clay-model stage before that cash crunch intervened. This foretelling clay was one of two GMX721 models sculpted, the other representing a midengine replacement for Cadillac's XLR.

d) Nice try (echoed even by some former GM people) but no cigar.
e) Not exactly how it happened, but we'll give Don this one just for playing our game.

In Sherman's defense, know that, in pro baseball, reaching safely *only* three tries out of 10 might just someday earn a plane ticket to Cooperstown. C'mon, cut the guy some slack. After taking so many swings over so many years, nearly leaving the yard in the C8's case appeared damn near as sweet as touching 'em all. If only Duntov had been there to high-five him at home.

Having left this world for automotive Valhalla in 1996, Zora Duntov is still well remembered—as mentioned more than once in earlier chapters—for his dedication to the midengine ideal, a respect he surely brought with him when he came to work at Chevy Engineering in May 1953. His full consideration, however, his devotion to the cause, didn't start taking real root until his ill-fated SS racer (leaf rearward to pages 36–40) dropped out of Sebring's 12-hour endurance run so, so early in March 1957. Back in the Eighties, Duntov told Sherman he first concluded that moving the engine to the middle had to be a better idea after the SS "cooked" driver John Fitch during its disappointing on-track debut down in Florida. "We had to put the heat source behind the driver," Zora said.

C8 2020–2025

Basically the same veteran design team on board for C7 Stingray development was responsible for drawing up the C8, including, among others, Hwasup Lee, Brad Kasper, and Vlad Kapitonov. According to GM global design v.p. Michael Simcoe, packaging studies for a next-gen midengine Corvette existed in the corporation's Studio X "skunkworks" as far back as 2007.

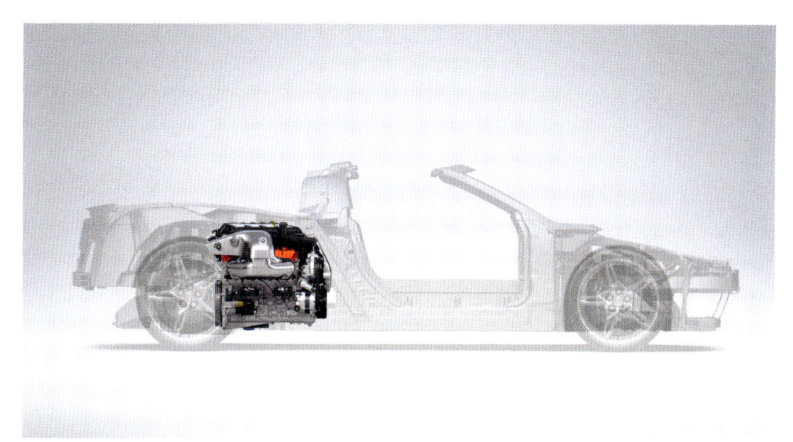

It was kinda like when the trooper pulled over the tandem-axle heavy hauler and asked, "What's behind the seat, sir?" "The whole rest of the truck, officer," offered the driver before being hauled off for impersonating a wiseguy across state lines. But seriously, the C8's main attraction did bring up its rear in the form of the midmounted 490-horsepower LT2 small-block (495 horses with optional performance exhausts). Curiously, though, one source claimed Chevrolet was sandbagging us with those ratings. In October 2019, *Motor Trend* published its own dyno test results: 656 horsepower, 558 at the wheels, leaving even some GM people wondering what they'd missed. Nothing. One month later, *Motor Trend* people published a detailed retraction. Bottom line: their math was flawed. But the sum of C8 parts still equaled a world-class Chevrolet sports car like America had never seen before.

Pushed back by Covid-19 restrictions to August 2, 2020, the 104th running of the Indianapolis 500 was paced by a Corvette for the fourth straight year. GM president Mark Reuss drove this 2020 Stingray coupe on race day. *Indianapolis Motor Speedway*

At some point before 2020 production rolled over into 2021, Chevrolet quietly treated Stingray customers to a couple Easter eggs in the form of humble commemorative windshield etchings. Discovered in September 2020 by the folks at Kerbeck Corvette in Atlantic City, New Jersey, this treatment included a Zora Arkus-Duntov silhouette on the lower passenger-side corner and a "Team Corvette" logo (complemented by the breed's familiar cross-flags) on the driver's side. A month or two later, a second Duntov image was discovered stamped into C8 undersides.

It was as simple as that. "Duntov always had some racing applications going on during all his projects," recalled Sherman in 2020. And, in Zora's mind early on, switching to midengine construction first and foremost represented the straightest, strongest course towards building more competitive competition vehicles. Hence came CERV I—which sure looked like an Indy car—in 1960, followed shortly by CERV II, another middie that appeared able to take any track by storm. Actual racing careers, however, escaped both, due primarily to evolving politics at GM.

Duntov's approach then refocused towards more production-friendly propositions, resulting in various other aforementioned midengine concepts (Astro II, XP-882, Aerovette, etc.) beginning in the late-Sixties. And it was this showcar stream that instantly inspired so many faulty press predictions.

Faulty due to various stumbling blocks, not the least of which involved a response that Duntov surely heard at most every turn: "Why repair something not in need of fixin'?" On their own, strong sales of conventionally constructed Corvettes served as stop signs hindering Zora's best efforts. Nonetheless, he continued pushing on, even after his 1975 retirement.

According to his successor, Dave McLellan, Duntov's parting words were, "you must do the midengine Corvette." The car's second engineering chief, however, fully recognized what its proud "father" apparently refused to see: both GM's executive hierarchy and Chevy's sales/marketing force at the time simply couldn't see reason to reinvent their highly popular niche-mobile.

And let's not forget those ever-present, always-enthusiastic customers, who also weren't necessarily

interested in shifting gears. As McLellan explained it to Sherman in an October 2018 *Hagerty* magazine interview, "when you consider that the move beyond chrome bumpers [during the Seventies] was initially a tough sell, it's clear how conservative Corvette owners can be." Market research as late as 1990 revealed that most of the fiberglass faithful still qualified as proverbial sticks-in-the-mud as far as trading engine positions was concerned.

Those same seemingly immovable market realities, those continuing upper-office cold shoulders, remained for David Hill, who took over for McLellan in 1992 in time to help shepherd the C5—with its Q-Corvette-style (see page 47) rear-mounted transmission—into production. But that was as close as the fifth gen came to fulfilling Duntov's dream, and no real consideration was given to making the sixth a middie, either.

287

CHAPTER NINE

In August 2019, GM donated two priceless pieces of C8 prehistory to the National Corvette Museum, one an actual running prototype, the other the first 40-percent clay model sculpted in 2014. On permanent display in the NCM's Design & Engineering exhibit, that scale-model (at right) was used for theme evaluation and initial aerodynamic tests. *Courtesy National Corvette Museum, Bowling Green, Kentucky*

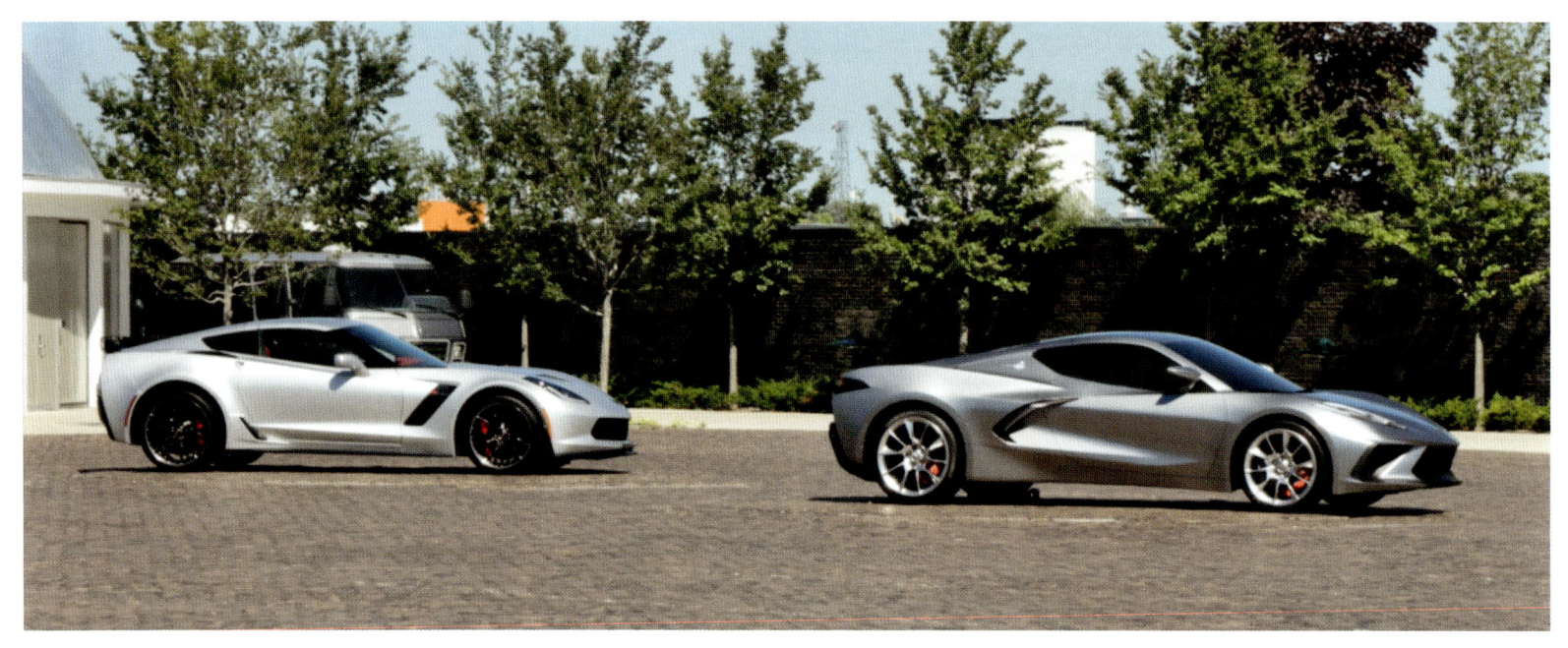

Exterior design manager Kirk Bennion startled more than a few seminar attendees during the National Corvette Museum's 25th anniversary celebration in August 2019 by announcing that C8 sketchwork indeed was well underway *before* the C7 was revealed to the world on January 13, 2013. Family ties between the two are evident here in this Design Center patio pose, a predictable result considering that the C7 was initially intended to emerge as Chevrolet's first midengine Corvette. While it ended up on a conventional front-engine platform, the 7th-gen Stingray still was treated to some of the lines and form that then evolved fully into the next-gen middie.

Instead, Hill's group concentrated on enhancing the status-quo to all-new heights, work that continued sensationally after Tom Wallace became the Corvette's fourth chief engineer in 2006. That year's Z06 and 2009's ZR1 each helped bolster the breed's newfound status as a true world-class supercar but at the same time didn't exactly help convince anyone that a radical change might be in order.

Maybe so. But Wallace and crew still put wheels back beneath the midship proposition. Picking up where Duntov left off was a new driving force, Tadge Juechter, who Hill had brought onboard as his assistant in 1993. On the team as well was designer Kirk Bennion, a veteran who was already drawing up midengine Corvettes during his college days before joining GM in 1984.

Then there was Corvette product manager Harlan Charles. If Juechter was the driving force, Charles was the guy with the hand crank who first got things firing. As he recounted in two public presentations in 2021 and 2022, he had gone right to work following the C6's January 2004 intro building a file identifying all-new strategies that could make the C7 far and away the best Vette yet.

Bowling Green, Kentucky, became the first location not part of GM's traveling 2020 Stingray road show to showcase the next-gen C8 in August 2019, this after a fully functional prototype was donated to the National Corvette Museum. Assembled in 2018, this test mule was fitted with data-recording equipment to analyze LT2 performance parameters, primarily on the track at GM's Milford Proving Ground in Michigan. *Courtesy National Corvette Museum, Bowling Green, Kentucky*

Topping the list, as one might guess, were all the benefits of a midengine swap.

This file remained secret at first. It wasn't until solid C7 discussions began in 2005 that Charles decided to man up, to come right out and say what Duntov had boldly claimed many times over: it was time for a relocated V-8. "Why wouldn't we at least study a mid-engine design?" asked Harlan early that year in a conference attended by everyone from David Hill on down. At first, however, it seemed he might've struck out.

"The meeting room got totally quiet, and everyone waited for Dave to react, some half snickering," remembered Charles. "Dave's reaction was 'if Chevrolet'—meaning me—'has a different proposal for C7 than the rest of the team, then you are welcome to bring it in next week.' I gulped hard and said, 'Okay, you are on.'"

About the same time, unknown to Harlan, advance architecture engineer Kim Lind had done a side study on a midengine concept and had asked Bennion to sketch up his ideas. Kirk (Harlan's "partner in crime") and Lind then teamed up with Charles to build a blockbuster presentation per Hill's direction. And being one to give ample credit where credit is due, Harlan explained that Lind was the "creative engineer who was instrumental in showing how we could make the package work with good passenger and cargo space."

As for their presentation, amidst Harlan's well-presented case was a chart adding all existing strengths to the exotic car attributes promised by a midengine structure. The resulting sum was the new C7, "a grand slam home run!"

But once more, many seated in that second meeting apparently missed the play. "There was silence in the room again, but Tadge said

The man who picked up where Arkus-Duntov's dreams left off, the driving force behind 2020's radically new midengine Corvette, was the breed's fifth chief engineer, Tadge Juechter, shown here introducing the C8 to the world during a global event livestreamed from Tustin, California, on July 18, 2019. Barely visible in the graphic displayed above Juechter is the 2020 Corvette's coil-over-shock suspension, another total revision along with its engine location. Also new for 2020 was an aluminum foundation created through die-casting, not hydroforming, a practice dating back to 1997's C5.

'nicely done' to me quietly," recalled Charles in November 2021. "Everyone else [was] looking at Dave to see how he would react. [He] said, 'I don't think we should change our winning formula, but I do think we should put some

CHAPTER NINE

Truckloads of 2020 Stingrays began leaving GM's Bowling Green assembly plant on Wednesday, March 4, 2020, with two of those haulers (carrying twelve Corvettes) simply cruising across the street to the National Corvette Museum. This batch consisted of one model slated for on-track drives at nearby NCM Motorsports Park plus eleven museum-delivery (R8C) cars, a group that included the eagerly awaited VIN 25 C8 raffled off during the NCM's 25th anniversary celebration in August 2019. The museum's first R8C 2020 Corvette delivery (a Torch Red coupe, VIN 29) was made to Texans Glenn and Andrea Johnson the following Tuesday, March 10. *Courtesy National Corvette Museum, Bowling Green, Kentucky*

packages together and have a look.' So that was the foot in the door."

Nonetheless, Hill still held the key. And, while the chief engineer was intrigued, he wasn't about to move beyond package studies. Neither was Juechter exactly on board early on, not until Charles made his presentation. And when Tom Wallace filled Hill's shoes, he initially thought Harlan "was crazy." Furthermore, the big boss was emphatically not amused. "When Bob Lutz found out we were [meddling with] a midengine, he gave an order that anyone working on that would be fired," added Charles.

Job security concerns aside, Chevrolet's determined product manager wouldn't be denied. Bolstered by his ample evidence, and aided by Juechter, he quickly swayed Wallace to his side. Then Tom and Tadge took another presentation up the chain during the winter of 2005–2006 to Lutz and GM chairman Rick Wagoner. Included this time was a clever "March 2011" *Car and Driver* cover, fabricated (in the best Don Sherman tradition) by Charles, announcing an "American Revolution!" "Ferrari? Porsche? Irrelevant!" read the subhead. Irrelevant because there would be a new sheriff in town, 2012's midengine C7 Stingray, armed with a 500-horsepower DOHC small-block capable of both 200 miles per hour and 30 mpg, all for about $50,000.

Top brass was finally convinced. Harlan kept his job, and approval to take the existing C7 program—coded GMX721—into the midengine realm immediately followed. Same for a new dual-clutch transmission. A customer clinic, executed in Los Angeles in 2007, added additional momentum after showing good public support for the idea. Two clay models appeared that year—one in Detroit, the other in England, to serve as the Cadillac XLR's replacement. As Bob Lutz later told *Road & Track* in January 2015, "They were both gorgeous, [and] we started working on [GMX721]."

But not for long. Charles was in Florida for Christmas 2007 when Juechter called him with the bad news: the game was over; all bets were off. GMX721 work never progressed beyond the clay-model stage, and no running midengine C7 prototypes came forth due to GM's monumental money woes.

Yet another roadblock? Not at all. In fact, 2008's oft-maligned federal involvement may have helped keep the dream alive even after the corporation's temporary failing fortunes appeared to quash it. As Harlan Charles concluded, "The one positive was that we had cemented the midengine direction with leadership for some time in the future." And that spark remained ready to reignite once corporate finances re-fired.

But not until the momentary darkness subsided. And Corvette progress got progressing again.

Initially in 2008, Lutz let Wallace know that there would be no funding for future Corvette development, only trucks and SUVs, GM's undeniable cash cows. Wallace then took early retirement later that November as opposed to riding ahead on what he perceived to be a lame horse—leaving Juechter to take over the reins just in time to benefit from a bit of unexpected luck. After reviewing Chevrolet's books, Washington's accountants discovered that Corvette indeed carried its weight, that it too brought in the bucks, making it a viable investment.

"Hence, the feds said to get a new platform off the back burner, to get the next generation up and rolling," explained Don Sherman. "Not necessarily a midengine [model], mind you, but a new car." Preferring not to dare stare a gift equine in the snout, and still limited budget-wise, Juechter moved ahead with a "conventional" redesign, albeit the thoroughly maximized, better balanced C7 Stingray that hit the ground galloping in January 2013.

Midengine machinations, meanwhile, went underground, or at least did once discussions resumed in 2011 or 2012 following GM's triumphant return to profitability. Only discovered by the pubic in the late-summer of 2019, a full-sized clay was on the football-field-sized patio outside GM's Design Center even before Juechter and gang had unveiled the C7 six years prior. "We were putting pencils down on our work on the midship [design] at the time [of the C7 reveal]," said Bennion in August 2019. "We [then] had close to a dozen scale-models." Full-bore C8 engineering development—starring an engine mounted behind the driver, not in front—was underway within the next year or so, leading to the latest wave of prophetic magazine covers, which for once rang true. At least partially.

Like Sherman, other sources also reported C8 debut dates that didn't exactly pan out. Even the hometown paper, *The Detroit News*, initially missed the point, claiming sales would start early in 2019. "The eighth-generation Corvette—code named 'Emperor'—is targeted for an unveiling in early 2018," stated an August 4, 2016, report penned by Henry Payne and Melissa Burden. Most opinions across the country then coalesced into a conclusion that the next-gen middie surely would debut, like its C7 forerunner, at Detroit's North American International Auto Show, in this case in January 2019. But denied again.

Various factors may or may not have contributed to this "delay." Or lack of such, as Tadge Juechter explained to *Road & Track*'s Travis Okulski in July 2019. "People are saying, 'Oh, this car's taking so long to come out.' Actually, my argument is that it's early. We're not late. [Development] was about five years."

From a bigger-picture perspective, Juechter spoke the truth. But apparently the original plan did involve the C8 following in the C7's introductory tire tracks. The wrench in the works? According to gmauthority.com's Alex Luft in December 2018, an electrical system glitch sent engineers back to the drawing board, resulting in a six-month postponement of the big ball.

Other outlets, including *USA Today* later in March 2019, also spoke of a structural integrity issue, which resulted in a fractured rear hatch during some heavy horsepower experiments. Not true at all, according to Juechter, who blamed this false rumor on a case of mistaken identity. As he again told *Road & Track*, a Cadillac test coinciding with C8 work resulted in that broken glass, and some witness mistakenly assumed a Stingray connection. Silly them.

Helping further thicken the plot were various refits at the Corvette's ol' Kentucky home. In May 2015, Chevrolet announced construction of a new

A C8 feature familiar to Corvette customers was the coupe's removable roof. And almost lost amidst the 2020 Stingray's long list of innovations was its base price, about $59K, making it by far the best supercar buy in this world. Or probably any other. "Most people thought when we moved the Corvette to mid-engine it would no longer be attainable," said Chevrolet U.S. v.p. Brian Sweeney in August 2019. "But we knew we couldn't mess with a winning formula and the 2020 Stingray proves it." How Chevy managed to offer so much world-class performance for so few bucks amazed onlookers even more than those unworldly 0–60 times.

CHAPTER NINE

paint facility, a 450,000-square-foot addition that itself would end up being about half the size of the existing works. Hot on the heels of this $439 million investment came word that December of an additional $44 million improvement of Bowling Green's Performance Build Center. And in June 2016, another $290 million was sent the plant's way "to upgrade and modify vehicle assembly operations with new technologies and processes."

A couple temporary shutdowns ensued, both for general maintenance and to help facilitate completion of those construction projects. A one-week closure in January 2017 was followed by a three-month interruption beginning that August, the latter also serving to take a major bite out of 2018 model-year production. Of course, this break was required to integrate the new assembly line process and complete the new paint shop, but rumor-mongers couldn't resist assuming that this downtime also might involve retooling for *2019* C8 production.

New for the C8's brakes was an electronic eBoost system that both did away with a venerable vacuum booster (that took up valuable space under the C7's hood) and allowed driver tuning of braking right along with all other performance parameters. Optional color choices returned for the C8's Brembo calipers.

2020 Stingray

Model availability	coupe (with removable roof panel) and convertible (with folding hardtop); front splitter and large rear spoiler (supplying upwards of 400 pounds of downforce) included with Z51 Performance Package
Construction	composite/carbon-fiber body panels, high-pressure die-cast aluminum center-tunnel frame with carbon-fiber structural components
Wheelbase	107.2 inches
Length	182.3 inches
Width (without mirrors)	76.1 inches
Height	48.6 inches
Track (front/rear, inches)	64.9/62.4
Curb weight	3,366 pounds (coupe), 3,467 pounds (convertible)
Wheels	Sterling Silver five-spokes **Size:** 19x8.5 inches front; 20x11 inches rear
Tires	Michelin Pilot Sport ALS (Michelin Pilot Sport 4S with Z51 package) **Size:** P245/35ZR19, front; P305/30ZR20, rear
Brakes	eBoost-assisted antilock discs with Brembo four-piston/two-piece calipers, front, and four-piston/monobloc calipers, rear **Z51:** Brembo four-piston/monobloc calipers, front and rear
Brake rotor diameter	12.6 inches front, 13.6 inches rear **Z51:** 13.3 inches front, 13.8 inches rear
Suspension (standard FE1)	short/long-arm (SLA) double wishbone, forged-aluminum upper control arms, cast-aluminum L-shaped lower control arms and coilover dampers, front and rear (FE4 Magnetic Selective Ride Control available with Z51 Performance Package in 2020, also available individually as FE2 beginning in 2021)
Steering	power-assisted Bosch/ZF variable-ratio rack and pinion
Ratio	15.7:1
Engine (LT2)	6.2-liter (376 ci) direct-injection overhead-valve V-8 with Active Fuel Management and Variable Valve Timing, cast-aluminum cylinder block/heads and dry-sump oiling
Compression	11.5:1
Bore and stroke	4.06x3.62 inches
Output (std)	490 horsepower at 6,450 rpm; 465 lb-ft of torque at 5,150 rpm
Output w/NPP exhausts*	495 horsepower at 6,450 rpm; 470 lb-ft of torque at 5,150 rpm *offered separately or included with Z51 package
Transmission	M1L Tremec-supplied 8-speed dual-clutch transaxle
Axle ratio	3.55:1 (electronic limited-slip differential included with Z51)

Plant officials didn't help matters by suspending public tours for 18 months beginning in July 2017—to prevent (in some bloggers' opinions) prying eyes having premature peeks at the revolutionary next-gen Stingray perhaps? More likely to keep tourists out of harm's way while all the dust settled. Or simply out of the way completely while all construction/reconstituting efforts wrapped up. Or maybe all of the above?

The last Corvette off the "old" assembly line (which dated back to the C5's 1997 introduction), a Torch Red 2018 Stingray, was completed in July 2017. And when 2018 production restarted on Bowling Green's thoroughly modernized, now "flexible" line that fall, word quickly came of an early model-year tradeoff. The already short 2018 run would end on January 22 and 2019 production would commence a week later. Why? As Tadge Juechter explained to Don Sherman in March 2020, "It wasn't that we wanted to shorten model-year '18, it was that we wanted to get the 2019 ZR1 out as soon as we could. We brought [out] the rest of the minor changes for 2019 on the other models to match so the plant change-over was easy."

Hence, the anticipated C7 ZR1 did debut as a 2019 Corvette, not a 2018 as predicted by various sources. And the C8? While some industry-watchers continued discussing the possibility that C7 and C8 production might coincide, in the end, Chevy's last seventh-gen Corvette left Bowling Green Assembly in November 2019—so late due to the United Auto Workers strike that had idled

C8 2020–2025

Above: C7 customers had already demonstrated a preference for paddle-shifted automatics, first the six-speed 6L80 followed by the superior 8L90 eight-speed in 2015. So it was that Chevrolet completely ditched a good ol' stick in 2020, relying solely on the incredibly quick, ultraprecise, supremely durable M1L dual-clutch transmission—a paddle-shift wonder that guaranteed supercar-strata for all C8s right out of the box. According to Chevrolet tests, a 2020 Stingray could run from rest to 60 miles per hour in a tidy 3.0 seconds and complete the quarter-mile run in 11.2 clicks, topping out at 123 miles per hour—easily the best numbers ever posted by a standard model. Breaking into the twos was simply a matter of adding the carryover Z51 equipment, resulting in a 2.9-second 0–60 dash. Yowza!

Right: At 495 horsepower (with optional performance exhausts), the C8's LT2 V-8 set yet another standard for base-model output. "[It] is one of our best efforts yet in Corvette's history of naturally aspirated high-performance small-block engines," said global small-block chief engineer Jordan Lee in November 2019. "Power is readily available when the driver needs it." The LT2 also featured dry-sump oiling, a first in base ranks, too.

293

CHAPTER NINE

GM plants for nearly six weeks from September to October. Meanwhile, back in April, Chevrolet finally had made it official: the *2020* C8 would make its long-awaited public debut that summer.

Reportedly, more than 470,000 Corvette fans worldwide were plugged in when GM livestreamed the C8 reveal at 7:30 p.m. Pacific time on July 18, 2019. So many that downloads stalled more than once due to heavy traffic.

And just when Stingray buyers figured it was safe to go back in the water, to finally start trolling for an eighth-gen model, they once again were left high and dry. Along with perpetuating 2019 production, that UAW strike also forced Chevrolet's hand concerning initial C8 deliveries, originally slated to begin by year's end. A lucky customer who made the first dealer lists in 2019 didn't see his/her midengine Corvette until March 2020. A moment some 50 years in the making.

2020

Rumor mills mentioned a possible dual-clutch transmission almost as much as a midengine makeover during the decade or so prior to the C8's debut. Not sure what DCT stands for? Don't feel alone. "[Corvette enthusiasts] have asked for DCTs for years not even knowing what they were," said Tadge Juechter in August 2019. "They just read the media raving about the latest Porsche or whatever, so they ask for DCTs."

To make a long story short, a DCT is an automatic manual. Or is it a manual automatic? More seriously (or not) to the point, it's an electrohydraulic-controlled manual sans clutch pedal. And it also might be considered a paddle-shifted automatic sans torque converter. Got it?

As those three letters implied, making the C8's Tremec-supplied eight-speed transaxle work were two concentric clutches—one servicing odd-numbered gearsets, the other the evens—operating in the wet to keep friction-fired temps down. But not too wet. Openings in the housing allowed the two clutches to remain merely moist, not submerged in lubricant, which translated into less parasitic drag.

Additional advantages included the way in which 2020's DCT, like the C7's eight-speed 8L90, helped Stingray drivers forget all about how cool it used to be to stir their souls with a stick. Know, however, that manual transmissions had already lost considerable favor before the high-horsepower-handling 8L90's 2015 introduction. An indication, at least in part, of the ever-increasing age of the typical Corvette customer perhaps? Whatever the case, after hitting 65 percent in 2014, convenience-conscious automatic purchases reached 68 percent in 2015, 77 in 2016, then stuck at about 78 annually up through the last C7 run. And now we have no clutch pedals at all.

Among those not lamenting that loss in the least was Tadge Juechter, who pointed out a few convenient pluses provided by the left pedal's departure. No worries about how to fit it down inside the C8's driver-side footwell, which was tighter than ever thanks to its newfound proximity to the left front wheel. No further fuss over how to create a complicated linkage, for both said pedal and its complementary stick shift. And no unwanted weakening of 2020's revised center-tunnel structure by cutting into it to provide rearward routing for those links.

As for performance, again like the 8L90, the DCT relied on microchip technology to manage what no human, not even drag racing legend Don Garlits back in his biggest, baddest daddy days, could. That is, make lightning-quick, ultraprecise gear changes each and every time, whether functioning in convenient automatic or manual-shift fashion.

Terri Schulke's transmission engineering team even did the 8L90 more than one better by improving the steering-wheel-mounted paddles' electronic pathways to the Tremec transaxle's control module for even more responsive manual shifts. And while the C7's familiar driver mode choices—Weather, Tour, Sport, and Track—carried over into the C8 era, they were joined by two new multi-configurable settings: the full-time (set it, forget it) MyMode and the single-use, more specific

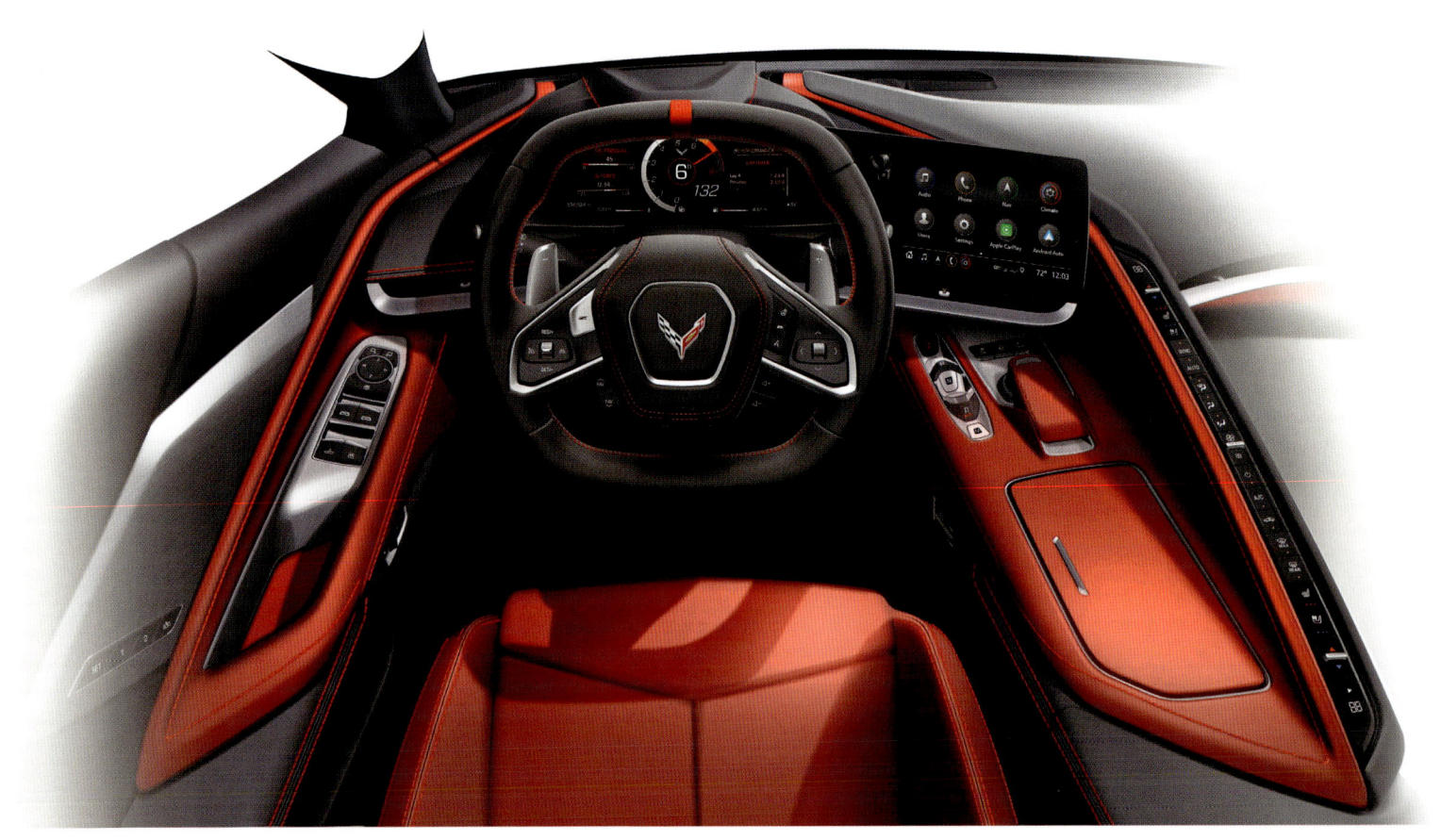

Like various exterior touches, the C8 interior demonstrated unmistakable inspiration from the aviation world. A rectangular steering wheel returned, but truly new were the console-mounted toggle controls for the LT2's dual-clutch transmission. A dial again appeared to the shift-change apparatus's right to set driver modes, now including a "MyMode" choice. A second new performance personification, "Z" mode, was controlled by an appropriately labeled button found just to the left of the steering wheel hub. Available 1LT, 2LT, and 3LT trim levels again were the norm for the latest generation Stingray.

"Z" mode, named in honor of the long-revered Z06, ZR1, and Z51 packages.

Making 2020's dual-clutch box appear even more advanced in contrast was the good ol' friend it bolted up to: Chevrolet's still-strong small-block. And still activated by traditional pushrods, not overhead cams as was also commonly rumored during the run-up to the latest best Vette yet. Now tabbed LT2, the C8's soldiering-on 6.2-liter V-8 again relied on direct injection and Active Fuel Management and, as expected, established yet another base-model output high, 495 horsepower when fitted with available performance exhausts. Another claim to fame came underneath in the form of standard dry-sump oiling, a Stingray first that at the same time enhanced on-track durability and helped the entire powertrain package fit so darned well within the eighth gen's revised confines.

Minimizing the LT2 oil pan below allowed a 1-inch-lower mounting position compared to the C7's LT1. While that may not sound like much, dropping a supercar's center of gravity (CG), however lightly, is always a good thing. As was, of course, relocating the LT2 behind the cockpit, which not only resulted in the desired rearward weight bias but also redefined the Corvette driving experience. In more ways than one.

With seating now positioned 16.5 inches farther forward, the C8's CG arrived just off a driver's inside hip, meaning the car literally turned right around its pilot, a confidence-boosting perception that

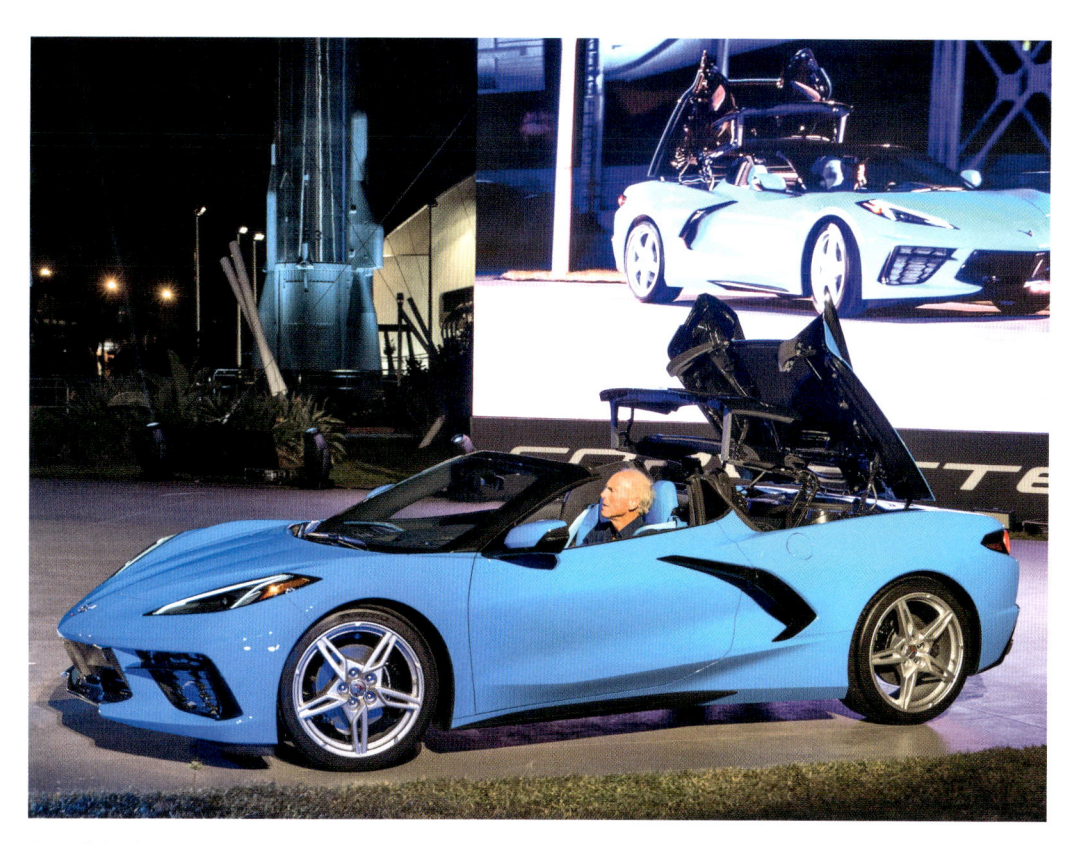

Above: Tadge Juechter rolled out the C8 convertible on October 2, 2019, at Florida's Kennedy Space Center. Another first for the breed, this high-powered hair-musser featured a two-piece retractable hardtop that could be activated even at speeds up to 30 miles per hour. Six electric motors helped that retraction happen in about 16 seconds. When raised, the roof simply refused to reveal this model's split personality. The 2020 convertible appeared every bit as weather-proof as it actually was. And did so with as much flair as, if not more than, its targa-top running mate.

Below: Introduced along with the C8 convertible in October 2019, the C8.R appears here during its debut at Daytona in January 2020. Like the C7.R it superseded, Chevrolet's eighth-gen competition Corvette relied on an IMSA-spec DOHC 5.5-liter small-block producing 500 horsepower. A production-based foundation went underneath, demonstrating the ready-to-race nature of a regular production 2020 Stingray.

CHAPTER NINE

benefited even the most novice Corvette owner. But track-tested or otherwise, all occupants surely appreciated the newfound forward visibility afforded by a nose that ended so much closer and lower now that all that aluminum arrayed in a V was no longer forcing front parameters up and away.

Supporting the C8's cab-forward body was an even stronger aluminum chassis based on the aforementioned backbone structure that Juechter preferred not to cut into for shift-linkage access. Now created using high-pressure die-casting instead of hydroforming, this foundation featured six cross beams and did away with the side rails located inside the rocker panels of previous generations. Totally new out back was an industry-first carbon-fiber rear bumper beam, added to cut pounds. And, along with various other carbon-fiber applications, a special fiberglass mix—used for the dash and both trunks—helped save additional weight. Yes, trunks, as in plural; one at the nose, one in back behind the LT2 small-block.

Gone, too, were conventional vacuum-assisted brakes and those long-familiar transverse leaf springs, the latter items replaced at both ends by coil-over dampers. The C8's new eBoost "brake-by-wire" system traded that big booster for a space-saving electronic module that additionally allowed drivers to personally tune brake pedal feel right along with all other performance variables.

Meanwhile, an established tradition, a removable roof panel, remained in place atop 2020 coupes. But the C8 convertible represented a true departure, crowned by the breed's first hideaway hardtop.

C8.R racer

Chevrolet's ground-breaking eighth-generation convertible was introduced at NASA's Kennedy Space Center in Florida on October 2, 2019, right along with another eye-opener: the next-generation competition Corvette, the C8.R. The company's first clean-sheet racing design since 1999's C5.R, the C8.R shared a higher percentage of parts with its regular-production counterpart than any of its predecessors. Save, once again, for its power source.

Like the C7.R, Chevy's latest track-ready two-seater was fitted with an IMSA-legal naturally aspirated 5.5-liter small-block that generated 500 horsepower and 480 lb-ft of torque. But this time those sixteen familiar pushrods finally were retired in favor of state-of-the-art dual overhead cams. And accompanying those four cams was a flat-plane crankshaft that further enhanced this V-8's rev-ready nature. Backing up this ground-breaking DOHC powerplant was a compact Xtrac six-speed sequential transmission, and holding up the works was a Bowling Green-built production chassis, a testament to the supreme ruggedness of the C8's die-cast aluminum foundation. Taking full advantage of the midengine platform's improved balance were 18-inch Michelin Pilot Sport GT competition tires created specifically for this application.

The C8.R's on-track debut came in January 2020 at Daytona's Rolex 24 endurance epic, where the yellow no. 3 car finished fourth in GTLM competition. Drivers Antonio Garcia, Jordan Taylor, and Nicky Catsburg completed 785 laps for 2,794.5 miles, a distance record for any Corvette racing team at the Florida event.

2021

Be they anxious buyer or frustrated seller, most members of the Corvette community just couldn't catch a break during the transformative years spanning tail-end C7 and early C8 production runs. On the customer side, you'll recall that 2018's supply was severely hampered by the much-needed modernization of the Bowling Green plant.

But then all seemed back on track the following year, when the Kentucky crew proved they still knew how to git 'er done by assembling almost 35,000 Stingrays, more than 3.5 times the previous model year's humble total. And this even after that late startup and the aforementioned 40-day UAW strike, the former coming in January 2019 and the latter concluding at the end of October. Helping compensate for these hiccups was the addition in April of a second shift and 400 extra workers, boosting Bowling Green's workforce to more than 1,300.

Such apparent good fortune quickly waned, however, as that fall's temporary plant paralysis (while labor/management negotiations droned on) predictably resulted in yet another order fulfillment extension—which in turn understandably delayed the next generation's arrival. Chevrolet's last C7—a black Z06 "pre-auctioned" for charity earlier in June—left the line on November 14, 2019. Enter the epochal C8 only a tad tardy, correct?

Not exactly. Pre-strike reports of a December 2019 kickoff for midengine production, of course, proved premature. Duh, right? Because there remained the deferred need to prepare both assembly line and assemblers to ably produce the uber-advanced best Vette yet, an atypical reboot that translated into a tantalizing wait for the first buyers on their blocks to order up. Not to mention the feverish flood that followed.

Early that December, GM exec Barry Engle told *Motor Trend* that 2020's allotment—at the time tabbed at 40,000 cars—could be considered sold out due to the piles of advance orders already on dealers' desks, all of which remained in seeming limbo for seemingly too long. Once the Corvette story's seventh chapter finally closed, Bowling Green didn't manage its first ready-to-ship C8 until February 3, 2020, after which time all was well, yes?

Again, not so fast. Why couldn't luck have been more ladylike, tonight or otherwise? Stingray-laden haulers had no more started rolling up and down

C8.R

Model availability	coupe (produced by Bowling Green Assembly and Pratt & Miller)
Construction	composite/carbon-fiber body panels, widened by 4.6 inches at the corners to house larger racing wheels/tires; stiffened and lightened diecast aluminum frame with aluminum and magnesium structural/chassis components; competition roll cage welded to frame
Wheelbase	107.2 inches
Length	182.3 inches
Width	80.7 inches
Height	45.2 inches (3.4 inches lower than production Corvette)
Weight	2,733 pounds
Wheels	BBS six-split-spoke with single center locking nut
	Size: 18x11.5 inches front, 18x12.5 inches rear
Tires	Michelin racing slicks
	Size: 300/33-R18 front, 310/41-R18 rear
Brakes	steel rotors (vented and slotted) with six-piston Alcon Racing monobloc calipers, front and rear
Brake rotor diameter	15.35 inches front, 14.0 inches rear
Suspension	short/long-arm (SLA) double wishbone fabricated steel upper/lower control arms and coil-over adjustable shock absorbers
Engine	5.5-liter (336 ci) naturally aspirated direct-injected DOHC V-8 with flat-plane crankshaft, four valves per cylinder, and dry-sump oiling
Compression	12.5:1
Bore and stroke	4.104x3.150 inches
Output	500 horsepower at 7,400 rpm; 480 lb-ft torque at 7,400 rpm
Transmission	six-speed Ztrac P529 sequential manual

Above: Red Mist Metallic Tintcoat (shown here) was one of two new paint choices for 2021, the other being Silver Flare Metallic. Two 2020 shades, Long Beach Red Metallic Tintcoat and Blade Silver Metallic, didn't roll over into 2021.

Right: This Sky Cool Gray/Strike Yellow interior was new for 2021 and served as the exclusive cabin treatment for 2022's C8.R Edition Stingray. Additional 2021 interior choices included Jet Black, Sky Cool Gray (without yellow complements), Adrenaline Red, Natural, Natural Dipped, Tension/Twilight Blue Dipped, and Morello Sky.

I-65 when the Covid-19 pandemic brought on a particularly painful plant closure, this one running from March 20 to May 26. Then, to inflict additional injury onto injury, pandemic-related parts shortages continued to infect the C8 bloodline even after our whole wide world—along with all those sets of 18 big wheels—began spinning again.

Left holding the bag by recovering hamstrung suppliers who couldn't hold up their end, Bowling Green officials first temporarily halted production for a couple weeks in October 2020, further inhibiting deliveries of Chevy's first midengine Corvette. Like the final C7, its initial C8 successor also ran long, this time into December.

Parts procurement snafus continued into 2021, temporarily halting work on the line for about two weeks in January/February, plus a few more days in early March, then again for another week the following March. While truant computer chips were the common culprits in so many manufacturing holdups witnessed around the globe in the pandemic's wake, GM officials more than once stated that this wasn't necessarily true in the new C8's case. Reportedly, the earliest headaches involved unavailable transmission components.

Further frustrating both buyer and builder was an occasional inability to deliver on some RPO selections, because those sub-assemblies also couldn't be readily sourced.

The plot thickened further on Saturday, December 11, 2021, when, of all things, a tornado started a fire that tore through a plant roof. No one was injured, though more than 90 lives were claimed elsewhere that week by various twisters that ravaged six states across the Southeast and Midwest. Far less tragic, albeit certainly sad, was news of more than 100 completed Corvettes ending up scrapped due to damage incurred during that heavy weather. And in other bad news, production was held up for nine days to allow for repair work to the factory.

Yet through it all, the C8 supply line soldiered on. Back on May 18, 2020, a partial line restart following the Covid clampdown had helped ease fears expressed earlier by at least one knowledgeable web reporter that the 2,700 models built prior to that closure just might've represented 2020's whole shebang, making this the rarest Vette since 1955. Not so, of course. Indefatigably

CHAPTER NINE

Above: Bowling Green Assembly began rolling out its first right-hand-drive (RHD) Corvettes in January 2021, delivering them initially only to Japan. Five of these pioneering RHD Stingrays appeared at this coming-out party held at Mt. Fuji Speedway on May 29 that year. RHD Japanese export production was 361 in 2021, 147 in 2022, and 575 in 2023, with that last batch including eight Z06s.

Left: From the get-go, the C8 design team made sure that some interior features were "mirror-imaged," with the goal being to make flipping their positions in RHD applications a relative piece of cake. RHD deliveries expanded to Australia/New Zealand and the United Kingdom in 2022. Production counts for the former were 212 in 2022 and 492 in 2023. Totals for the latter were 83 in 2022 and 245 in 2023.

Opposite: In May 2021, Chevrolet announced a renewed export push into Europe for the C8. Both left-hand- and right-hand-drive Stingrays were sent across the Atlantic depending on the needs of the targeted populations, and all were fitted with Z51 equipment. While most also were 2LT models, a few better-equipped Launch Edition Stingrays were treated to 3LT upgrades, plus additional carbon-flash exterior details. Optional 3LT trim was available in all applications, but no 1LT exports were delivered.

defying heavy odds, Bowling Green Assembly built slightly more than 20,000 Stingrays by the time the convoluted 2020 run ended.

A gallant effort, yes, but do the math. Remember that biggo stack of advance orders? Something had to give.

In March, word leaked out that GM had directed dealers to stop taking 2020 orders, leading more than one internet reporter to play Captain Obvious: no way would we see 40,000 2020 Corvettes. At the same time, it became abundantly clear that many eager customers with their hearts set on owning the first of a new breed would eventually end up coming home with the second.

So it was that a hunk of first-year C8 orders went unfulfilled, leaving some buyers no choice but to return to their dealers to reorder and purchase 2021s, which went into production on December 14, 2020. In a goodwill gesture, Chevrolet initially rolled over all pricing for 2021 customers who had signed on the dotted line prior to March 3 that year. From that day forward, however, a new-model hike of $1,000 appeared on all bottom lines, meaning no more Stingray coupes "suggestedly" priced at less than 60 grand.

That typical hike aside, 2021's Corvette nonetheless remained every bit the same world-class, best-performance buy as its predecessor. All aspects predictably rolled over from 2020 as well, with the most noticeable revisions typically coming to the paint palette. Two 2020 exterior colors (Long Beach Red Metallic Tintcoat and Blade Silver Metallic) didn't return, while two new ones (Red Mist Metallic Tintcoat and Silver Flare Metallic) were added—and quickly emerged as crowd-pleasers. Out of a dozen available finishes, the former shade was 2021's third most popular choice; the latter came in at number five.

More notable in many minds was a familiar mechanical adjustment to the RPO list, which again (as in 2016) morphed Magnetic Selective Ride Control into an individual option, now available either alone (FE2) or along with the Z51 Performance Package (FE4). In keeping with pre-2016 practices, adding FE4 equipment to a 2020 Stingray had required anteing up an additional $5,000 for the Z51 deal. Meanwhile, not costing an extra penny was some new wireless technology, Apple CarPlay and Android Auto, now standard on all 2021 models. On the outside, various remade striping options also appeared this year, and that was pretty much that as far as all the latest news fit to print was concerned. Local news, that is. There remained more to report from a bigger-picture perspective—historically more.

Though GM basically soft-pedaled the idea during the C8's July 2019 intro, a renewed plan to promote Corvette sales more aggressively worldwide—per what Ford had implemented in 2015 with its new-and-improved globe-trotting Mustang—was mentioned. Only in small type. Most newsworthy was a tidy, easy-to-miss press release line announcing 2020 Stingray "availability in both left- and right-hand-drive variants."

Never before had Chevrolet offered a factory-built Corvette capable of comfortably and/or confidently traveling the thoroughfares where drivers sit on the right while their vehicles roll down the left. But here it was in black and white, another major first for Chevy's midshipman—just not a first for 2020's model nor for overseas destinations initially announced by various sources.

In July 2019, Dave Buttner, top dingo at Holden, GM's longtime subsidiary Down Under, insinuated (along with various media outlets) that an Aussie would be the first to own an RHD Corvette delivered direct from the Bluegrass State. Then, the following

June, *Car and Driver*'s Mike Duff wrote that GM to that point had released only British pricing for this upcoming modified milestone, explaining his magazine's headline concluding that a UK bloke (or bloke-ess) would end up a history-maker. But all bets were off two months later after Tadge Juechter officially identified the grand-prize winner during a virtual appearance at Pennsylvania's annual "Corvettes at Carlisle" event: Japan.

Surely influencing this decision was the rampant enthusiasm exhibited by future Far East customers after seeing the new C8 at the Tokyo Auto Salon in January 2020—and furthermore hearing that import versions would be built with right-hand steering. More than 300 preorders were placed within 60 hours, about three times the *total* number of 2019 Corvettes delivered to Japan, the last seen prior to that moment. Yes, all those 2019 Japanese exports were left-handers, just like all Corvettes delivered up until then to other right-hand-driven markets. Problematic for sure.

But a problem no more, at least for now in Japan. The first of 361 right-handers eventually shipped west across the Pacific in 2021 left the Bowling Green line on January 21. A "private preview" starring five of these followed on May 29 at Japan's Fuji Speedway, located, as one might guess, in the shadow of stunning Mt. Fuji. Inviting selected live celebrants to this coming-out gala was the original plan, but Covid restrictions intervened, forcing party plotters to livestream their two-hour on-track event instead, welcoming the entire world to watch in the process.

Back stateside, earlier in January more than one astute Corvette-watcher had noticed a progressive slowdown at Bowling Green Assembly, inspiring a fully logical conclusion: unprecedented RHD production surely had started, albeit at a lessened, more careful pace to allow line workers to familiarize themselves with what initially was a foreign (pun intended) process.

This downward trend also might've motivated a wide-eyed Vette whisperer to further infer, with similar confidence, why GM waited a year to build its first right-handed Corvette. Fully necessary in January 2021, a line slowdown never would've worked 12 months prior, when stalled supply had already fallen behind overwhelming domestic demand, requiring Chevrolet to focus all-in on that humungous preorder backlog. America first? Certainly so—or, more correctly, conventional LHD production first. GM did export 1,791 first-year C8s, all with steering wheels where they "belonged."

Surely, still other quizzical witnesses additionally wondered why GM had not rewritten this history during previous generations. Easy: the time finally was right. Working with more conventional, less cooperative chassis, engineers had briefly experimented with right-hand conversions years before, then quietly shelved the idea. Too labor-intensive? Damn straight. But this time the radically reimagined C8 platform lent itself more willingly to this surgery and undoubtedly would have simply due to basic design, regardless of any extra forethought given to possible pilot repositioning during initial planning.

First and foremost, moving the engine out of the picture up front seriously (not to mention predictably) reduced the fuss/muss inherent to this transformation. Same for deleting a possible clutch pedal installation, along with its linkage. Further aiding this cause, but completely hidden from view, was the C8's simplified dash wiring, made possible by all the latest/greatest Vette's cutting-edge digital hardware. Also not readily apparent, at least to casual viewers, were various clever interior touches, which *were* specifically drawn up right out of the blocks to facilitate a steering wheel reshuffle.

As Juechter explained during his own streamed presentation at Carlisle in August 2020, "In [the C8], everything is driver-focused, everything is angled towards the driver. The cockpit wraps around you, and so when [we] do a right-hand-drive, we didn't want to dumb that down. We wanted those customers to have the same exact experience whether it's Japan, U.K., Australia. So, what we actually did was tool up all those unique parts that are kinda mirror shaped so that we can flip them over to the other side, and it would just be an exact [reflection] of the rest of the world's [Stingrays], the left-hand-drive cars."

Beyond that, making sure the C8 platform was RHD friendly wasn't all that tough, at least according to former chief engineer Ed Piatek. "I don't remember too many [issues] besides the obvious, like ensuring packaging space for the steering intermediate shaft and e-boost brake system," he said in August 2024. "A slightly bigger challenge [involved] packaging space for the heads-up display. That module isn't exactly mirrored to minimize tooling costs, so the space required to fit it was something we protected [from the beginning.] The pedal box was interesting, too, as we had to ensure adequate spacing for the accelerator pedal near the front body hinge-pillar structure. Then, putting the dead pedal against the tunnel, that wasn't terribly difficult."

Don't you just love it when a plan comes together?

CHAPTER NINE

2022

Releasing a RHD C8 represented the first shot fired in a multipronged assault on underserved far-off markets, again an updated strategy initially hinted at back in 2019. "Updated" because, of course, it wasn't like Corvettes hadn't infiltrated international boundaries before. That year, for example, GM shipped 11.6 percent of final C7 production to five foreign destinations: Japan, as mentioned, plus Canada, Mexico (a Corvette customer dating back to 1956), Europe, and the Middle East. Another 8.8 percent followed in 2020, this time only to Canada, Mexico, and the Middle East. These three then were rejoined on the export list the next year by Japan, thanks to those spearheading right-handers.

Helping set 2022's offensive apart were two new tactics. Additional attention (detailed in a moment) was given to making America's sports car fit in better behind friendly lines beyond the switch to RHD, yes, but that represented only half the plan. The lesser half. Making real news was the re-targeting of a wide-open front that had been left all too quiet for two years: Europe. Official word of this particular push came from GM's Zurich press offices in May 2021—after US media outlets had spent those two years speculating about this movement like it was massing immediately on the border. Massing, yes. On the border? "Restricted to barracks" perhaps better served the truth.

Once more, why the delay? For starters, GM surely needed those ground-breaking Japanese right-handers to hit the beaches first. Furthermore, following up with a reformed European incursion probably was held off a bit longer to keep things "tidy and ship-shape," to save the big barrage until the traditional outset of the next new model year.

After debuting at England's annual Goodwood Festival of Speed in July 2021, Euro-aimed 2022 Corvettes began showing up on dealer lots eastward across the pond that October, with appropriate RHD in nations accustomed to this practice. Cranking things up a notch—reportedly to better serve European tastes—these export models came standard with swankier, more convenient 2LT equipment, not to mention the bottom-line-busting Z51 performance package. Toss in predictably heavier delivery costs, and the pricier nature of these deals was no surprise. Example: Initial suggested stickers in the U.K. read (in contemporary Yankee equivalents) $86,150 for a coupe and $91,850 for a convertible.

Note that these examples were even-better-equipped Launch Edition Stingrays, which GM specifically created to proudly lead the charge across the Atlantic. Along with Z51 armament, Launch Editions featured upscale 3LT trim, plus Carbon Flash exterior upgrades, all the better to turn more European heads. Aforementioned 2LT models cost less, of course, but remained high on the wallet-wilting scale, whether measured

Left: Right-hand-drive export production (coupe and convertible) to the United Kingdom was 83 in 2022 and 241 in 2023, with that latter total including 4 Z06s. Another 610 Stingrays were exported to continental Europe in 2022, followed by 1,912 more, plus another 38 Z06s, in 2023.

Below: When in Rome . . .? Or Frankfurt perhaps? Switching the speedometer from "mph" to "kph" was only the beginning in the export C8's case. All that digital readout wizardry was not much good if drivers couldn't decipher it, hence the appropriate translation seen here. No, American C8 instrumentation did not specially compensate for Boston, Brooklyn, Chicago, or New Orleans dialects.

Above: Again, revised paint choices made up the most noticeable Stingray changes for 2022. Left to right here: 3LT coupe, 3LT convertible with Z51 package, and 3LT Z51 coupe. At 17,727, Z51 production made up 68.6 percent of the 2022 run. In other completely easy-to-miss news, 2022's LT2 small-block was updated (higher pressure injection, broader rpm range for the Active Fuel Management system) to keep up with federal emissions standards, resulting in no output variances. Official fuel economy ratings didn't waver either for what was the cleanest-running Vette V-8 yet. Some sources even claimed mpg performance improved, regardless of the advertised EPA-approved figure.

300

Above: In June 2021, Chevrolet introduced its latest special-edition Corvette at Detroit's Raceway at Belle Isle Park, this one honoring the Corvette Racing team's dominating C8.R competition Stingrays. While this 2022 options package (RPO ZCR) was recognized on paper as the "IMSA-GTLM Championship C8.R Edition," the translation on the street was simply "C8.R Edition." Posed from right to left here are a Stingray, a C8.R Edition coupe, an actual C8.R racer, and a C8.R Edition convertible. What journalists probably didn't know that day was that the two yellow ZCR cars were in truth 2021 models mocked up to represent the upcoming 2022 C8. Editions. The "experimental" coupe at right sold at a Barrett-Jackson auction in July 2022. Its convertible counterpart at far left remained under wraps as these words went to print in September 2024.

Right: Mimicking the actual #3 and #4 C8.R racers, 2022's C8.R Edition Stingrays were available in only two colors: Accelerate Yellow or Hypersonic Gray. Special wheels, a high-wing spoiler, and a Sky Cool Gray/Strike Yellow interior also were included in the deal. As was a numbered "C8.R Edition" plaque between the seats. Those numbers ran from 0001 to 1007 in production models, while the two pre-production models shown off at Belle Isle Park in June 2021 featured plaques engraved with "01EX" and "02EX."

in pounds, dollars, or euros. Or . . . per earlier predictions, Australia, along with neighboring New Zealand, also made the list of new export objectives in 2022, as did Southeast Asia and Israel the following year.

Here on the domestic front, basically all reporting remained fully familiar, from essentially each nut and bolt up to final-product availability. Coupe with removable roof or retractable-hardtop convertible. Check. Base 1LT, plus optional 2LT or 3LT equipment packages. Ditto. Mid-mounted LT2 small-block producing either 490 or 495 horses depending on the optional installation of those ever-popular performance exhausts. That's a big 10-4, good buddy. And certainly not last, discriminating customers once more could install any transaxle they liked out back—as long as it was Tremec's eight-speed, dual-clutch box. Done and done, Henry Ford style.

Paint picks again, unsurprisingly, represented 2022's most notable revisions. Into the archives went two metallics (Zeus Bronze and Shadow Grey), plus Sebring Orange Tintcoat. In their place came Caffeine Metallic, Hypersonic Gray Metallic, and Amplify Orange Tintcoat. Option list changes included a binned wheel choice (LPO 5DG) and a couple spoiler/splitter/ground effects additions.

Understandably unnoticed were fuel-delivery upgrades made to 2022's LT2 to comply with ever-tightening federal emissions standards. Note that not one pony was harmed during the remaking of this cleaner, greener small-block, nor did those wizards in engineering lose a single pound of torque. Eat your heart out, Ozempic.

But not all 2022 updates were so easy to miss. In June 2021, Chevrolet introduced the latest in a long line of special-edition Corvettes, this one honoring—in no uncertain terms—those dominating C8.R competition cars. Priced at $6,595, the IMSA GTLM Championship Edition package (RPO ZCR) included a lengthy list of enhancements meant to help Stingray drivers imagine they really were out there on track, say, at the Detroit Grand Prix—or Le Mans. And they didn't even have to make those burbling engine noises with their lips. Because the Z51 package was mandatory with RPO ZCR, meaning in turn that those growling NPP pipes were part of the deal as well, again along with those 5 extra horses.

The ZCR package was available for coupes or convertibles, and top-shelf 3LT trim also was mandated. On the other hand, customers were free to choose between two finishes, Accelerate Yellow Metallic or Hypersonic Gray Metallic, each chosen to mimic the #3 and #4 C8.R racers that had run at competition venues around the world. And if that wasn't enough, additionally included was a car cover artistically rendered to mimic the look of either #3 or #4. Just one last peek before switching off the garage lights, then dream the dream, no? Either that or flick on *SportsCenter*. Your call.

Requisite exterior identification announced the arrival of a "C8.R Edition" 2022 Stingray and was joined by a corresponding numbered plaque inside, where the Sky Cool Gray/Strike Yellow upholstery introduced the year before was the sole choice. Further complementing the image on the outside was a competition-style high-wing spoiler done in Carbon Flash, with more Carbon Flash on the side mirrors, blackened side rockers and splash guards, yellow brake calipers, and black Trident wheels. No hiding this baby in the corner.

Overall, it was a desirable concoction, made even more attractive by its limited-edition status. Initial announcements put a 1,000-unit lid on the ZCR run, inspiring potential customers to get while the gettin' was good. Did they ever. Chevrolet received 412 preorders between opening day (July 1, 2021) and production startup early in September. The dreaded "sold-out" sign appeared two months later. In the end, 2022's C8.R Edition tally read 1,007: 470 coupes, 537 convertibles.

That, however, wasn't that. A single pre-production C8.R Edition coupe—actually a 2021 model—curiously made it into the wild, a marketing vehicle mocked up for publicity purposes prior to the real thing's release. While this loner featured nearly everything that made a 2022 ZCR a ZCR,

CHAPTER NINE

it wasn't technically a ZCR, at least not on paper. Nowhere on its build sheet was that three-letter code mentioned—perhaps because Chevy had yet to codify it when this mockup was mocked up?

Reportedly put to work, among other things, promoting the upcoming C8.R Edition models at racetracks where actual competition Corvettes were wreaking havoc in 2021, this Accelerate Yellow oddity then showed up on the block at a Barrett-Jackson auction in July 2022. According to a November 2022 hagerty.com interview, the winning bidder noticed adhesive residue where, evidently, "C8.R Edition" graphics had previously stuck to each rear quarter. Though those exterior tells apparently were peeled off before that sale, a ZCR interior plaque remained to reveal this mystery machine's true identity. In place of 2022's 0001–1007 engravings was a simple "01EX."

Know, too, that a yellow convertible, "02EX," also was mocked up in 2021 to work alongside 01EX. So, keep your eyes open.

Top: Chevrolet gave America a sneak peek at its next best Vette yet on May 29, 2022, when a 2023 Z06 coupe paced the 106th Indianapolis 500. Adorned in 70th Anniversary garb, this 670-horsepower supercar was driven by former Indy racer Sarah Fisher. Another 2023 Z06, a convertible, paced "The Greatest Spectacle in Racing" in May 2023, making it the 34th time a Chevrolet product lead the way around the Brickyard on race day. *Indianapolis Motor Speedway*

Above: Ever throw a party and have no one come? Perhaps the oddest subplot in the 70-year telling of the Corvette tale came in June 2022, when GM announced its "Own the Color" auction. The idea involved bidding on NFT artwork depicting a Z06 Corvette specially done in a unique color: Minted Green. Bids began at about $210,000 in real currency for both the NFT plus the actual model depicted, a fully loaded Z06 that, once built at a later date, would be the only Corvette ever painted in that shade. And the winner was? No one. Not one bidder showed, even after the auction was extended an additional 24 hours. It was then left to real-world Z06s to bring customers running. Done and done.

2023

Seventy and fifty-three, two big numbers that really stood out this year. In the latter's case, that was as in an eye-popping 53,000 a towering production plateau reached only once before during the Corvette's long history. After rolling out 20,368 C8s in 2020, 26,216 in 2021, and 25,831 in 2022, Bowling Green Assembly really fired things up in 2023, delivering 53,785 two-seaters, a sharp spike that fell 23 cars short of the legacy's all-time record set in St. Louis way back in 1979.

"Missed it by *that* much" might've gone Maxwell Smart's take here. (Yeah, googling will help you get smart about this.) Or, as Hall of Fame Chicago Cubs broadcaster Harry Caray used to say whenever a promising long drive was caught right at the wall, "One more biscuit for breakfast . . ." Either way, ain't it funny how much assembly a thoroughly modern assembly plant can manage when it isn't inhibited or interrupted and when supply can simply coexist with demand? Holy cow!

"[Yes], it wasn't until 2023 that we were finally able to effectively match supply to demand," said Ed Piatek in August 2024. "Sports cars tend to have what our marketing teams call 'fashion/trend' lifestyles, with intense demand [rising] initially, and then [they] have interest fall off quite quickly." But that clearly wasn't the case here. Added Piatek, "It was exciting to see so much demand even into the fourth and now fifth years of [C8] production."

Record build rates were still exciting Bowling Green watchers as these words went to press in the fall of 2024.

As for that long history mentioned moments ago, make that *really* long. Say, 70 years, the most for any American nameplate remaining on the road in 2023. Still sounds hard to believe, doesn't it? Even more so considering the various times Corvettes have been Mark Twained during that span. Translated: Past accounts of upcoming demises always were unfounded. Geez, don't we all wish we could be as resilient and at a similarly advanced age to boot?

Still proving it all night, all 2023 models were treated to a 70th Anniversary interior plaque, located on the center speaker grille, plus additional commemorative graphics imprinted at the base of the rear glass. Nice touches, sure, but even more celebratory upgrades awaited buyers with an extra $5,995 burning its way out of their pockets.

On January 24, 2022, GM announced yet another special-edition Vette, this one not simply paying prominent tribute to seven decades on the road but at the same time showcasing how much the C8 has done to keep this venerable legacy looking, as well as feeling, fresh and youthful, if not downright hot-to-trot.

"Passion for the Corvette runs deep at Chevrolet, and this anniversary is extra special because of the excitement we've achieved with the eighth generation of America's iconic sports car," said Chevy marketing VP Steve Majoros while presenting 2023's birthday gift to young buyers of all ages. "Even after 70 years, Corvette still makes hearts race and kids dream of the open road." We could've only hoped those pumpers could still stand the strain. As for the kids, surely they were alright.

FYI, at the other end of that nearly $6,000 price tag was RPO Y70, available in 2023 for Stingray coupes and convertibles, along with the new Z06's two bodies. You'll meet that latter rocket ship soon enough, promise. In the meantime, recognize further that taking delivery of the resulting 70th Anniversary Edition Corvette required shelling out considerably more cash for a fully trimmed model, the 3LT Stingray or 3LZ Z06.

From there, Y70 customers were given a choice of two new finishes and two only—White Pearl Tintcoat or Carbon Flash Metallic—each limited to 70th Anniversary models. As you might guess, commemorative exterior badges further enhanced the image, along with Edge Red treatment for both the LT2 cover (in coupes) and brake calipers, plus 70th logo center caps for the forged-aluminum wheels, which differed slightly per application. Y70 Stingrays rolled on new-for-2023 Midnight Gray 20-spoke wheels, while their Z06 counterparts were fitted with Satin Graphite Spider–design units. Red striping complemented impressions for both types.

A black bumper protector and black trunk cover brought up the rear, and dual racing stripes could've been added for another $995. Stripe colors were Satin Grey for White Pearl cars, Satin Black for their Carbon Flash running mates.

Anniversary logos repeated inside (seats, steering wheel, and sill plates), where upholstery was done in two-tone Ceramic White leather stitched up in red. Either GT2 or Competition Sport seats were installed, each type featuring sueded microfiber inserts. Red seatbelts also were part of the deal, as was a custom set of luggage adorned with additional red stitching and more commemorative identification.

Yes, the Y70 package dominated Corvette news for 2023, at least in Stingray ranks. Save for those two shades reserved for 70th Anniversary models, even the paint lineup repeated this year. There were a few optional additions (including a new

Above: Prior to the Z06's return for 2023, the hottest Stingray available was the Z51-equipped model, which among other things included bigger brakes, a stiffer suspension with Magnetic Selective Ride Control, and the NPP performance exhausts that upped the output ante to 495 horses. Z51 sales remained strong in 2023, making up 57.5 percent of that year's Stingray run.

Right: Corvette production reached the 50,000 plateau for the second time in 2023, topping out at 53,785, a scant 23 cars short of breaking the breed's record set in 1979. On the Stingray side of the fence, 28,834 were coupes and 22,538 were convertibles. Demonstrating perhaps how discriminating Corvette buyers have long been, the most popular model this year (12,524 built) was a loaded 3LT Stingray convertible.

CHAPTER NINE

A 70th Anniversary Edition package was offered for all 2023 models but limited to top-shelf 3LT (Stingray) and 3LZ (Z06) coupes or convertibles. Exclusive paint choices were White Pearl Metallic Tincoat or Carbon Flash Metallic. The Satin Matrix Gray dual racing stripes seen here were optional. Satin Black Metallic stripes were available on Carbon Flash cars. Total production was 5,601: 1,569 Stingray coupes, 3,010 Stingray convertibles (shown here), 389 Z06 coupes, and 633 Z06 convertibles. Of the Stingrays, 1,186 coupes and 2,415 convertibles also were fitted with Z51 equipment, while 135 Z06 coupes and 211 Z06 convertibles were additionally equipped with the Z07 Performance Package.

20-spoke wheel, available in three finishes), and that was it. Of course, next to no one cared about this particular shortage, certainly not once that aforementioned second model made the stage.

Make that supermodel.

2023 Z06

Prior to this year, Z51-equipped C8s had inspired all the headlines. First came news in 2020 of this RPO-enhanced model's entry into the 2-second club, joining the C7's Z06/ZR1 duo. Translated, eighth-gen Z51s were able to scream from rest to 60 miles per hour in fewer than 3 seconds, a still-startling achievement considered impossible not all that long beforehand.

Carrying over from C7 to C8, RPO Z51 once more added most everything a midengine man or woman needed to truly put the sting in their Stingrays. Cooler cooling for the LT2, boosted to 495 horsepower thanks to the inclusion of NPP performance exhausts. Track-ready suspension with Magnetic Ride Control. Typically beefed brakes. A performance axle ratio in an electronic limited-slip differential. And a rear spoiler/front splitter combo that supplied 400 pounds of extra downforce. What more could Walter or Wilma Mitty ask? (C'mon, no googling required here, right?)

Wait, don't answer—the Mitty question, that is. Because along with all the noise made by a Z51, Tadge Juechter's team just couldn't resist turning the volume up to 11 with a little help from their new friend, Chevy's latest Z06. As Rich Ceppos recounted in *Car and Driver*, "The Z06 sounds like a race car, [its] engine blaring, howling, yawping out a soundtrack that was equal parts Gatling gun and circular saw."

Sounds like a race car? Progressively thinned during C7 years, the once indelibly inked line between road and road course was all but erased by this supreme supercar, which rolled on a chassis borrowed to a great degree from Corvette Racing's proven C8.R track star—along with much of its exterior parameters, aero treatments, and engine design. Ah, so that's why Chevrolet insiders started calling the C8.R "a Z06 hiding in plain sight."

That street/strip engine was the LT6, an ultimate small-block that, like the C8.R's V-8, displaced 5.5 liters, featured dual overhead cams and four valves per cylinder, was naturally aspirated, and contained that flat-plane crankshaft that everyone had been talking about dating back to the eighth-gen competition Corvette's October 2019 unveiling. Missing, however, were the air restrictors that limited the C8.R's heart to "only" 500 horsepower, per sanctioning body mandate.

With nothing clogging its airways, the daily driven (sorta) LT6 produced 670 horsepower, making this the strongest non-blown, non-turbo V-8 to ever find its way, planet-wide, into a regular-production vehicle. Two words: "yow" and "zuh." Note that achieving that un-force-fed record was made possible by the LT6's unique lightweight crankshaft. Thanks to its flat-plane layout, it was capable of ridiculous rev counts previously unimagined in street-machine guise just the year before. At least not on any street in our neighborhood.

To call this mind-blowing small-block a high-winder was akin to describing Stephen Hawking as bright. Those 670 max horses arrived at 8,400 revs.

That's not a typo: Eighty. Four. Hundred. Enough rpm to leave pretty much every other powerplant in Detroit history melting its way downward to China. On top of that, the LT6 generated 460 maximum pounds of torque at an also-dizzying 6,300 revolutions, meaning a Z06 might've done double-duty pulling stumps on the side.

When transmitted to the rear wheels, all that rotational muscle translated into 0 to 60 in a tiny 2.6 ticks and the quarter-mile in 10.6 seconds at 131 miles per hour. Again, no kidding. Even more all-time bests. For now.

Z06 wheels made news too. Measuring 20 inches across and 10 inches wide in front, 21 and 13 in back, these steamrollers were the largest ever included on a Corvette's standard-equipment list. Of course it only followed that protecting these massive hunks of aluminum from rough roads would be left to some serious rubber. To that end, Michelin's Pilot Sport 4S ZP tires (275/30ZRs fronts, 345/25ZRs rears) were standard too.

Stickier Michelin Sport Cup 2 R ZP tires—developed specifically for this application—were optional, but only as part of the Z06-exclusive Z07 Performance Package, an $8,995 option that, among other things, helped deplete our supply of exclamation points big time. To begin at the bottom, Z07 equipment included even tighter FE7 underpinnings, also created exclusively for this package. Bigger Brembo brakes were predictably present as well and featured carbon-ceramic rotors, yet another Corvette first. Compared to standard Z06 rotors, which measured 14.6 inches across up front and 15 out back, Z07 counterparts were truly huge: 15.7/15.4, front/rear. Z06 front calipers used six pistons, two more than their Stingray cousins.

On top, the Z07 deal added carbon-fiber aero treatments (a larger front splitter, front-corner dive planes, a rear wing, and downforce-enhancing underbody strakes) that took the already heavily modified Z06 body to even greater heights. A recounting of all the standard Z06 shell's functional upgrades and aesthetic enhancements would practically fill a page or two in their own right, so please refer to captions and spec boxes for more specific info. But do recognize that the body was widened by 3.6 inches, and unique front/rear fascias were added, with the rear panel incorporating center-exiting exhaust bezels. Along with looking way cool, those four conjoined trumpets helped produce the exhaust note that Rich Ceppos described earlier, which truly had to be heard to be appreciated.

In other news we can't hold back on, also standard for the Z06 was the Corvette's ever-present stability technology, Magnetic Ride Control 4.0, which was recalibrated exclusively in Z07 installations. And let's not forget one last first: carbon-fiber wheels that shaved off 41 pounds of unwanted unsprung weight. Available in two

2023–2025 Z06

Model availability	coupe (with removable roof panel) and convertible (with folding hardtop)
Construction	composite/carbon-fiber body panels, widened by 3.6 inches at the corners to house larger wheels/tires; high-pressure diecast aluminum center-tunnel frame with carbon-fiber structural components
Body features	standard front splitter; unique front/rear fascias (front with revised venting for maximum cooling, rear incorporating center-floating exhaust bezels); enlarged rear-quarter cooling intakes; unique reconfigurable standard rear spoiler with adjustable wickerbill elements; available Carbon Aero Package (RPO T0F* or T0G*, Z06 only in 2023) added larger front splitter, front-corner dive planes, a pedestal-mounted high rear wing, and underbody aero strakes * T0F also included carbon-fiber-painted ground effects (RPO CFZ), T0G included visible-carbon-fiber ground effects (RPO CFV); CFZ and CFV options also Z06-only in 2023
Z07 Performance Package	added T0F or T0G aero packages, plus carbon-ceramic brakes, unique FE7 suspension with specially calibrated Magnetic Selective Ride Control 4.0, and Michelin Sport Cup 2 R ZP tires developed specifically for Z06; carbon-fiber wheels (RPO ROY or ROZ) available with Z07
Wheelbase	107.2 inches
Length	184.6 inches (standard) 185.9 inches with carbon-fiber ground effects (CFV or CFZ)
Width (without mirrors)	79.7 inches
Height	48.6 inches
Track (front/rear)	66.3/66.1 inches
Curb weight	3,494 pounds (coupe), 3,593 pounds (convertible)
Wheels	Titanium Satin–painted forged-aluminum Spider-design (10-spoke), standard in 2023–24; restyled 10-spoke design debuted in 2025; ROY or ROZ five-spoke carbon-fiber wheels, optional (with carbon-flash paint or visible carbon-fiber finish, respectively, limited to Z07 applications) **Size:** 20x10 inches front, 21x13 inches rear
Tires	Michelin Pilot Sport 4S ZP summer-only, standard **Size:** 275/30ZR20 front, 345/25ZR21 rear
Brakes	eBoost-assisted antilock discs with Brembo six-piston/two-piece front calipers and four-piston monobloc rears, standard; larger carbon-ceramic rotors included with Z07
Brake rotor diameter	**Standard:** 14.6 inches front, 15.0 inches rear **Optional:** 15.7 inches front, 15.4 inches rear (carbon-ceramic brakes)
Suspension	short/long-arm (SLA) double wishbone, forged-aluminum upper control arms, cast-aluminum L-shaped lower control arms and coilover dampers, front and rear; Magnetic Selective Ride Control 4.0, standard; special calibration included with Z07 package
Steering	electric power-assisted Bosch/ZF variable-ratio rack and pinion
Ratio	15.7:1
Engine	5.5-liter (336 ci) direct-injection DOHC V-8 with Variable Valve Timing, cast-aluminum cylinder block/heads, dry-sump oiling, flat-plane crankshaft, and stainless-steel four-into-two tubular headers
Compression	12.5:1
Valvetrain	dual overhead cams activating four valves per cylinder via mechanical finger-followers and dual valve springs
Valves	1.654-inch titanium intakes, 1.378-inch sodium-filled exhausts
Bore and stroke	4.104x3.150 inches
Output	670 horsepower at 8,400 rpm, 460 lb-ft torque at 6,300 rpm
Transmission	M1L Tremec-supplied eight-speed dual-clutch transaxle
Axle ratio	5.56:1 (electronic limited-slip differential standard)

CHAPTER NINE

Above: On October 26, 2021, Chevrolet announced the return of the Z06 Corvette, "a new American supercar that puts the world on notice." Hiding beneath its bulging body—widened by 3.6 inches to house the largest wheel/tire combo ever bolted up to a regular-production Corvette—was the breed's most powerful naturally aspirated V-8 yet, the 670-horsepower LT6 small-block. Enlarged vents, both in that unique front fascia and on the rear quarters, allowed ample cooling breezes to reach the brakes, engine, and transaxle. Aiding that cause further was a center-mounted heat exchanger in the nose.

Left: Like its C7 forerunner, the aluminum C8 platform proved tough enough to allow the strongest Corvette models to run around topless with no worries about enhancing support. Z06 production for 2023 included 3,109 coupes and 3,304 convertibles. The Z06's rear fascia was unique, too, and incorporated center-floating quad exhausts, which were key to the Z06's unforgettable exhaust note. European export Z06s used the Stingray's outboard exhausts due to the mandated installation of gas particulate filters (GPF) in those countries, which slightly lowered outputs for both the LT6 and its LT2 sibling.

Below: The Z06's record-setting wheels measured 20x10 inches at the nose, 21x13 in the rear. Five different forged-aluminum split-spoke wheels were available, plus two truly new carbon-fiber designs, one painted Carbon Flash, the other (shown here) visibly demonstrating its true carbon-fiber makeup.

Above: The 5.5-liter LT6 small-block's 670 horses not only established a new high for unblown Corvette engines, but they also represented the most produced to date by a naturally aspirated production-car V-8. In keeping with Chevrolet's maximum-performance manufacturing tradition, each LT6 was hand-assembled, start to finish, by a master technician at Bowling Green's Performance Build Center. On each intake manifold was a signed plaque honoring the artists responsible for these masterpieces.

Below left: Standard Z06 brakes were big enough: 14.6-inch rotors up front, 15 in back. Adding the Z07 Performance Package traded those serious stoppers for Brembo carbon-ceramic rotors measuring 15.7 inches in front, 15.4 out back. Shut the front door!

Below right: Typical V-8 crankshafts have long been of 90-degree, cross-plane design, meaning each throw is rotated 90 degrees around the reciprocating circumference, with the goal being to inhibit unwanted secondary vibrations. Much more rare are flat-plane cranks that, as the name implies, feature throws opposed 180 degrees from each other. While shaky operation can be an issue, flat-plane cranks offer better exhaust scavenging compared to cross-planers and also are generally much lighter because they don't require the heavy counterweighting needed to balance a 90-degree crank. Less mass, less weight means flat-plane cranks can wind up quicker and higher than cross-plane counterparts—like, say, to 8,400 rpm. Chevy engineers minimized vibrations in their flat-plane-crank small-block by relying on an oversquare layout to achieve those 5.5 liters. Oversquare means bore is larger than stroke, quite a bit in the LT6's case: 4.104 by 3.150 inches. That oh-so-short stroke meant less piston speed, which in turn translated into minimal piston shake. Note the weight-saving short-skirt forged-aluminum pistons, which also contributed to the LT6's high-winding willingness. Connecting rods were tough forged-titanium pieces.

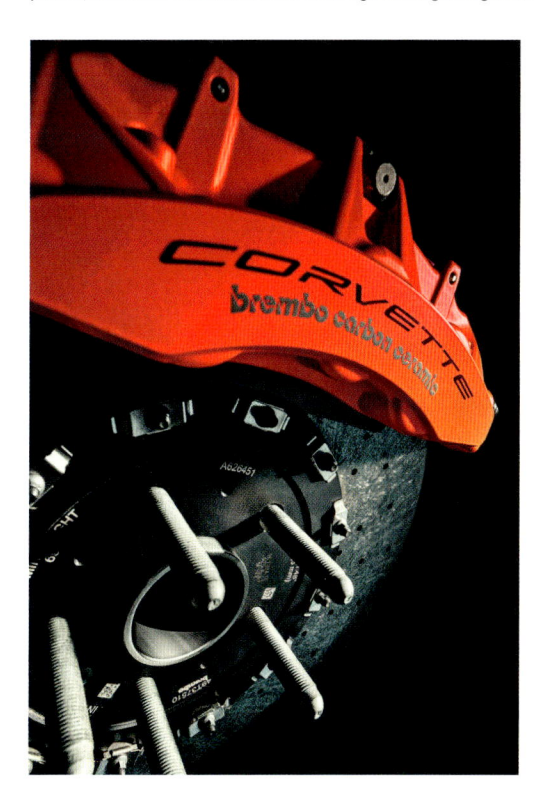

finishes—visible carbon fiber and carbon-fiber painted—these "pricey" ($11,995 and $9,995, respectively) options could only be installed on Z07-equipped Corvettes.

Repeating raves would be a full-time job, so let's listen to GM president Mark Reuss: "The new Z06 defines the American supercar," he said in October 2021. "It builds on the distinctive design and groundbreaking dynamics introduced with the midengine Corvette and elevates them to deliver refined but uncompromising track capability with world-class performance."

Or, as *Car and Driver* put it a bit more succinctly, the Z06 was "an American Ferrari and then some."

2024

Corvette chief engineer. In the beginning, that title didn't even exist. Remember that the first to wear this hat, Zora Arkus-Duntov, didn't officially receive the new chief designation until 1967. You might further recall that Dave McLellan followed Duntov in 1975, then passed the baton on to David Hill in 1992. Hill retired in 2006, leaving the top engineering office to Tom Wallace, who stayed there only two years, after which time Tadge Juechter took the main stage.

Ah, but what some observers may have overlooked is that the story kinda complicated itself from there. If you're one of these, now hear this: the Corvette's chain of command was reshuffled about this time to include a new leadership arrangement made up of chief engineer *and* executive chief engineer. But also know that this revised pecking order required a couple years to evolve following some earlier management restructuring across the board at GM dating back to 1995.

New that year were 12 vehicle line executive (VLE) positions, each dedicated to a specific model group. One of these went to Tom Wallace, who was initially tasked with shepherding midsized trucks through design stages all the way into the sales/service realm. Meanwhile, along with his Corvette chief engineer chores, David Hill also became a VLE, in this case in charge of GM's performance machines.

And so it was that Wallace assumed Hill's dual role on January 1, 2006. Tom then solely became performance-group VLE in July, when Juechter was promoted from assistant to full chief, a position by that point answering to the performance VLE. Following Wallace's early retirement in November 2008, Tadge was promoted again, this time to *executive* chief engineer. Ed Piatek eventually became chief engineer after Juechter's rise to exec, and Josh Holder later succeeded Piatek in 2020 once Ed moved over to chief engineer of future product—that is, performance electric vehicles.

Clear as mud? If that's not convoluted enough, it was about this time that GM further spread the wealth around by creating vehicle line director (VLD) and vehicle chief engineer (VCE) positions, making

CHAPTER NINE

Above: Wheel choices made new-model news for 2024 as 2023's five-trident spoke design was superseded on the RPO list by a revised five-split-spoke wheel, offered in Satin Graphite (with machine edge) or machine-face Sterling Silver. Various safety-conscious driver-assist features also became standard inside all Corvettes this year.

Below: Like 2024's Stingray, Chevrolet's second-edition Z06 rolled over essentially unchanged, save for varying wheel choices and a few trade-outs on the available color palette. New exclusively in the 2024 Z06's case was a third carbon-fiber wheel option, this one adorned with a red stripe.

for overcrowded office parties that remained in place worldwide only briefly. In 2012, General Motors announced a quick guillotining of this three-headed beast, opting instead to let able generals like Juechter stand tall on their own. No more VLE, VLD, or VCE to potentially complicate the process.

"GM charges the new executive chief engineers with complete responsibility for their vehicle groups, from the initial design sketch through production," wrote *Ward's AutoWorld* senior editor James Amend in July that year. According to a corporate statement, the executive chief job description included "defining the requirements of GM's new vehicles to ensure they win in the marketplace, as well as understanding the competitive landscape and managing cost, quality and performance targets."

Juechter didn't just accept this rebuilt role, he put it in a headlock and made it his own. But even he wasn't shy about admitting he had strong support all along. Rarely told completely, let alone accurately, the power progression during C7/C8 development of course involved other important players, including various engineers who stepped up temporarily into the VLE void following Tom Wallace's departure.

"There was a transition period between Tom and Tadge where we had Gene Stefanyshyn, Bill Shaw, Randy Schwartz, [and] Dave Leone, but they had other programs too as VLEs," explained Corvette product manager Harlan Charles in August 2024. "Publicly it was Tadge after Tom in the eye of the customer, so we didn't really introduce the others in that role. A little confusing." Definitely. The result was some not-so-correct reporting on this subject that occasionally repeats itself to this day.

And there remains another behind-the-scenes name that deserves mention: Jim Danahy. According to Ed Piatek, "[his] role was to provide 'executive launch support' during the challenging days of getting C7 into high-volume production. He had a brief stint as Corvette VCE when C7 was potentially going to be mid-engine, but when the decision was made to do [the seventh-gen] as an 'evolutionary' model, he was redeployed and took [a GM position] in Korea. He jokingly said I stole 'his job' after he repatriated."

Bringing this somewhat oversimplified telling to a close was Tadge's announcement in April 2024 that he would be retiring that summer after 47 hard-rockin' years at GM. More official words followed in June stating that Tony Roma, yet another former Cadillac man, would assume the executive role come July 1. Just in time to see the final feather in Juechter's cap, 2025's Earth-shaking ZR1, make its "unthinkable" (GM's words) debut, which we'll revisit in a few pages.

"It's been the honor of a lifetime to work at this company, leading the men and women who have brought to life one of the most iconic and recognizable vehicles in recent American history," read Juechter's April 2024 farewell statement. "Their tenacity and ability to push what is possible with every variant and generation of Corvette was inspiring to see. I know the future of the nameplate is in the right hands."

Calling Juechter "a whirlwind of quiet competence and astute leadership," veteran *Car and Driver* scribe Rich Ceppos added the following: "The measure of a great chief engineer isn't just a thorough understanding of physics and thermodynamics—or even a detailed knowledge of the desires of customers. It's their ability to get their company to produce great cars. Tadge got the huge and often recalcitrant GM organization to do just that, which makes him the most important Corvette chief engineer since Zora Arkus-Duntov."

From all of us watching in August 2024, here's to ya, Mr. Juechter. Hope you finally get some well-earned rest, not to mention a break from all those pesky auto writers and authors. Ain't they the worst? And, yes, Zora certainly would be proud.

The main man behind the machine, the "godfather of the C7 and C8" (according to *Car and Driver*) may have left the scene, but his beloved ward rolled on, once again exhibiting nearly no changes in 2024. No major changes, that is, at least not as far as Stingray and Z06 were concerned. Recognize that this ever-potent duo was joined this year by yet another history-maker, the electrified E-Ray, which we'll shine a spotlight on soon enough. Patience.

In the meantime, 2024's list of deletions/additions, though once again humble, was as long as it had been in some time. Leading the way were various safety features added in along with familiar 1LT/1LZ equipment: following distance indicator, forward collision alert, lane-keep assist with lane-departure warning, Intellibeam auto high beam, automatic emergency braking, and front pedestrian/bicyclist braking. A rear camera mirror became standard in all applications too, and the performance data recorder (included in 2LT/2LZ and 3LT/3LZ packages) was now an option for 1LT/1LZ models.

In other RPO news, two new engine appearance packages—one for coupes, another for convertibles—debuted for 2024. And at the corners, 2023's two five-spoke Trident wheel RPOs (painted silver or Spectra Gray) didn't return but were ably superseded by three new Stingray choices: two five-split-spokers (done in Satin Graphite or Sterling Silver), plus a bright-polished 15-spoker wheel, available via LPO 5DO. Z06 buyers, meanwhile, were treated to a second visible carbon-fiber option (RPO STZ, joining ROZ, introduced in 2023), this one complemented by a red stripe.

Up topside, new wider dual racing stripes joined 2023's rollover striping on 2024's RPO list, in this case done in Carbon Flash trimmed by one of five accent colors: blue, yellow, orange, red, or silver. And after seemingly resting for a year, the Corvette art department got back in the act full-brush, first draining away three 2023 paints: Elkhart Lake Blue Metallic, Caffeine Metallic, and the 70th Anniversary package's exclusive White Pearl Metallic. Contrarily, RPO Y70's Carbon Flash Metallic option flowed over from 2023 (available this year for all models) and was joined by three fresh shades: Riptide Blue Metallic, Cacti Green, and Sea Wolf Gray Tintcoat.

Truly fresh inside was an Artemis Dipped interior, named after the Greek goddess of, among other things, the hunt and vegetation. Hence the great-outdoors-inspired color behind this handle: dark olive green, complemented by green seatbelts. Artemis, Schmartemis. In this case, St. Patrick might've been the one chock full o' pride. You know him, no? The god of assuming an Irish identity (by donning green socks) while drinking yourself into a mug shot?

2024 E-Ray

Prognosticating about future product had long been a full-time job in the Corvette world by the time the C8 landed in 2019. Yet the rumor mill still managed to find another gear from there. Midengine first impressions had yet to fully make their way into the mainstream, and journalists already were pouring out type by the bucket load about the next new Z06, the next new ZR1, and even a so-called "Zora Edition," proposed in more than one form in years to come.

Then there was the electric vehicle (EV) question. Kibitzers also couldn't help themselves as far as this trending subject was concerned, repeatedly asking when would Chevrolet make some genuinely historic history? This time by throwing the switch on an electric C8.

Among others, a *Detroit Free Press* account in July 2021 echoed earlier predictions that the upcoming Z06 could be that milestone, basing its conclusion on a short video promoting Chevrolet's most powerful free-breathing (AKA: unblown) Vette yet. Along with calling this new model "the supercar that will put the world on notice," that vid further proclaimed that "the streets will never sound the same." *Detroit Free Press* reporter Jamie LaReau was inspired to humbly speculate that such talk *might* represent "a hint at an all-electric Corvette."

Of course, those words also could've simply signaled that this loud, street-legal race car was ready, willing, and able, with its internal-combustion engine (ICE), to take on all comers from all points around the globe. Which, per previous reporting here, actually was the translation. As *Saturday Night Live*'s malaprop-prone Emily Litella (Gilda Radner) always used to say, "nevermind."

It was then left to GM president Mark Reuss to set Emily, among others, straight. "In addition to the amazing new Z06 and other gas-powered variants coming, we will offer an electrified and fully electric Corvette in the future," he wrote on LinkedIn in April 2022. "In fact, we will offer [the former] as early as next year." The latter, that full-bore EV, remains a mystery as these keys were clicking away in August 2024. The "electrified" model Reuss spoke of, however, definitely was waiting right around the corner while he was punching up his words.

On January 17, 2023, GM announced a second addition to 2024's Corvette lineup, the aptly named (and aforementioned) E-Ray. And, man, did it shake up the Chevy saga. For starters, all four wheels did the driving, an obvious first for the breed. But more notably, while a 495-horsepower LT2 remained in control of the rear wheels, the fronts were now powered, as one might guess, by an electric motor, a compact three-phase, radial-flux unit energized by a 1.9-kWh lithium-ion battery located in the tunnel between the seats. This second power source added another 160 horses into the mix, bringing the combined count to 655. Total torque was 595 lb-ft, more than enough to rotate our planet back on its heels.

CHAPTER NINE

2024–2025 E-Ray

Model availability	all-wheel-drive hybrid-electric coupe (with removable roof panel) and all-wheel-drive hybrid-electric convertible (with folding hardtop)
Construction	composite/carbon-fiber body panels, widened by 3.6 inches at the corners to house larger wheels/tires; high-pressure diecast aluminum center-tunnel frame with carbon-fiber structural components; rear spoiler, standard
ZER Performance Package	added Michelin Pilot Sport 4 S ZP summer-only tires and specially calibrated FE0 suspension with Magnetic Selective Ride Control 4.0
Wheelbase	107.2 inches
Length	184.6 inches
Width (without mirrors)	79.7 inches
Height	48.6 inches
Track (front/rear)	66.3/66.1 inches
Curb weight	3,774 pounds (coupe), 3,856 pounds (convertible)
Wheels	five-spoke forged-aluminum Pearl Nickel Star-design, exclusive to E-Ray (three extra-cost finishes available) **Size:** 20x10 inches front, 21x13 inches rear **Optional:** ROY, ROZ, or STZ five-spoke carbon-fiber wheels (with carbon-flash paint, visible carbon-fiber finish, or red-striped visible carbon fiber, respectively)
Tires	Michelin Pilot Sport all-season, standard; Michelin Pilot Sport 4S ZP summer-only, optional with ZER package **Size:** 275/30ZR20 front, 345/25ZR21 rear
Brakes	eBoost-assisted antilock discs with Brembo six-piston/two-piece front calipers and four-piston monobloc rears, carbon-ceramic rotors
Brake rotor diameter	15.7 inches front, 15.4 inches rear
Suspension	short/long-arm (SLA) double wishbone, forged-aluminum upper control arms, cast-aluminum L-shaped lower control arms and coilover dampers, front and rear; half-shafts with constant velocity joints, front; Magnetic Selective Ride Control, standard
Steering	electric power-assisted Bosch/ZF variable-ratio rack and pinion
Ratio	15.7:1
Hybrid powertrain	LT2 small-block powering rear wheels, electric motor driving fronts (battery located between seats)
Engine (LT2)	6.2-liter (376 ci) direct-injection overhead-valve V-8 with Active Fuel Management, Variable Valve Timing, and Stop/Start; cast-aluminum cylinder block/heads; dry-sump oiling; and NPP performance exhausts
Compression	11.5:1
LT2 bore and stroke	4.06x3.62 inches
LT2 output	495 horsepower at 6,450 rpm, 470 lb-ft torque at 6,150 rpm
LT2 valves	2.13-inch hollow intakes, 1.59-inch sodium-filled exhausts
Electrification system	magnesium-cased three-phase radial-flux electric motor (mounted below front trunk) energized by a 1.9 kWh lithium-ion battery located in the chassis tunnel between the seats
Electric motor output	160 horsepower, 125 lb-ft torque
Total output	655 horsepower, 595 lb-ft torque
Transmission	M1L Tremec-supplied eight-speed dual-clutch transaxle

Additional electro-wizardry made sure no one could mistake this truly new 2024 Corvette for just *another* hybrid. For starters, there was no plug. Recharging was automatic via regenerative energy provided by coasting and braking, as well as by borrowing some torque now and again from that generous gas-fed V-8 behind the seats. There also was a "Charge+" feature that allowed a driver to specifically control battery revitalization in cases of more urgent need.

Furthermore, both power-providers were joined together in cooperative fashion by the E-Ray's intelligent eAWD system, which literally learned its way down the road. These cutting-edge electronics recognized when the time was right to amp up the front drive more vigorously to enhance traction and/or stability in moments of imperative necessity. Summed up in Tadge Juechter's words, this "electrification technology enhances the feeling of control in all conditions, adding an unexpected degree of composure."

On the other side of the chip, eAWD incorporated a Stealth Mode, which also put a human back (mostly) in control. Selected at startup, this operational option was meant to keep things quiet, say, while slowly leaving your sleepy neighborhood first thing in the morning. Or hers even earlier that morn. It stuck with soft-spoken electric front-wheel-drive alone until speeds reached 45 miles per hour, after which time the LT2 awoke to begin barking its way up to highway limits. Stealth Mode automatically relinquished its hold on the rear wheels whenever it sensed low-speed torque demand or a battery in need of charging.

Initial design work on this innovative drive system dated back to 2016, just two years into the C7 run. So it only followed that, similar to those RHD export models, the game-changing E-Ray hybrid was developed right along with the eighth-gen platform.

"It was absolutely part of the [original C8] plan," said Ed Piatek. "A proof point that it was considered from the start is that the E-Ray has a front trunk that can still swallow an airline carry-on roller bag. It would have been easy to just take that from the [E-Ray] customer, but we found a way to make it all work together." After considerable consideration, some consternation, and the immolation of mucho midnight oil.

"Exactly what the hardware and output would be were not obvious from the start," added Piatek. "[But] we ultimately decided to use a motor decoupled from the ICE hardware and use it to power the front axle. There are many ways to incorporate electrification, and many of our competitors have gone different routes than we have chosen: attaching the electric motor to the engine, or between engine and transmission, or after the transmission, or a combination of these. We did a vast amount of computer simulation around optimizing the performance gains while trying to minimize mass, cost and the packaging challenges [involved] with finding space, in what was primarily an ICE architecture, for the high-voltage battery and motor drive unit."

Accordingly, much attention was paid to making the E-Ray's electric motor so markedly compact, as well as amazingly light. Aluminum hardware proliferated, and inside a weight-saving magnesium case were two sets of reduction gears and a differential that all appeared quite toy-like yet were well up to the task of handling those 160 horses. The sum of these diminutive parts? A 2024 E-Ray coupe only weighed 260 pounds more than a comparable Z06—quite an achievement in hybrid terms.

Above: After working so hard to move the C8's power source behind the seat, engineers put one right back up front in 2024. But this time it was an electric motor driving the front wheels. Hence, the aptly named E-Ray not only became the first gas-electric hybrid in the Corvette lineage, but it also set a precedent with its all-wheel-drive (AWD) layout. E-Ray wheels were the same sizes as the Z06's but featured their own exclusive five-spoke star design. Properly housing those big wheels meant sharing the Z06's widened body too. The Z06's optional carbon-fiber wheels also were available for E-Ray coupes and convertibles.

Below: While the E-Ray used the widened (by 3.6 inches) Z06 body, the hybrid's rear fascia retained the Stingray's outboard exhausts. An exclusive body-length stripe package, done in Electric Blue, was optional for the 2024 E-Ray.

CHAPTER NINE

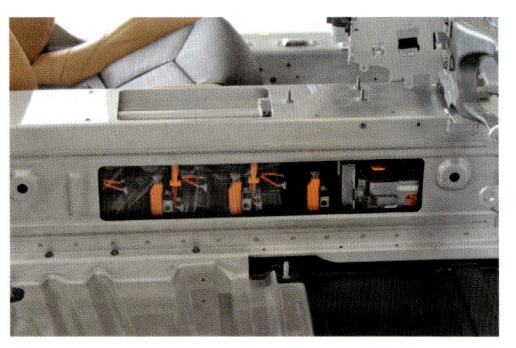

Top left: Note the digital readouts at right for both of the E-Ray hybrid's power sources. Including the NPP performance exhausts meant the E-Ray's LT2 small-block produced 495 horsepower. Adding 160 electrified horses up front brought the E-Ray total to 655 horsepower, which translated into a 0–60 mph run of 2.5 seconds—a new Corvette record. *Wright's Media*

Top right: Much effort went into keeping the E-Ray's electric motor both compact and lightweight to preserve cargo space in the "frunk" and limit weight gain. Note that the motor's case was magnesium. *Wright's Media*

Bottom left: Major League Baseball Hall of Famer Ken Griffey Jr. led the 2024 Indianapolis 500 field in this E-Ray coupe on May 26 that year, making it eight straight pace lap appearances for America's Sports Car. *Indianapolis Motor Speedway*

Center right: Located between the seats, the E-Ray's 1.9 kWh battery pack consisted of eight 10-cell modules, with each cell able to supply as many as 525 amps. In normal driving, individual output was more like 400 amps. *Wright's Media*

Bottom right: The E-Ray's electrical hardware relied on various cooling tricks to help keep heat down in the battery packs. Electronics for both the battery and motor relied on a separate glycol-based cooling system, and this radiator (located behind the left front wheel) kept temps down inside the drive unit. *Wright's Media*

Also vital was keeping things relatively simple with modular construction to minimize repair/maintenance efforts. Equally compact, the batteries and accompanying electronics fit nicely together in a single composite-encased housing that all but fell out the bottom of the car if removal was required. Can you say *not* labor-intensive?

Understandably not quite as simple, however, was locating that little electric power source beneath that still-spacious "frunk" (for "front trunk"), as Piatek further explained. "As far as packaging the front motor in the E-Ray, it was absolutely a challenge and influenced things like suspension link design, steering gear packaging, electrical harnesses, coolant plumbing, and the front cradle structure." But talk about sweat investments paying off.

Raves in the automotive press were rampant, with Rich Ceppos representing one of this historic hybrid's biggest fans. "Unlike the shrieking, feral Z06, E-Ray is a domesticated beast, manifesting much of the over-the-road sophistication and refinements that we've marveled at in the base C8 with the Z51 package," he praised in *Car and Driver* in November 2023.

Domesticated or not, Chevrolet's headline-grabbing AWD Corvette could run toe-to-toe with its uncivilized 670-horsepower cousin. Or should that read "could stomp all over" those Z06 digits? E-Ray's 0–60 performance came in at a time-tripping 2.5 seconds, making it the quickest Corvette ever. Ouch! And if that wasn't hot enough for you, consider quarter-mile performance: 10.5 seconds at 130 miles per hour. As Dr. Zachary Smith used to exclaim, "Oh the pain!" (Once again, google 'em if ya got 'em.)

How could've GM possibly stepped up from this one, you say? Give us a couple pages.

312

GT3.R Racer

Dominating simply wasn't a big enough word in the C8.R's case. Recording names while booting butt was more like it. As of 2023, eighth-gen competition Corvettes had scored 22 wins worldwide, leading to two manufacturer's championships, plus three driver's titles. Let's not forget that class victory at Le Mans in June 2023, the work of Nicky Catsburg, Ben Keating, and Nico Varrone in their #33 C8.R, a suitable exclamation point to four years' worth of on-track successes.

The ninth for Corvettes all-time, and the first since a C7.R turned the trick in 2015, 2023's celebrated triumph in France represented an honorable send-off because the C8.R retired not long afterward. Beating progress, that was where Father Time drew the line. But there was no need to grieve, especially after the next best-racing-Vette-yet took over.

Officially confirmed by GM sources in November 2021, the upcoming Z06 GT3.R eventually picked up where its forerunner left off, turning its first lap in anger—like the C8.R—during the Rolex 24 at Daytona, this time in January 2024. No glory down in Florida then, but this career was still young, capeesh? Initial computerized design work dated back to early 2021, followed by actual on-track testing in September 2022. Chevrolet then threw a formal coming-out party for the GT3.R, also in Daytona, on January 27, 2023.

That moniker was the result of a refocused effort in Corvette terms. First and foremost came news in June 2023 of the impending closure of Corvette Racing by the end of the year. This famed factory team had campaigned the various .R models for 24 years, primarily on the International Motor Sports Association (IMSA) circuit. Not that Chevy was getting out of the game, it just no longer would invest whole-hog at the track.

"We're not going to have a fully factory-funded Corvette race program," explained GM motorsport director Mark Stielow. "But we're going to have a pool of Corvette drivers [with customer teams]. We're going to support different teams to a degree but it's not going to be like Cadillac, which is a fully funded factory effort."

Another change involved the tracks Corvettes had begun visiting in 2022, when C8.Rs started battling for both IMSA's Weathertech SportsCar Championship and the Federation Internationale de l'Automobile's (FIA) World Endurance Championship, a two-pronged global strategy that then carried on after the C8.R's retirement. Various endurance classes existed in both leagues, but one, the FIA's Le Mans GT3 (LMGT3), was off-limits to the C8.R, which didn't meet its specifications. But its successor did, so it was off around the globe to GT3 events for the racing Corvette of the same name.

Like its predecessors, the GT3.R was a product of collaboration between Bowling Green Assembly and the Pratt & Miller works in Michigan. But this time, the platform they started with was the new Z06, by far the closest thing to a street-legal race car that American drivers had ever seen. Save for adding track-ready springs and brakes, race-spec wheels and tires, and ample crash-proof reinforcement, most of the work was already done before the cars left Kentucky.

No need to fix something that ain't broke, right? That adage especially applied in the engine compartment, where the GT3.R's 5.5-liter flat-plane-crank DOHC small-block shared more than 70 percent of its makeup with its street-going counterpart.

"[Competition] continues to play a key role in the development of our production engines," said Performance and Racing Propulsion Team director Russ O'Blenes while announcing the new GT3.R in January 2023. "The flow of information from Corvette Racing to production engineering and back has helped us build race and street-car engines that are fast, reliable and efficient."

"Racing improves the breed," anyone?

The C8.R retired along with the Corvette Racing team, but Chevrolet wasn't about to leave competition-minded customers empty handed. New for 2024 was the GT3.R, a Z06-based racer now targeted at "independent" teams. From left to right here during the GT3.R's unveiling at Daytona International Speedway on January 27, 2023, are Chevrolet global design executive Phil Zak, Chevy motorsports engineering director Mark Stielow, and GM sports car racing program manager Laura Wontrop Klauser.

GT3.R

Model availability	Z06 coupe (produced by Bowling Green Assembly and Pratt & Miller)
Construction	composite/carbon-fiber body panels, widened by 4.6 inches at the corners to house larger racing wheels/tires; stiffened and lightened diecast aluminum frame with aluminum and magnesium structural/chassis components; competition roll cage welded to frame
Wheelbase	107.2 inches
Length	182.3 inches
Width	80.7 inches
Height	45.2 inches (3.4 inches lower than production Corvette)
Weight	2,733 pounds
Wheels	BBS six-split-spoke with single center locking nut **Size:** 18x11.5 inches front, 18x12.5 inches rear
Tires	Michelin racing slicks **Size:** 300/33-R18 front, 310/41-R18 back
Brakes	steel rotors (vented and slotted) with six-piston Alcon Racing monobloc calipers front and rear
Brake rotor diameter	15.35 inches front, 14.0 inches rear
Suspension	short/long-arm (SLA) double wishbone fabricated steel upper/lower control arms and coilover adjustable shock absorbers
Engine	5.5-liter (336 ci) naturally aspirated direct-injected DOHC V-8 with flat-plane crankshaft, four valves per cylinder, and dry-sump oiling
Bore and stroke	4.104x3.150 inches
Output	500 horsepower at 7,400 rpm
Transmission	six-speed Ztrac P529 sequential manual

CHAPTER NINE

2025

Yes, the Corvette song did remain the same more than ever this year, at least at the outset. But who could stop tuning in to these greatest hits? Still-sensational Stingray with 490 or 495 horsepower. Strongest-ever Z06 with 670 horsepower. Electrified AWD E-Ray with those familiar 495 horses at the rear wheels, plus another 160 up front. Can you dig it? We knew that you could.

Then, just when WVET's listening audience thought they'd heard it all, Tadge Juechter and his band had to go and twist the knob up again. Right off the dial. "Eleven" this time was left well behind in the rearview after word hit the streets in July 2024 of the upcoming release of yet another new chart-buster, a history-mutilating hard-rocker that had long needed no introduction. Even those living beneath stones knew what was coming: ZR1, the supreme Corvette that Bowling Green watchers had been looking out for, day and night, dating back to the first C8 off the line in February 2020. Bollocks! Now it was Nigel Tufnel cranking the pride-o-meter over the limit. (Once again, Google is right there.)

With far more muscle than even the most learned prophets had predicted, Chevrolet's latest, undeniably greatest ZR1 shocked the senses like nothing on four wheels before. However, we've said it before, and we'll say it again: you gotta wait to meet this reborn mondo-model. Sure, you could jump ahead a couple column inches, but then we'd have to kill ya. Or at least deliver a harsh letter of complaint.

Back here on Earth, 2025's updates list typically included new exterior finishes: Competition Yellow Tintcoat Metallic and blue-leaning Hysteria Purple

314

Metallic. A third eye-opening shade, Sebring Orange Tintcoat Metallic, returned after a three-year hiatus, and Velocity Yellow paint joined the optional brake caliper coloring book.

Meanwhile, two fashionable interior choices made the latest mix: Habanero (a black/orange two-tone design) and blue-stitched Jet Black, each limited to 3LT/3LZ cars. Formerly also available only with 3LT/3LZ packages, a leather airbag cover this year graced steering wheels across the board as part of an effort to enhance the premium feel in all Corvette cockpits.

Optional revisions included a redesigned spoiler for the Z51 package that both looked more muscular and better cheated the wind compared to 2024's unit. Furthermore, Z06 buyers this year could choose from four new finishes for their 10-spoke wheels: Pearl Nickel, gloss black, polished, and bright polished.

So there you have the most notable appetizers. Now for the main course.

The Corvette lineup expanded to four in 2025 with the reintroduction of the ZR1, seen here second from right. At far left is a 3LZ E-Ray hybrid convertible done in Riptide Blue, an exterior color introduced the year before. Second from left is a Sebring Orange Metallic 3LZ Z06 coupe fitted with the Z07 Performance Package. The Hysteria Purple 3LZ ZR1 coupe is equipped with the Carbon Fiber Aero Package and the ZTK Performance Package. The Torch Red 3LT Stingray convertible at far right is maxed out with Z51 equipment. Its rear spoiler was new for 2025.

CHAPTER NINE

CORVETTE 2025

SEBRING

Returning for 2025 was Sebring Orange Tintcoat paint, a color last seen in 2021.

316

C8 2020–2025

CORVETTE 2025

COMPETITION YELLOW

Two new choices, Competition Yellow Tintcoat Metallic and Hysteria Purple Metallic also showed up in 2025's paint locker.

317

CHAPTER NINE

2025 ZR1

Beamon Edition. Perhaps that's a tag GM shoulda considered for Chevy's newest born-again Z-car. What the hey, you say? Remember when US Olympian Bob Beamon long-jumped 29 feet, 2.5 inches at the Mexico City games in 1968? No? Well, take our word for it. And understand further that Beamon's gloriously golden moment didn't simply *make* track and field history, it murdilated the previous world record by an incomprehensible 21.75 inches! Nearly 2 inhuman feet, a measure equaling an off-the-charts 6.6 percent leap up the stat-o-meter.

Bob himself was so shaken, so seemingly intoxicated by what he'd just wrought, he almost immediately collapsed in shock. "Almost" because, along with being a bit metric-challenged, he simply couldn't wrap his head around the exact result until a coach verified it in no uncertain imperial terms. Then doink! Once back on his feet, he remained this planet's long-jump standard bearer for almost 23 years. It's still one of sporting annals most jaw-dropping moments, if not humankind's.

Your verdict aside, let us now flash-forward to 2025's ZR1, an unquestionably mean, momentous machine that rocked this world wide awake via web video on the night of July 25, 2024.

Anyhoo, those considered faint of heart by this country's surgeon general were cautioned the previous April that viewer discretion might be advised. "The Unthinkable Is Coming This Summer," warned a 21-second statement announcing GM's upcoming YouTube event. Even though "they told you so," who knows how many viewers also crumpled to Earth in a heap upon glimpsing this vid.

As nearly all earthlings with eyes and/or ears now accept as a stone-cold fact, Chevy's 2025 ZR1 inspired the most ink, the heaviest headlines, with its reality-racking 1,064-horse LT7 small-block. Yes, it's still true, folks. Four. Frickin'. Digits. Add your own exclamation points if you like—we're all out.

In an attempt to help us get our noggins straight here, Chevrolet explained it like this: the LT7 produced more power than two of the C6's LS7 V-8s combined, akin to one LS7 engine per LT7 cylinder bank. Glad they cleared that up.

Arguably the web's top source for Corvette breaking news, gmauthority.com could only imagine a mere 850 horsepower in April 2024 while rumoring about the ZR1-to-come. Silly them? Not really, considering that next to no one with a right mind could've forecasted such a bodacious bound upward into the output stratosphere. This frightening (to some) full-figured figure topped the Corvette legacy's existing, six-year-old all-purpose (both naturally aspirated and otherwise) standard by 40 percent.

Not to beat an analogous horse to death, but if Bob Beamon 57 years prior had raised the long-jump bar by 40 percent, he would've traveled 38 feet, arriving at his distant destination (undoubtedly sans luggage) after a nonstop flight that surely would've required a stewardess. Maybe even two. (Remember those?)

So how did Juechter's team do it? How'd they out-Beamon Beamon?

First, they copped the Z06's proven bomb, the naturally aspirated, 32-valve, 5.5-liter LT6 with its rev-crazy flat-plane crank. But recognizing that 670 puny ponies would never do, nor 850, they blew away the megatonnage by incorporating what gmauthority.com *had* spot-on predicted (along with various rivals) well in advance: twin turbochargers. Force-feeding this doomsday device were two Borg-Warner pressure cookers that shoved in upwards of 20 pounds of extra ambient atmosphere.

Remember back in the day when confounding turbo lag effectively negated much of the performance gain promised by this installation? Back when disco would never die? Well, a lot has changed since then, including the reduction of polyester in our wardrobes. Not to mention the deletion of *confounding* from our lexicon. Furthermore, while inherent lag has yet to be rubbed out completely, today's turbo wizards have come a long way, baby.

In the ZR1's case, modern computerization working in concert with familiar wastegate technology more effectively allowed turbo pressure to remain at the ready even when a driver's foot lifted off the go-pedal. As small-block program assistant chief engineer Dustin Gardner told *Road and Track*'s Mike Dustin in July 2024, "When you come out of that braking zone, that corner, and you roll back into it, the [LT7's] turbos [are] at speed, you have boost ready to go. In this situation, there's virtually no lag because we maintain turbo speed during those off-throttle excursions." What will they think of next, gosh-dernit?

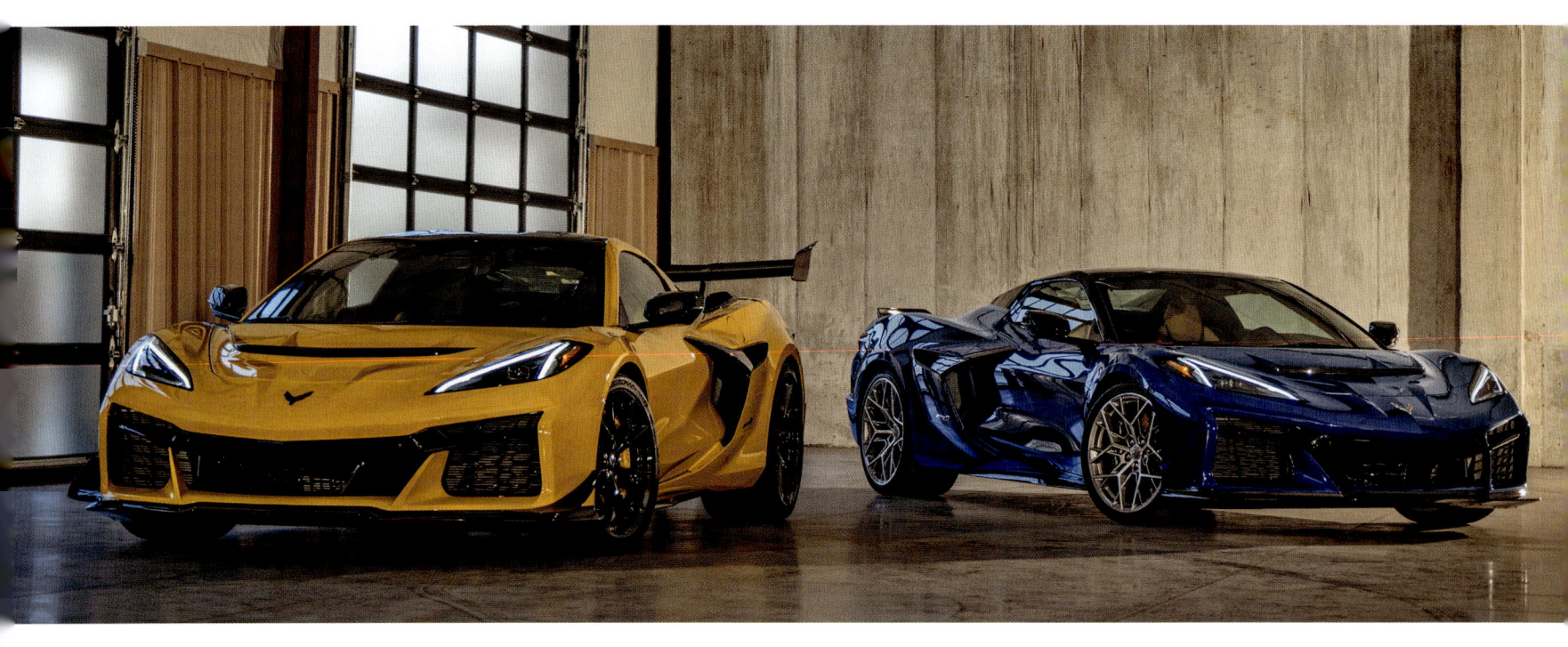

In April 2024, GM officials announced via internet video that the ZR1 would be officially introduced to the world in July. "The Unthinkable Is Coming This Summer" was all that vid said. No kidding. Along with shattering the Corvette output record beyond all recognition, the 1,064-horsepower, twin-turbo ZR1 stood tall as the most powerful V-8 car ever released by an American manufacturer. While it shared the Z06's widened shell, the ZR1 body went one better with even more attention given to cooling all internal components. An available aero package (showcased on the coupe at left) also created more than 1,200 pounds of downforce at top speed.

2025 ZR1

Model availability	coupe (with removable roof panel) and convertible (with folding hardtop)
Construction	composite/carbon-fiber body panels, widened by 3.6 inches at the corners to house larger wheels/tires; high-pressure diecast aluminum center-tunnel frame with carbon-fiber structural components; weight-saving visible carbon-fiber roof panel standard for both coupe and convertible; front trunk space deleted to allow for installation of two large heat exchangers, one for engine, the other for turbo intercoolers
Body features, standard	front splitter with Gurney deflectors; Z06-style front/rear fascias (front featuring revised venting for maximum cooling, rear incorporating center-floating exhaust bezels); enlarged rear-quarter air intakes with additional brake-cooling ductwork incorporated above; turbo-cooling intakes added to coupe's rear hatch; flow-through hood that allowed airflow to exit heat exchangers beneath and amplified front aerodynamic downforce; carbon-fiber rear window spine; reconfigurable standard rear spoiler with adjustable wickerbill elements
Carbon Aero Package, optional	added larger front splitter, front-corner dive planes, a pedestal-mounted high rear wing, underbody aero strakes (in place of standard Gurney deflectors), and tall Gurney lip to leading edge of hood vent, all constructed from woven carbon-fiber (included with ZTK Performance Package and available separately for ZR1, this aero option upped overall downforce to more than 1,200 pounds at top speed)
ZTK Performance Package, optional	along with Carbon Aero Package, this option added stiffer springs with specially calibrated Magnetic Selective Ride Control 4.0 and Michelin Pilot Sport Cup 2R tires
Wheelbase	107 inches
Length	185.9 inches
Width (without mirrors)	79.7 inches
Height	48.6 inches
Track (front/rear)	66.3/66.1 inches
Curb weight	3,670 pounds (coupe), 3,758 pounds (convertible)
Wheels	forged-aluminum split-spoke design, standard; Carbon Revolution carbon-fiber 10-spoke (exclusive to ZR1), optional **Size:** 20x10 inches front, 21x13 inches rear
Tires	Michelin Pilot Sport 4S, standard; Michelin Pilot Sport Cup 2R, optional with ZTK Performance Package **Size:** 275/30ZR20 front, 345/25ZR21 rear
Brakes	eBoost-assisted antilock discs with six-piston monbloc front calipers and four-piston monobloc rears; carbon-ceramic rotors, standard
Brake rotor diameter	15.7 inches front, 15.4 inches rear
Suspension	short/long-arm (SLA) double wishbone, forged-aluminum upper control arms, cast-aluminum L-shaped lower control arms and coilover dampers, front and rear; Magnetic Selective Ride Control 4.0, standard; adjustable front lift with memory, optional
Steering	electric power-assisted Bosch/ZF variable-ratio rack and pinion with Active Steer Stops
Ratio	15.7:1
Engine (LT7)	5.5-liter (336 ci) twin-turbocharged direct-injection DOHC V-8 with Variable Valve Timing, cast-aluminum cylinder block/heads, dry-sump oiling, and flat-plane crankshaft; various internal modifications made to handle extra power and maximize turbo oiling/cooling
Compression	9.8:1
Forced induction	twin 76mm mono-scroll ported shroud-ball-bearing turbochargers with electronic E-Waste gate system and dual engine-mounted water-to-air charge coolers
Valvetrain	dual overhead cams activating four valves per cylinder via mechanical finger-followers and dual valve springs
Valves	1.77-inch titanium intakes, 1.37-inch sodium-filled Nimonic exhausts
Bore and stroke	4.104x3.150 inches
Output	1,064 horsepower at 7,000 rpm, 828 lb-ft torque at 6,000 rpm
Transmission (LT7)	modified M1L eight-speed dual-clutch transaxle

CHAPTER NINE

The Z06's center-mounted quad exhausts also carried over for the ZR1, but unique to 2025's super Corvette was a split rear window layout that honored 1963's new Stingray sport coupe. The "stinger" this time also served to help vent engine heat into the atmosphere. The competition-style high rear wing seen here was included in the ZTK performance package.

320

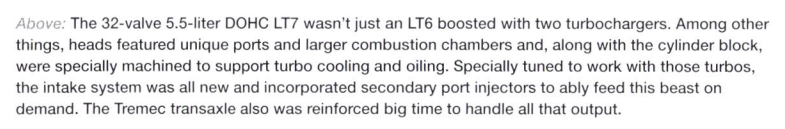

Above: The 32-valve 5.5-liter DOHC LT7 wasn't just an LT6 boosted with two turbochargers. Among other things, heads featured unique ports and larger combustion chambers and, along with the cylinder block, were specially machined to support turbo cooling and oiling. Specially tuned to work with those turbos, the intake system was all new and incorporated secondary port injectors to ably feed this beast on demand. The Tremec transaxle also was reinforced big time to handle all that output.

Top right: Like the Z06, the ZR1 was fitted with a 9,000-rpm tach, but the redline didn't go quite as high. Those 1,064 horses maxed out at 7,000 revs. Note the leather airbag cover in the steering wheel. Previously limited to 3LT and 3LZ packages, this upgrade became standard inside all 2025 Corvettes.

Bottom right: Called "the godfather of the C7 and C8 Corvette" by *Car and Driver*, Tadge Juechter retired just before the ZR1 was introduced to the world in July 2024. The former executive chief engineer then was honored with a glass engraving similar to the Zora Duntov silhouette seen on C8 windshields dating back to 2020. Juechter's bust first appeared at the top left of the ZR1's rear window.

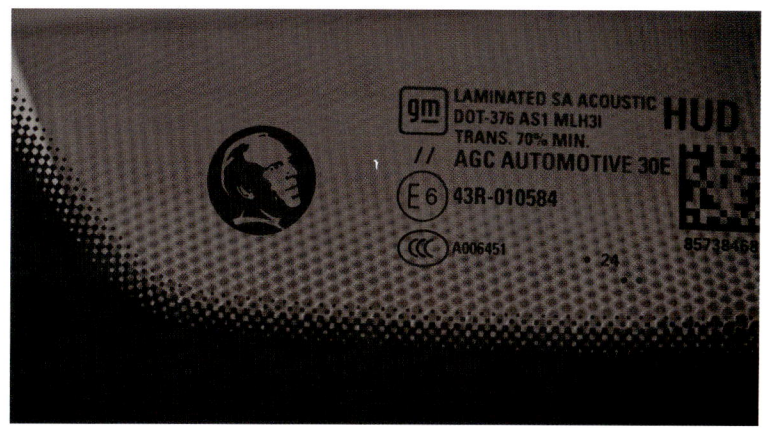

As for all-out throttle-mashing, official numbers for 2025's ZR1 (both performance parameters and pricing) remained unpublished as these words were being smithed. But Chevrolet officials did claim, rather half-heartedly, that in-house test drives had produced quarter-mile excursions of fewer than 10 seconds. Remember when that realm was reserved for Pro Stock drag cars? Beyond that, GM estimates put top-end at "more than 215 mph." Again, what part of "street car" did Juechter's people not understand?

As was the case a couple subheads ago concerning the Z06's beautiful body, attempting to tell the epic LT7 tale in 25 words or less is beyond impossible. Again, you'll have to look elsewhere amidst these pages for additional facts and/or figures detailing the transformation from LT6 to 7. For now, highlights included revised heads with larger combustion chambers and unique ports; a retuned intake system coupled with secondary port fuel injectors, added to better feed this hungry beast; and remachined block and heads to better facilitate turbo cooling and oiling.

Lastly, the ZR1's Tremec transaxle also was treated to various upgrades to better handle, in Chevy's words, that "dramatic increase in power." Calling Stephen Hawking "bright" a second time? You be the judge.

Much of the ZR1's remaining makeup carried over from the Z06 paddock. But in the braking department, the Z06's optional carbon-ceramic stoppers were standard in the ZR1's case. And the Z07 Performance Package was repackaged slightly and labeled ZTK in ZR1 terms.

Unique to ZR1 models was the rear glass, which revived an almost forgotten Sting Ray feature. "Sting Ray," because this blast from the past dated back to 1963. Yes, the split window was back. And in this application, the carbon-fiber "stinger," or "spine" in ZR1-speak, not only honored Chevy's original Sting Ray, but it also improved engine heat extraction. Clever, no?

"Not only does this element provide function, but we were able to integrate passionate design into the form and do it in a way that paid homage to Corvette's history," explained executive design director Phil Zak. "ZR1 felt like the right time to bring the split-window back."

Well, in truth, this design actually debuted earlier on the GT3.R competition Corvette. But no biggie, impressions remained fond regardless of which came first.

At issue, more importantly, is how much more familiar can Corvette road and track identities become? Where will the breed go from here? If the envelope gets rammed out any further, Tony Roma's team just might be entering Michael J. Fox DeLorean territory.

Only the future will tell. Come back then.

Epilogue

While the future of performance car design probably won't involve time travel—at least not the *near* future—it will unswervingly mandate fully electric drive. And not because the Feds say so, but due simply to the fact that our world *will* run out of dinosaur squeezings much sooner than later. Couple with that an undeniable need to continue cleaning up our transportation act or suffer additional, surely horrific consequences, also before we know it. Our point? No ifs, ands, or big buts about it, we will see not "just" a hybrid but a full-fledged Corvette EV, once again not all that far down the road.

How far and in what form/forms? Take your pick. Press rumors surrounding this subject date almost all the way back to the C8's debut and have flourished ever since. This buzz really started singing after Ford's all-electric Mustang Mach-E began charging onto Main Street USA late in 2020.

Raising the raucous further in August that year was then-presidential-candidate Joe Biden, who announced in a campaign ad that—no malarkey about it, Jack—he'd been told of an upcoming all-electric Corvette. A vintage Sting Ray (a '67) owner himself, President Biden then backed up that claim 12 months later. "I've got a commitment from [GM CEO] Mary Barra," he said in a 2021 press conference focusing on the fate of our automotive industry. "When they make the first electric Corvette, I get to drive it." Just leave some for us, Joe.

Early media speculation covered the gamut, including reports of future SUVs, crossovers, and/or four-door sport sedans, all EVs adorned with Corvette badges. And most of these accounts were immediately shot down by GM. Until April 2022, when, as mentioned a few pages ago, corporate president Mark Reuss went on record to at least partially confirm some basics. But there are still no hard/fast facts, per GM's ever-present "no comment on future product" protocol.

That apparently not-so-hard/fast rule notwithstanding, various outlets in 2024 continued to claim they had been told of more specific plans by credible sources inside GM. Most notable to this point was a *Motor Trend* mention of an all-electric concept car, a Stingray, slated to appear before 2024's end. Still other sources claimed an actual production EV—utility vehicle or sedan—would be seen wearing Corvette identification before the calendar rolled over into 2025.

Might they all be founded? That is, could Chevrolet introduce the Corvette faithful to the EV ideal in prototypical, fondly familiar Stingray form, then go into production with one of those "badge-engineered" models to better target, to better penetrate, the electrified performance market? Some mouthful, huh?

Whatever the case, never fear. Your beloved ICE Stingray isn't going anywhere soon. Apparently all odds seem (in August 2024) to favor a Ford Mach-E response or some other secondary platform rebadging, basically due to packaging issues. As performance EV chief engineer Ed Piatek explained, "It is fairly clear in the [case of] EVs that [already] have come to market with even a modicum of performance that using the same architecture for both pure ICE and pure EV models comes with extensive compromise and sub-optimization."

One particular challenge involves driver safety. Or, more to the point, the "safety" of the voltage source. "On an ICE product, the primary [crash-proofing] concern involves delivering acceptable passenger performance results," continued Piatek. "But an EV [design further demands] little or zero intrusion into the battery, which is typically nearly the full width of the vehicle. This [requirement] can be onerous for EVs and results in significantly more structure along the rocker than you would find in a gasoline-only product. There are a few examples where manufacturers have made EVs from ICE-derived architectures, but I don't think they are highly regarded by consumers or third parties."

Piatek's account helps explain why still other prognosticators have claimed that a regular-production EV Stingray probably will have to wait until a better-suited C9 platform shows up. Better suited to full electric conversion, that is. Putting together a single concept for the auto show stage—actually functional or not—is one thing. Going to all the trouble of morphing a partly uncooperative ICE package into a cutting-edge EV for full production, now that's a whole 'nother matzo ball for sure.

Stand by and stay tuned.

If any of you do know a way to traverse time, if you've perhaps been hiding a flux capacitor from us, please come back and let us know how this all turns out. Right now.

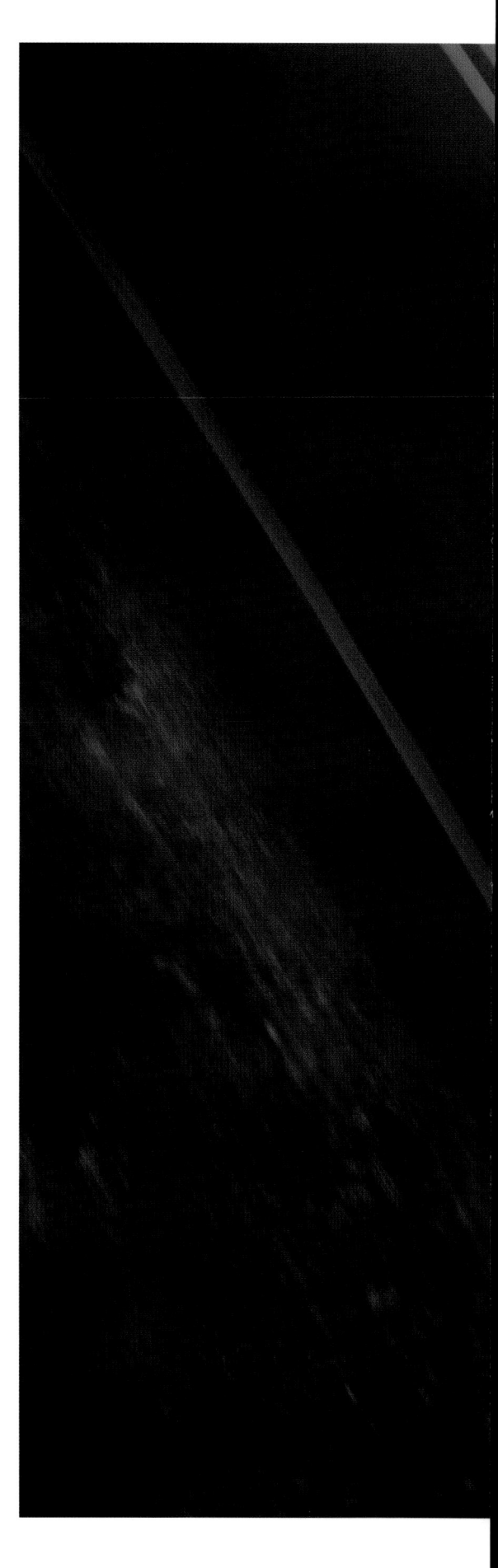

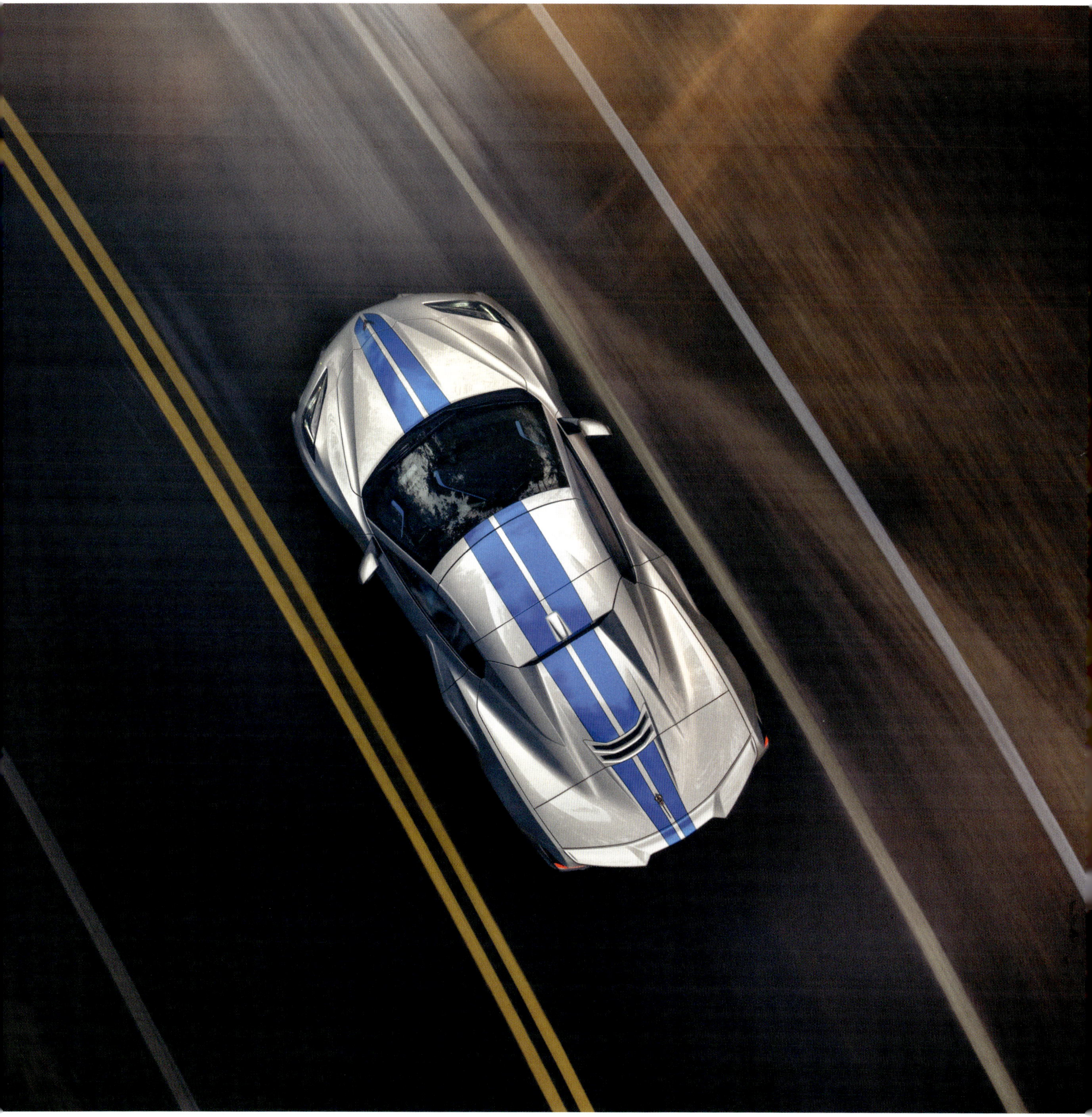

Appendix
Options

OPTIONS

1953 Options

CODE	DESCRIPTION	QTY	RETAIL $
2934	Base Corvette Roadster	300	$3,498.00
101A	Heater	300	91.40
102B	AM Radio, signal seeking	300	145.15

1954 Options

CODE	DESCRIPTION	QTY	RETAIL $
2934	Base Corvette Roadster	3,640	$2,774.00
100	Directional Signal	3,640	16.75
101A	Heater	3,640	91.40
102A	AM Radio, signal seeking	3,640	145.15
290B	Whitewall Tires, 6.70x15	3,640	26.90
313M	Powerglide Automatic Transmission	3,640	178.35
420A	Parking Brake Alarm	3,640	5.65
421A	Courtesy Lights	3,640	4.05
422A	Windshield Washer	3,640	11.85

1955 Options

CODE	DESCRIPTION	QTY	RETAIL $
2934-6	Base Corvette Roadster, six-cylinder	7	$2,774.00
2934-8	Base Corvette Roadster, V8	693	2,909.00
100	Directional Signal	700	16.75
101	Heater	700	91.40
102A	AM Radio, signal seeking	700	145.15
290B	Whitewall Tires, 6.70x15	—	26.90
313	Powerglide Automatic Transmission	—	178.35
420A	Parking Brake Alarm	700	5.65
421A	Courtesy Lights	700	4.05
422A	Windshield Washers	700	11.85

1956 Options

CODE	DESCRIPTION	QTY	RETAIL $
2934	Base Corvette Convertible	3,467	$3,120.00
101	Heater	—	123.65
102	AM Radio, signal seeking	2,717	198.90
107	Parking Brake Alarm	2,685	5.40
108	Courtesy Lights	2,775	8.65
109	Windshield Washers	2,815	11.85
290	Whitewall Tires, 6.70x15	—	32.30
313	Powerglide Automatic Transmission	—	188.50
419	Auxiliary Hardtop	2,076	215.20
426	Power Windows	547	64.60
440	Two-Tone Paint Combination	1,259	48.45
684	Heavy Duty Racing Suspension	51	780.10
685	4-Speed Manual Transmission	664	188.30

1957 Options

CODE	DESCRIPTION	QTY	RETAIL $
2934	Base Corvette Convertible	6,339	$3,176.32
101	Heater	5,373	118.40
102	AM Radio, signal seeking	3,635	199.10
107	Parking Brake Alarm	1,873	5.40
108	Courtesy Lights	2,849	8.65
109	Windshield Washers	2,555	11.85
276	Wheels, 15x5.5 (5)	51	15.10
290	Whitewall Tires, 6.70x15	5,019	31.60
303	3-Speed Manual Transmission, close ratio	4,282	0.00
313	Powerglide Automatic Transmission	1,393	188.30
419	Auxiliary Hardtop	4,055	215.20
426	Power Windows	379	59.20
440	Two-Tone Paint Combination	2,794	19.40
469A	283ci, 245hp Engine (2x4 carburetors)	2,045	150.65
469C	283ci, 270hp Engine (2x4 carburetors)	1,621	182.95
473	Power Operated Folding Top	1,336	139.90
579A	283ci, 250hp Engine (fuel injection)	182	484.20
579B	283ci, 283hp Engine (fuel injection)	713	484.20
579C	283ci, 250hp Engine (fuel injection)	102	484.20
579E	283ci, 283hp Engine (fuel injection)	43	726.30
677	Positraction Rear Axle, 3.70:1	327	48.45
678	Positraction Rear Axle, 4.11:1	1,772	48.45
679	Positraction Rear Axle, 4.56:1	—	48.45
684	Heavy Duty Racing Suspension	51	780.10
685	4-Speed Manual Transmission	664	188.30

1958 Options

CODE	DESCRIPTION	QTY	RETAIL $
867	Base Corvette Convertible	9,168	$3,591.00
101	Heater	8,014	96.85
102	AM Radio, signal seeking	6,142	144.45
107	Parking Brake Alarm	2,883	5.40
108	Courtesy Light	4,600	6.50
109	Windshield Washers	3,834	16.15
276	Wheels, 15x5.5 (5)	404	0.00
290	Whitewall Tires, 6.70x15	7,428	31.55
313	Powerglide Automatic Transmission	2,057	188.30
419	Auxiliary Hardtop	5,607	215.20
426	Power Windows	649	59.20
440	Two-Tone Exterior Paint	3,422	16.15
469	283ci, 245hp Engine (2x4 carburetors)	2,436	150.65
469C	283ci, 270hp Engine (2x4 carburetors)	978	182.95
473	Power Operated Folding Top	1,090	139.90
579	283ci, 290hp Engine (fuel injection)	504	484.20
579D	283ci, 290hp Engine (fuel injection)	1,007	484.20
677	Positraction Rear Axle, 3.70:1	1,123	48.45
678	Positraction Rear Axle, 4.11:1	2,518	48.45
679	Positraction Rear Axle, 4.56:1	370	48.45
684	Heavy Duty Brakes and Suspension	144	780.10
685	4-Speed Manual Transmission	3,764	215.20

1959 Options

CODE	DESCRIPTION	QTY	RETAIL $
867	Base Corvette Convertible	9,670	$3,875.00
101	Heater	8,909	102.25
102	AM Radio, signal seeking	7,001	149.80
107	Parking Brake Alarm	3,601	5.40
108	Courtesy Light	3,601	6.50
109	Windshield Washers	7,929	16.15
121	Radiator Fan Clutch	67	21.55
261	Sunshades	3,722	10.80
276	Wheels, 15x5.5 (5)	214	0.00
290	Whitewall Tires, 6.70x15	8,173	31.55
313	Powerglide Automatic Transmission	1,878	199.10
419	Auxiliary Hardtop	5,481	236.75
426	Power Windows	587	59.20
440	Two-Tone Exterior Paint	2,931	16.15
469	283ci, 245hp Engine (2x4 carburetors)	1,417	150.65
469C	283ci, 270hp Engine (2x4 carburetors)	1,846	182.95

APPENDIX

CODE	DESCRIPTION	QTY	RETAIL $
473	Power Operated Folding Top	661	139.90
579	283ci, 250hp Engine (fuel injection)	175	484.20
579D	283ci, 290hp Engine (fuel injection)	745	484.20
675	Positraction Rear Axle	4,170	48.45
684	Heavy Duty Brakes and Suspension	142	425.05
685	4-Speed Manual Transmission	4,175	188.30
686	Metallic Brakes	333	26.90
1408	Blackwall Tires, 6.70x15 nylon	—	—
1625	24 Gallon Fuel Tank	—	—

1960 Options

CODE	DESCRIPTION	QTY	RETAIL $
867	Base Corvette Convertible	10,261	$3,872.00
101	Heater	9,808	102.25
102	AM Radio, signal seeking	8,166	137.75
107	Parking Brake Alarm	4,051	5.40
108	Courtesy Light	6,774	6.50
109	Windshield Washers	7,205	16.15
121	Temperature Controlled Radiator Fan	2,711	21.55
261	Sunshades	5,276	10.80
276	Wheels, 15x5.5 (5)	246	0.00
290	Whitewall Tires, 6.70x15	9,104	31.55
313	Powerglide Automatic Transmission	1,766	199.10
419	Auxiliary Hardtop	5,147	236.75
426	Power Windows	544	59.20
440	Two-Tone Exterior Paint	3,312	16.15
469	283ci, 245hp Engine (2x4 carburetors)	1,211	150.65
469C	283ci, 270hp Engine (2x4 carburetors)	2,364	182.95
473	Power Operated Folding Top	512	139.90
579	283ci, 250hp Engine (fuel injection)	100	484.20
579D	283ci, 290hp Engine (fuel injection)	759	484.20
675	Positraction Rear Axle	5,231	43.05
685	4-Speed Manual Transmission	5,328	188.30
686	Metallic Brakes	920	26.90
687	Heavy Duty Brakes and Suspension	119	333.60
1408	Blackwall Tires, 6.70x15 nylon	—	15.75
1625A	24 Gallon Fuel Tank	—	161.40

1961 Options

CODE	DESCRIPTION	QTY	RETAIL $
867	Base Corvette Convertible	10,939	$3,934.00
101	Heater	10,671	102.25
102	AM Radio, signal seeking	9,316	137.75
242	Positive Crankcase Ventilation	—	5.40
276	Wheels, 15x5.5 (5)	337	0.00
290	Whitewall Tires, 6.70x15	9,780	31.55
313	Powerglide Automatic Transmission	1,458	199.10
353	283ci, 275hp Engine (fuel injection)	118	484.20
354	283ci, 315hp Engine (fuel injection)	1,462	484.20
419	Auxiliary Hardtop	5,680	236.75
426	Power Windows	698	59.20
440	Two-Tone Exterior Paint	3,351	16.15
468	283ci, 270hp Engine (2x4 carburetor)	2,827	182.95
469	283ci, 245hp Engine (2x4 carburetor)	1,175	150.65
473	Power Operated Folding Top	442	161.40
675	Positraction Rear Axle	6,915	43.05
685	4-Speed Manual Transmission	7,013	188.30
686	Metallic Brakes	1,402	37.70
687	Heavy Duty Brakes and Suspension	233	333.60
1408	Blackwall Tires, 6.70x15 nylon	—	15.75
1625	24 Gallon Fuel Tank	—	161.40

1962 Options

CODE	DESCRIPTION	QTY	RETAIL $
867	Base Corvette Convertible	14,531	$4,038.00
102	AM Radio, signal seeking	13,076	137.75
203	Rear Axle, 3.08:1 ratio	—	0.00
242	Positive Crankcase Ventilation	—	5.40
276	Wheels, 15x5.5 (5)	561	0.00
313	Powerglide Automatic Transmission	1,532	199.10
396	327ci, 340hp Engine	4,412	107.60
419	Auxiliary Hardtop	8,074	236.75
426	Power Windows	995	59.20
441	Direct Flow Exhaust System	2,934	0.00
473	Power Operate Folding Top	350	139.90
488	24 Gallon Fuel Tank	65	118.40
582	327ci, 360hp Engine (fuel injection)	1,918	484.20
583	327ci, 300hp Engine	3,294	53.80
675	Positraction Rear Axle	14,232	43.05
685	4-Speed Manual Transmission	11,318	188.30
686	Metallic Brakes	2,799	37.70
687	Heavy Duty Brakes and Steering	246	333.60
1832	Whitewall Tires, 6.70x15	—	31.55
1833	Blackwall Tires, 6.70x15 nylon	—	15.70

1963 Options

RPO #	DESCRIPTION	QTY	RETAIL $
837	Base Corvette Sport Coupe	10,594	$4,252.00
867	Base Corvette Convertible	10,919	4,037.00
898	Genuine Leather Seats	1,114	80.70
941	Sebring Silver Exterior Paint	3,516	80.70
A01	Soft Ray Tinted Glass, all windows	629	16.15
A02	Soft Ray Tinted Glass, windshield	470	10.80
A31	Power Windows	3,742	59.20
C07	Auxiliary Hardtop (for convertible)	5,739	236.75
C48	Heater and Defroster Deletion (credit)	124	-100.00
C60	Air Conditioning	278	421.80
G81	Positraction Rear Axle, all ratios	17,554	43.05
G91	Special Highway Axle, 3.08:1 ratio	211	2.20
J50	Power Brakes	3,336	43.05
J65	Sintered Metallic Brakes	5,310	37.70
L75	327ci, 300hp Engine	8,033	53.80
L76	327ci, 340hp Engine	6,978	107.60
L84	327ci, 360hp Engine (fuel injection)	2,610	430.40
M20	4-Speed Manual Transmission	17,973	188.30
M35	Powerglide Automatic Transmission	2,621	199.10
N03	36 Gallon Fuel Tank (for coupe)	63	202.30
N11	Off Road Exhaust System	—	37.70
N34	Woodgrained Plastic Steering Wheel	130	16.15
N40	Power Steering	3,063	75.35
P48	Cast Aluminum Knock-Off Wheels (5)	—	322.80
P91	Blackwall Tires, 6.70x15, (nylon cord)	412	15.70
P92	Whitewall Tires, 6.70x15 (rayon cord)	19,383	31.55
T86	Back-up Lamps	318	10.80
U65	Signal Seeking AM Radio	11,368	137.75
U69	AM-FM Radio	9,178	174.35
Z06	Special Performance Equipment	199	1,818.45

OPTIONS

1964 Options

RPO #	DESCRIPTION	QTY	RETAIL $
837	Base Corvette Sport Coupe	8,304	$4,252.00
867	Base Corvette Convertible	13,925	4,037.00
—	Genuine Leather Seats	1,334	80.70
A01	Soft Ray Tinted Glass, all windows	6,031	16.15
A02	Soft Ray Tinted Glass, windshield	6,387	10.80
A31	Power Windows	3,706	59.20
C07	Auxiliary Hardtop (for convertible)	7,023	236.75
C48	Heater and Defroster Deletion (credit)	60	-100.00
C60	Air Conditioning	1,988	421.80
F40	Special Front and Rear Suspension	82	37.70
G81	Positraction Rear Axle, all ratios	18,279	43.05
G91	Special Highway Axle, 3.08:1 ratio	2,310	2.20
J50	Power Brakes	2,270	43.05
J56	Special Sintered Metallic Brake Package	29	629.50
J65	Sintered Metallic Brakes, power	4,780	53.80
K66	Transistor Ignition System	552	75.35
L75	327ci, 300hp Engine	10,471	53.80
L76	327ci, 365hp Engine	7,171	107.60
L84	327ci, 375hp Engine (fuel injection)	1,325	538.00
M20	4-Speed Manual Transmission	19,034	188.30
M35	Powerglide Automatic Transmission	2,480	199.10
N03	36 Gallon Fuel Tank (for coupe)	38	202.30
N11	Off Road Exhaust System	1,953	37.70
N40	Power Steering	3,126	75.35
P48	Cast Aluminum Knock-Off Wheels (5)	806	322.80
P91	Blackwall Tires, 6.70x15 (nylon cord)	372	15.70
P92	Whitewall Tires, 6.70x15 (rayon cord)	19,977	31.85
T86	Back-up Lamps	11,085	10.80
U69	AM-FM Radio	20,934	176.50

1965 Options

RPO #	DESCRIPTION	QTY	RETAIL $
19437	Base Corvette Sport Coupe	8,186	$4,321.00
19467	Base Corvette Convertible	15,378	4,106.00
—	Genuine Leather Seats	2,128	80.70
A01	Soft Ray Tinted Glass, all windows	8,752	16.15
A02	Soft Ray Tinted Glass, windshield	7,624	10.80
A31	Power Windows	3,809	59.20
C07	Auxiliary Hardtop (for convertible)	7,787	236.75
C48	Heater and Defroster Deletion (credit)	39	-100.00
C60	Air Conditioning	2,423	421.80
F40	Special Front and Rear Suspension	975	37.70
G81	Positraction Rear Axle, all ratios	19,965	43.05
G91	Special Highway Axle, 3.08:1 ratio	1,886	2.20
J50	Power Brakes	4,044	43.05
J61	Drum Brakes (substitution credit)	316	-64.50
K66	Transistor Ignition System	3,686	75.35
L75	327ci, 300hp Engine	8,358	53.80
L76	327ci, 365hp Engine	5,011	129.15
L78	396ci, 425hp Engine	2,157	292.70
L79	327ci, 350hp Engine	4,716	107.60
L84	327ci, 375hp Engine (fuel injection)	771	538.00
M20	4-Speed Manual Transmission	21,107	188.30
M35	Powerglide Automatic Transmission	2,021	199.10
N03	36 Gallon Fuel Tank (for coupe)	41	202.30
N11	Off Road Exhaust System	2,468	37.70
N14	Side Mount Exhaust System	759	134.50
N32	Teakwood Steering Wheel	2,259	48.45
N36	Telescopic Steering Column	3,917	43.05
N40	Power Steering	3,236	96.85
P48	Cast Aluminum Knock-Off Wheels (5)	1,116	322.80
P91	Blackwall Tires, 7.75x15 (nylon cord)	168	15.70
P92	Whitewall Tires, 7.75x15 (rayon cord)	19,300	31.85
T01	Goldwall Tires, 7.75x15 (nylon cord)	989	50.05
U69	AM-FM Radio	22,113	203.40
Z01	Comfort and Convenience Group	15,397	16.15

1966 Options

RPO #	DESCRIPTION	QTY	RETAIL $
19437	Base Corvette Sport Coupe	9,958	$4,295.00
19467	Base Corvette Convertible	17,762	4,084.00
—	Genuine Leather Seats	2,002	79.00
A01	Soft Ray Tinted Glass, all windows	11,859	15.80
A02	Soft Ray Tinted Glass, windshield	9,270	10.55
A31	Power Windows	4,562	57.95
A82	Headrests	1,033	42.15
A85	Shoulder Belts	37	26.35
C07	Auxiliary Hardtop (for convertible)	8,463	231.75
C48	Heater and Defroster Deletion (credit)	54	-97.85
C60	Air Conditioning	3,520	412.90
F41	Special Front and Rear Suspension	2,705	36.90
G81	Positraction Rear Axle, all ratios	24,056	42.15
J50	Power Brakes	5,464	42.15
J56	Special Heavy Duty Brakes	382	342.30
K19	Air Injection Reactor	2,380	44.75
K66	Transistor Ignition System	7,146	73.75
L36	427ci, 390hp Engine	5,116	181.20
L72	427ci, 425hp Engine	5,258	312.85
L79	327ci, 350hp Engine	7,591	105.35
M20	4-Speed Manual Transmission	10,837	184.35
M21	4-Speed Man Trans, close ratio	13,903	184.35
M22	4-Speed Man Trans, close ratio, heavy duty	15	237.00
M35	Powerglide Automatic Transmission	2,401	194.85
N03	36 Gallon Fuel Tank (for coupe)	66	198.05
N11	Off Road Exhaust System	2,795	36.90
N14	Side Mount Exhaust System	3,617	131.65
N32	Teakwood Steering Wheel	3,941	47.40
N36	Telescopic Steering Column	3,670	42.15
N40	Power Steering	5,611	94.80
P48	Cast Aluminum Knock-Off Wheels (5)	1,194	316.00
P92	Whitewall Tires, 7.75x15, (rayon cord)	17,969	31.30
T01	Goldwall Tires, 7.75x15 (nylon cord)	5,557	46.55
U69	AM-FM Radio	26,363	199.10
V74	Traffic Hazard Lamp Switch	5,764	11.60

1967 Options

RPO #	DESCRIPTION	QTY	RETAIL $
19437	Base Corvette Sport Coupe	8,504	$4,388.75
19467	Base Corvette Convertible	14,436	4,240.75
—	Genuine Leather Seats	1,601	79.00
A01	Soft Ray Tinted Glass, all windows	11,331	15.80
A02	Soft Ray Tinted Glass, windshield	6,558	10.55
A31	Power Windows	4,036	57.95
A82	Headrests	1,762	42.15
A85	Shoulder Belts	4,426	26.35
C07	Auxiliary Hardtop (for convertible)	6,880	231.75
C08	Vinyl Covering (for auxiliary hardtop)	1,966	52.70

APPENDIX

CODE	DESCRIPTION	QTY	RETAIL $
C48	Heater and Defroster Deletion (credit)	35	-97.85
C60	Air Conditioning	3,788	412.90
F41	Special Front and Rear Suspension	2,198	36.90
G81	Positraction Rear Axle, all ratios	20,308	42.15
J50	Power Brakes	4,766	42.15
J56	Special Heavy Duty Brakes	267	342.30
K19	Air Injection Reactor	2,573	44.75
K66	Transistor Ignition System	5,759	73.75
L36	427ci, 390hp Engine	3,832	200.15
L68	427ci, 400hp Engine	2,101	305.50
L71	427ci, 435hp Engine	3,754	437.10
L79	327ci, 350hp Engine	6,375	105.35
L88	427ci, 430hp Engine	20	947.90
L89	Aluminum Cylinder Heads for L71	16	368.65
M20	4-Speed Manual Transmission	9,157	184.35
M21	4-Speed Man Trans, close ratio	11,015	184.35
M22	4-Speed Man Trans, close ratio, heavy duty	20	237.00
M35	Powerglide Automatic Transmission	2,324	194.35
N03	36 Gallon Fuel Tank (for coupe)	2	198.05
N11	Off Road Exhaust System	2,326	36.90
N14	Side Mount Exhaust System	4,209	131.65
N36	Telescopic Steering Column	2,415	42.15
N40	Power Steering	5,747	94.80
N89	Cast Aluminum Bolt-On Wheels (5)	720	263.30
P92	Whitewall Tires, 7.75x15	13,445	31.35
QB1	Redline Tires, 7.75x15	4,230	46.65
U15	Speed Warning Indicator	2,108	10.55
U69	AM-FM Radio	22,193	172.75

1968 Options

RPO #	DESCRIPTION	QTY	RETAIL $
19437	Base Corvette Sport Coupe	9,936	$4,663.00
19467	Base Corvette Convertible	18,630	4,320.00
—	Genuine Leather Seats	2,429	79.00
A01	Soft Ray Tinted Glass, all windows	17,635	15.80
A02	Soft Ray Tinted Glass, windshield	5,509	10.55
A31	Power Windows	7,065	57.95
A82	Headrests	3,197	42.15
A85	Custom Shoulder Belts (std with coupe)	350	26.35
C07	Auxiliary Hardtop (for convertible)	8,735	231.75
C08	Vinyl Covering (for auxiliary hardtop)	3,050	52.70
C50	Rear Window Defroster	693	31.60
C60	Air Conditioning	5,664	412.90
F41	Special Front and Rear Suspension	1,758	36.90
G81	Positraction Rear Axle, all ratios	27,008	46.35
J50	Power Brakes	9,559	42.15
J56	Special Heavy Duty Brakes	81	384.45
K66	Transistor Ignition System	5,457	73.75
L36	427ci, 390hp Engine	7,717	200.15
L68	427ci, 400hp Engine	1,932	305.50
L71	427ci, 435hp Engine	2,898	437.10
L79	327ci, 350hp Engine	9,440	105.35
L88	427ci, 430hp Engine	80	947.90
L89	Aluminum Cylinder Heads with L71	624	805.75
M20	4-Speed Manual Transmission	10,760	184.35
M21	4-Speed Man Trans, close ratio	12,337	184.35
M22	4-Speed Man Trans, close ratio, heavy duty	80	263.30
M40	Turbo Hydra-Matic Automatic Transmission	5,063	226.45
N11	Off Road Exhaust System	4,695	36.90
N36	Telescopic Steering Column	6,477	42.15

CODE	DESCRIPTION	QTY	RETAIL $
N40	Power Steering	12,364	94.80
P01	Bright Metal Wheel Cover	8,971	57.95
PT6	Red Stripe Tires, F70x15, nylon	11,686	31.30
PT7	White Stripe Tires, F70x15, nylon	9,692	31.30
UA6	Alarm System	388	26.35
U15	Speed Warning Indicator	3,453	10.55
U69	AM-FM Radio	24,609	172.75
U79	AM-FM Radio, stereo	3,311	278.10

1969 Options

RPO #	DESCRIPTION	QTY	RETAIL $
19437	Base Corvette Sport Coupe	22,129	$4,781.00
19467	Base Corvette Convertible	16,633	4,438.00
—	Genuine Leather Seats	3,729	79.00
A01	Soft Ray Tinted Glass, all windows	31,270	16.90
A31	Power Windows	9,816	63.20
A82	Headrests	38,762	17.95
A 85	Custom Shoulder Belts (std with coupe)	600	42.15
C07	Auxiliary Hardtop (for convertible)	7,878	252.80
C08	Vinyl Covering (for auxiliary hardtop)	3,266	57.95
C50	Rear Window Defroster	2,485	32.65
C60	Air Conditioning	11,859	428.70
F41	Special Front and Rear Suspension	1,661	36.90
G81	Positraction Rear Axle, all ratios	36,965	46.35
J50	Power Brakes	16,876	42.15
J56	Special Heavy Duty Brakes	115	384.45
K05	Engine Block Heater	824	10.55
K66	Transistor Ignition System	5,702	81.10
L36	427ci, 390hp Engine	10,531	221.20
L46	350ci, 350hp Engine	12,846	131.65
L68	427ci, 400hp Engine	2,072	326.55
L71	427ci, 435hp Engine	2,722	437.10
L88	427ci, 430hp Engine	116	1,032.15
L89	Aluminum Cylinder Heads with L71	390	832.05
MA6	Heavy Duty Clutch	102	79.00
M20	4-Speed Manual Transmission	16,507	184.80
M21	4-Speed Man Trans, close ratio	13,741	184.80
M22	4-Speed Man Trans, close ratio, heavy duty	101	290.40
M40	Turbo Hydra-Matic Automatic Transmission	8,161	221.80
N14	Side Mount Exhaust System	4,355	147.45
N37	Tilt-Telescopic Steering Column	10,325	84.30
N40	Power Steering	22,866	105.35
P02	Deluxe Wheel Covers	8,073	57.95
PT6	Red Stripe Tires, F70x15, nylon	5,210	31.30
PT7	White Stripe Tires, F70x15, nylon	21,379	31.30
PU9	White Letter Tires, F70x15, nylon	2,398	33.15
TJ2	Front Fender Louver Trim	11,962	21.10
UA6	Alarm System	12,436	26.35
U15	Speed Warning Indicator	3,561	11.60
U69	AM-FM Radio	33,871	172.75
U79	AM-FM Radio, stereo	4,114	278.10
ZL1	Special L88 (aluminum block)	2	4,718.35

1970 Options

RPO #	DESCRIPTION	QTY	RETAIL $
19437	Base Corvette Sport Coupe	10,668	$5,192.00
19467	Base Corvette Convertible	6,648	4,849.00
—	Custom Interior Trim	3,191	158.00
A31	Power Windows	4,813	63.20

OPTIONS

CODE	DESCRIPTION	QTY	RETAIL $
A85	Custom Shoulder Belts (std with coupe)	475	42.15
C07	Auxiliary Hardtop (for convertible)	2,556	273.85
C08	Vinyl Covering (for auxiliary hardtop)	832	63.20
C50	Rear Window Defroster	1,281	36.90
C60	Air Conditioning	6,659	447.65
G81	Optional Rear Axle Ratio	2,862	12.65
J50	Power Brakes	8,984	47.40
L46	350ci, 350hp Engine	4,910	158.00
LS5	454ci, 390hp Engine	4,473	289.65
LT1	350ci, 370hp Engine	1,287	447.60
M21	4-Speed Man Trans, close ratio	4,383	0.00
M22	4-Speed Man Trans, close ratio, heavy duty	25	95.00
M40	Turbo Hydra-Matic Automatic Transmission	5,102	0.00
NA9	California Emissions	1,758	36.90
N37	Tilt-Telescopic Steering Column	5,803	84.30
N40	Power Steering	11,907	105.35
P02	Deluxe Wheel Covers	3,467	57.95
PT7	White Stripe Tires, F70x15, nylon	6,589	31.30
PU9	White Letter Tires, F70x15, nylon	7,985	33.15
T60	Heavy Duty Battery (std with LS5)	165	15.80
UA6	Alarm System	6,727	31.60
U69	AM-FM Radio	14,529	172.75
Y79	AM-FM Radio, stereo	2,462	278.10
ZR1	Special Purpose Engine Package	25	968.95

1971 Options

RPO #	DESCRIPTION	QTY	RETAIL $
19437	Base Corvette Sport Coupe	14,680	$5,496.00
19467	Base Corvette Convertible	7,121	5,259.00
—	Custom Interior Trim	2,602	158.00
A31	Power Windows	6,192	79.00
A85	Custom Shoulder Belts (std with coupe)	677	42.00
C07	Auxiliary Hardtop (for convertible)	2,619	274.00
C08	Vinyl Covering (for auxiliary hardtop)	832	63.00
C50	Rear Window Defroster	1,598	42.00
C60	Air Conditioning	11,481	459.00
ZQ1	Optional Rear Axle Ratio	2,395	13.00
J50	Power Brakes	13,558	47.00
LS5	454ci, 365hp Engine	5,097	295.00
LS6	454ci, 425hp Engine	188	1,221.00
LT1	350ci, 330hp Engine	1,949	483.00
M21	4-Speed Man Trans, close ratio	2,387	0.00
M22	4-Speed Man Trans, close ratio, heavy duty	130	100.00
M40	Turbo Hydra-Matic Automatic Transmission	10,060	0.00
N37	Tilt-Telescopic Steering Column	8,130	84.30
N40	Power Steering	17,904	115.90
P02	Deluxe Wheel Covers	3,007	63.00
PT7	White Stripe Tires, F70x15, nylon	6,711	28.00
PU9	White Letter Tires, F70x15, nylon	12,449	42.00
T60	Heavy Duty Battery (std with LS5, LS6)	1,455	15.80
UA6	Alarm System	8,501	31.60
U69	AM-FM Radio	18,078	178.00
U79	AM-FM Radio, stereo	3,431	283.00
ZR1	Special Purpose LT1 Engine Package	8	1,010.00
ZR2	Special Purpose LS6 Engine Package	12	1,747.00

1972 Options

RPO #	DESCRIPTION	QTY	RETAIL $
19437	Base Corvette Sport Coupe	20,496	$5,533.00
19467	Base Corvette Convertible	6,508	5,296.00
—	Custom Interior Trim	8,709	158.00
AV3	Three Point Seat Belts	17,693	—
A31	Power Windows	9,495	85.35
A85	Custom Shoulder Belts (std with coupe)	749	42.15
C07	Auxiliary Hardtop (for convertible)	2,646	273.85
C08	Vinyl Covering (for auxiliary hardtop)	811	158.00
C50	Rear Window Defroster	2,221	42.15
C60	Air Conditioning	17,011	464.50
ZQ1	Optional Rear Axle Ratio	1,986	12.65
J50	Power Brakes	18,770 4	7.40
K19	Air Injection Reactor	3,912	—
LS5	454ci, 270hp Engine (n/a California)	3,913	294.90
LT1	350ci, 255hp Engine	1,741	483.45
M21	4-Speed Manual Trans, close ratio	1,638	0.00
M40	Turbo Hydra-Matic Automatic Transmission	14,543	0.00
N37	Tilt-Telescopic Steering Column	12,992	84.30
N40	Power Steering	23,794	115.90
P02	Deluxe Wheel Covers	3,593	63.20
PT7	White Stripe Tires, F70x15, nylon	6,666	30.35
PU9	White Letter Tires, F70x15, nylon	16,623	43.65
T60	Heavy Duty Battery (std with LS5)	2,969	15.80
U69	AM-FM Radio	19,480	178.00
U79	AM-FM Radio, stereo	7,189	283.35
YF5	California Emission Test	1,967 1	5.80
ZR1	Special Purpose LT1 Engine Package	20	1,010.05

1973 Options

RPO #	DESCRIPTION	QTY	RETAIL $
1YZ37	Base Corvette Sport Coupe	25,521	$5,561.50
1YZ67	Base Corvette Convertible	4,943	5,398.50
—	Custom Interior Trim	13,434	154.00
A31	Power Windows	14,024	83.00
A85	Custom Shoulder Belts (std with coupe)	788	41.00
C07	Auxiliary Hardtop (for convertible)	1,328	267.00
C08	Vinyl Covering (for auxiliary hardtop)	323	62.00
C50	Rear Window Defroster	4,412	41.00
C60	Air Conditioning	21,578	452.00
—	Optional Rear Axle Ratio	1,791	12.00
J50	Power Brakes	24,168	46.00
LS4	454ci, 275hp Engine	4,412	250.00
L82	350ci, 250hp Engine	5,710	299.00
M21	4-Speed Manual Trans, close ratio	3,704	0.00
M40	Turbo Hydra-Matic Automatic Transmission	17,927	0.00
N37	Tilt-Telescopic Steering Column	17,949	82.00
1YZ37	Base Corvette Sport Coupe	25,521	5,561.50
1YZ67	Base Corvette Convertible	4,943	5,398.50
—	Custom Interior Trim	13,434	154.00
A31	Power Windows	14,024	83.00
A85	Custom Shoulder Belts (std with coupe)	788	41.00
C07	Auxiliary Hardtop (for convertible)	1,328	267.00
C08	Vinyl Covering (for auxiliary hardtop)	323	62.00
C50	Rear Window Defroster	4,412	41.00
C60	Air Conditioning	21,578	452.00
—	Optional Rear Axle Ratio	1,791	12.00
J50	Power Brakes	24,168	46.00
LS4	454ci, 275hp Engine	4,412	250.00
L82	350ci, 250hp Engine	5,710	299.00
M21	4-Speed Manual Trans, close ratio	3,704	0.00
M40	Turbo Hydra-Matic Automatic Transmission	17,927	0.00

APPENDIX

CODE	DESCRIPTION	QTY	RETAIL $
N37	Tilt-Telescopic Steering Column	17,949	82.00
N40	Power Steering	27,872	113.00
P02	Deluxe Wheel Covers	1,739	62.00
QRM	White Stripe Steel Belted Tires, GR70x15	19,903	32.00
QRZ	White Letter Steel Belted Tires, GR70x15	4,541	45.00
T60	Heavy Duty Battery (standard with LS4)	4,912	15.00
U58	AM-FM Radio, stereo	12,482	276.00
U69	AM-FM Radio	17,598	173.00
UF1	Map Light (on rearview mirror)	8,186	5.00
YF5	California Emission Test	3,008	15.00
YJ8	Cast Aluminum Wheels (5)	4	175.00
Z07	Off Road Suspension and Brake Package	45	369.00

1974 Options

RPO #	DESCRIPTION	QTY	RETAIL $
1YZ37	Base Corvette Sport Coupe	32,028	$6,001.50
1YZ67	Base Corvette Convertible	5,474	5,765.50
—	Custom Interior Trim	19,959	154.00
A31	Power Windows	23,940	86.00
A85	Custom Shoulder Belts (std with coupe)	618	41.00
C07	Auxiliary Hardtop (for convertible)	2,612	267.00
C08	Vinyl-Covered Auxiliary Hardtop	367	329.00
C50	Rear Window Defroster	9,322	43.00
C60	Air Conditioning	29,397	467.00
FE7	Gymkhana Suspension	1,905	7.00
—	Optional Rear Axle Ratios	1,219	12.00
J50	Power Brakes	33,306	49.00
LS4	454ci, 270hp Engine	3,494	250.00
L82	350ci, 250hp Engine	6,690	299.00
M21	4-Speed Manual Trans, close ratio	3,494	0.00
M40	Turbo Hydra-Matic Automatic Transmission	25,146	0.00
N37	Tilt-Telescopic Steering Column	27,700	82.00
N41	Power Steering	35,944	117.00
QRM	White Stripe Steel Belted Tires, GR70x15	9,140	32.00
QRZ	White Letter Steel Belted Tires, GR70x15	24,102	45.00
U05	Dual Horns	5,258	4.00
U58	AM-FM Radio, stereo	19,581	276.00
U69	AM-FM Radio	17,374	173.00
UA1	Heavy Duty Battery (std with LS4)	9,169	15.00
UF1	Map Light (on rearview mirror)	16,101	5.00
YF5	California Emission Test	—	20.00
Z07	Off Road Suspension and Brake Package	47	400.00

1975 Options

RPO #	DESCRIPTION	QTY	RETAIL $
1YZ37	Base Corvette Sport Coupe	33,836	$6,810.10
1YZ67	Base Corvette Convertible	4,629	6,550.10
—	Custom Interior Trim	—	154.00
A31	Power Windows	28,745	93.00
A85	Custom Shoulder Belts (std with coupe)	646	41.00
C07	Auxiliary Hardtop (for convertible)	2,407	267.00
C08	Vinyl Covered Auxiliary Hardtop (conv)	279	350.00
C50	Rear Window Defroster	13,760	46.00
C60	Air Conditioning	31,914	490.00
FE7	Gymkhana Suspension	3,194	7.00
—	Optional Rear Axle Ratios	1,969	12.00
J50	Power Brakes	35,842	50.00
L82	350ci, 205hp Engine	2,372	336.00
M21	4-Speed Manual Trans, close ratio	1,057	0.00

CODE	DESCRIPTION	QTY	RETAIL $
M40	Turbo Hydra-Matic Automatic Transmission	28,473	0.00
N37	Tilt-Telescopic Steering Column	31,830	82.00
N41	Power Steering	37,591	129.00
QRM	White Stripe Steel Belted Tires, GR70x15	5,233	35.00
QRZ	White Letter Steel Belted Tires, GR70x15	30,407	48.00
U05	Dual Horns	22,011	4.00
U58	AM-FM Radio, stereo	24,701	284.00
U69	AM-FM Radio	12,902	178.00
UA1	Heavy Duty Battery	16,778	15.00
UF1	Map Light (on rearview mirror)	21,676	5.00
YF5	California Emission Test	3,037	20.00
Z07	Off Road Suspension and Brake Package	144	400.00

1976 Options

RPO #	DESCRIPTION	QTY	RETAIL $
1YZ37	Base Corvette Sport Coupe	46,558	$7,604.85
—	Custom Interior Trim	36,762	164.00
A31	Power Windows	38,700	107.00
C49	Rear Window Defogger	24,960	78.00
C60	Air Conditioning	40,787	523.00
FE7	Gymkhana Suspension	5,368	35.00
—	Optional Rear Axle Ratios	1,371	13.00
J50	Power Brakes	46,558	59.00
L82	350ci, 210hp Engine	5,720	481.00
M21	4-Speed Manual Trans, close ratio	2,088	0.00
M40	Turbo Hydra-Matic Automatic Transmission	36,625	0.00
N37	Tilt-Telescopic Steering Column	41,797	95.00
N41	Power Steering	46,385	151.00
QRM	White Stripe Steel Belted Tires, GR70x15	3,992	37.00
QRZ	White Letter Steel Belted Tires, GR70x15	39,923	51.00
U58	AM-FM Radio, stereo	34,272	281.00
U69	AM-FM Radio	11,083	187.00
UA1	Heavy Duty Battery	25,909	16.00
UF1	Map Light (on rearview mirror)	35,361	10.00
YF5	California Emission Test	3,527	50.00
YJ8	Aluminum Wheels (4)	6,253	299.00

1977 Options

RPO #	DESCRIPTION	QTY	RETAIL $
1YZ37	Base Corvette Sport Coupe	49,213	$8,647.65
A31	Power Windows	44,341	116.00
B32	Color Keyed Floor Mats	36,763	22.00
C49	Rear Window Defogger	30,411	84.00
C60	Air Conditioning	45,249	553.00
D35	Sport Mirrors	20,206	36.00
FE7	Gymkhana Suspension	7,269	38.00
G95	Optional Rear Axle Ratios	972	14.00
K30	Speed Control	29,161	88.00
L82	350ci, 210hp Engine	6,148	495.00
M21	4-Speed Manual Trans, close ratio	2,060	0.00
M40	Turbo Hydra-Matic Automatic Transmission	41,231	0.00
NA6	High Altitude Emission Equipment	854	22.00
N37	Tilt-Telescopic Steering Column	46,487	165.00
QRZ	White Letter Steel Belted Tires, GR70x15	46,227	57.00
UA1	Heavy Duty Battery	32,882	17.00
U58	AM-FM Radio, stereo	8,483	281.00
U69	AM-FM Radio	4,700	187.00
UM2	AM-FM Radio, stereo with 8-track tape	24,603	414.00
V54	Luggage and Roof Panel Rack	16,860	73.00

OPTIONS

CODE	DESCRIPTION	QTY	RETAIL $
YF5	California Emission Certification	4,084	70.00
YJ8	Aluminum Wheels (4)	12,646	321.00
ZN1	Trailer Package	289	83.00
ZX2	Convenience Group	40,872	22.00

1978 Options

RPO #	DESCRIPTION	QTY	RETAIL $
1YZ87	Base Corvette Sport Coupe	40,274	$9,351.89
1YZ87/78	Limited Edition Corvette (pace car)	6,502	13,653.21
A31	Power Windows	36,931	130.00
AU3	Power Door Locks	12,187	120.00
B2Z	Silver Anniversary Paint	15,283	399.00
CC1	Removable Glass Roof Panels	972	349.00
C49	Rear Window Defogger	30,912	95.00
C60	Air Conditioning	37,638	605.00
D35	Sport Mirrors	38,405	40.00
FE7	Gymkhana Suspension	12,590	41.00
G95	Optional Rear Axle Ratio	382	15.00
K30	Cruise Control	31,608	99.00
L82	350ci, 220hp Engine	12,739	525.00
M21	4-Speed Manual Trans, close ratio	3,385	0.00
MX1	Automatic Transmission	38,614	0.00
NA6	High Altitude Emission Equipment	260	33.00
N37	Tilt-Telescopic Steering Column	37,858	175.00
QBS	White Letter SBR Tires, P255/60R15	18,296	216.32
QGR	White Letter SBR Tires, P225/70R15	26,203	51.00
UA1	Heavy Duty Battery	28,243	18.00
UM2	AM-FM Radio, stereo with 8-track tape	20,899	419.00
UP6	AM-FM Radio, stereo with CB	7,138	638.00
U58	AM-FM Radio, stereo	10,189	286.00
U69	AM-FM Radio	2,057	199.00
U75	Power Antenna	23,069	49.00
U81	Dual Rear Speakers	12,340	49.00
YF5	California Emission Certification	3,405	75.00
YJ8	Aluminum Wheels (4)	28,008	340.00
ZN1	Trailer Package	972	89.00
ZX2	Convenience Group	37,222	84.00

1979 Options

RPO #	DESCRIPTION	QTY	RETAIL $
1YZ87	Base Corvette Sport Coupe	53,807	$10,220.23
A31	Power Windows	20,631	141.00
AU3	Power Door Locks	9,054	131.00
CC1	Removable Glass Roof Panels	14,480	365.00
C49	Rear Window Defogger	41,587	102.00
C60	Air Conditioning	47,136	635.00
D35	Sport Mirrors	48,211	45.00
D80	Spoilers, front and rear	6,853	265.00
FE7	Gymkhana Suspension	12,321	49.00
F51	Heavy Duty Shock Absorbers	2,164	33.00
G95	Optional Rear Axle Ratio	428	19.00
K30	Cruise Control	34,445	113.00
L82	350ci, 225hp Engine	14,516	565.00
M21	4-Speed Manual Trans, close ratio	4,062	0.00
MX1	Automatic Transmission	41,454	0.00
NA6	High Altitude Emission Equipment	56	35.00
N37	Tilt-Telescopic Steering Column	47,463	190.00
N90	Aluminum Wheels (4)	33,741	380.00
QBS	White Letter SBR Tires, P255/60R15	17,920	226.20

CODE	DESCRIPTION	QTY	RETAIL $
QGR	White Letter SBR Tires, P225/70R15	29,603	54.00
U58	AM-FM Radio, stereo	9,256	90.00
UM2	AM-FM Radio, stereo with 8-track	21,435	228.00
UN3	AM-FM Radio, stereo with cassette	12,110	234.00
UP6	AM-FM Radio, stereo with CB	4,483	439.00
U75	Power Antenna	35,730	52.00
U81	Dual Rear Speakers	37,754	52.00
UA1	Heavy Duty Battery	3,405	21.00
YF5	California Emission Certification	3,798	83.00
ZN1	Trailer Package	1,001	98.00
ZQ2	Power Windows and Door Locks	28,465	272.00
ZX2	Convenience Group	41,530	94.00

1980 Options

RPO #	DESCRIPTION	QTY	RETAIL $
1YZ87	Base Corvette Sport Coupe	40,614	$13,140.24
AU3	Power Door Locks	32,692	140.00
CC1	Removable Glass Roof Panels	19,695	391.00
C49	Rear Window Defogger	36,589	109.00
FE7	Gymkhana Suspension	9,907	55.00
F51	Heavy Duty Shock Absorbers	1,695	35.00
K30	Cruise Control	30,821	123.00
LG4	305ci, 180hp Engine (required in California)	3,221	-50.00
L82	350ci, 230hp Engine	5,069	595.00
MM4	4-Speed Manual Transmission	5,726	0.00
MX1	Automatic Transmission	34,838	0.00
N90	Aluminum Wheels (4)	34,128	407.00
QGB	White Letter SBR Tires, P225/70R15	26,208	62.00
QXH	White Letter SBR Tires, P255/60R15	13,140	426.16
UA1	Heavy Duty Battery	1,337	22.00
U58	AM-FM Radio, stereo 6,	138	46.00
UM2	AM-FM Radio, stereo with 8-track	15,708	155.00
UN3	AM-FM Radio, stereo with cassette	15,148	168.00
UP6	AM-FM Radio, stereo with CB	2,434	391.00
U75	Power Antenna	32,863	56.00
UL5	Radio Delete	201	-126.00
U81	Dual Rear Speakers	36,650	52.00
V54	Roof Panel Carrier	3,755	125.00
YF5	California Emission Certification	3,221	250.00
ZN1	Trailer Package	796	105.00

1981 Options

RPO #	DESCRIPTION	QTY	RETAIL $
1YY87	Base Corvette Sport Coupe	40,606	$16,258.52
AU3	Power Door Locks	36,322	145.00
A42	Power Driver Seat	29,200	183.00
CC1	Removable Glass Roof Panels	29,095	414.00
C49	Rear Window Defogger	36,893	119.00
DG7	Electric Sport Mirrors	13,567	117.00
D84	Two-Tone Paint	5,352	399.00
FE7	Gymkhana Suspension	7,803	57.00
F51	Heavy Duty Shock Absorbers	1,128	37.00
G92	Performance Axle Ratio	2,400	20.00
K35	Cruise Control	32,522	155.00
MM4	4-Speed Manual Transmission	5,757	0.00
N90	Aluminum Wheels (4)	36,485	428.00
QGR	White Letter SBR Tires, P225/70R15	21,939	72.00
QXH	White Letter SBR Tires, P255/60R15	18,004	491.92
UL5	Radio Delete	315	-118.00

331

APPENDIX

CODE	DESCRIPTION	QTY	RETAIL $
UM4	AM-FM Radio, etr* stereo with 8-track	8,262	386.00
UM5	AM-FM Radio, etr* stereo with 8-track/CB	792	712.00
UM6	AM-FM Radio, etr* stereo with cassette	22,892	423.00
UN5	AM-FM Radio, etr* stereo with cassette/CB	2,349	750.00
	* electronic-tuned receiver		
U58	AM-FM Radio, stereo	5,145	95.00
U75	Power Antenna	32,903	55.00
V54	Roof Panel Carrier	3,303	135.00
YF5	California Emission Certification	4,951	46.00
ZN1	Trailer Package	916	110.00

1982 Options

RPO #	DESCRIPTION	QTY	RETAIL $
1YY87	Base Corvette Sport Coupe	18,648	$18,290.07
1YY07	Corvette Collector Edition Hatchback	6,759	22,537.59
AG9	Power Driver Seat	22,585	197.00
AU3	Power Door Locks	23,936	155.00
CC1	Removable Glass Roof Panels	14,763	443.00
C49	Rear Window Defogger	16,886	129.00
DG7	Electric Sport Mirrors	20,301	125.00
D84	Two-Tone Paint	4,871	428.00
FE7	Gymkhana Suspension	5,457	61.00
K35	Cruise Control	24,313	165.00
N90	Aluminum Wheels	16,844	458.00
QGR	White Letter SBR Tires, P225/70R15	5,932	80.00
QXH	White Letter SBR Tires, P255/60R15	19,070	542.52
UL5	Radio Delete	150	-124.00
UM4	AM-FM Radio, etr* stereo with 8-track	923	386.00
UM6	AM-FM Radio, etr* stereo with cassette	20,355	423.00
UN5	AM-FM Radio, etr* stereo with cassette/CB	1,987	755.00
	* electronic-tuned receiver		
U58	AM-FM Radio, stereo	1,533	101.00
U75	Power Antenna	15,557	60.00
V08	Heavy Duty Cooling	6,006	57.00
V54	Roof Panel Carrier	1,992	144.00
YF5	California Emission Certification	4,951	46.00

1984 Options

RPO #	DESCRIPTION	QTY	RETAIL $
1YY07	Base Corvette Sport Coupe	51,547	$21,800.00
AG9	Power Driver Seat	48,702	210.00
AQ9	Sport Seats, cloth	4,003	625.00
AR9	Base Seats, leather	40,568	400.00
AU3	Power Door Locks	49,545	165.00
CC3	Removable Transparent Roof Panel	15,767	595.00
D84	Two-Tone Paint	8,755	428.00
FG3	Delco-Bilstein Shock Absorbers	3,729	189.00
G92	Performance Axle Ratio	410	22.00
KC4	Engine Oil Cooler	4,295	158.00
K34	Cruise Control	49,832	185.00
MM4	4-Speed Manual Transmission	6,443	0.00
QZD	P255/50VR16 Tires/16" Wheels	51,547	561.20
UL5	Radio Delete	104	-331.00
UM6	AM-FM Stereo Cassette	6,689	153.00
UN8	AM-FM Stereo, Citizens Band	178	215.00
UU8	Stereo System, Delco-Bose	43,607	895.00
V01	Heavy-Duty Radiator	12,008	57.00
YF5	California Emission Requirements	6,833	75.00

CODE	DESCRIPTION	QTY	RETAIL $
Z51	Performance Handling Package	25,995	600.20
Z6A	Rear Window+Side Mirror Defoggers	47,680	160.00

1985 Options

RPO #	DESCRIPTION	QTY	RETAIL $
1YY07	Base Corvette Sport Coupe	39,729	$24,403.00
AG9	Power Driver Seat	37,856	215
AQ9	Sport Seats, leather	—	1,025
AR9	Base Seats, leather	—	400
—	Sport Seats, cloth	5,661	625
AU3	Power Door Locks	38,294	170
CC3	Removable Transparent Roof Panel	28,143	595
D84	Two-Tone Paint	6,033	428
FG3	Delco-Bilstein Shock Absorbers	9,333	189
G92	Performance Axle Ratio	5,447	22
K34	Cruise Control	38,369	185
MM4	4-Speed Manual Transmission	9,576	0
NN5	California Emission Requirements	6,583	99
UL5	Radio Delete	172	-256
UM6	AM-FM Stereo Cassette	2,958	122
UN8	AM-FM Stereo, Citizens Band	16	215
UU8	Stereo System, Delco-Bose	35,998	895
V08	Heavy-Duty Cooling	17,539	225
Z51	Performance Handling Package	14,802	470
Z6A	Rear Window+Side Mirror Defoggers	37,720	160

1986 Options

RPO #	DESCRIPTION	QTY	RETAIL $
1YY07	Base Corvette Sport Coupe	27,794	$27,027
1YY67	Base Corvette Convertible	7,315	32,032
AG9	Power Driver Seat	33,983	225
AQ9	Sport Seats, leather	13,372	1,025
AR9	Base Seats, leather	—	400
AU3	Power Door Locks	34,215	175
B4P	Radiator Boost Fan	8,216	75
B4Z	Custom Feature Package	4,832	195
C2L	Dual Removable Roof Panels	6,242	895
24S	Removable Roof Panel, blue tint	12,021	595
64S	Removable Roof Panel, bronze tint	7,819	595
C68	Electronic Air Conditioning Control	16,646	150
D84	Two-Tone Paint (coupe)	3,897	428
FG3	Delco-Bilstein Shock Absorbers	5,521	189
G92	Performance Axle Ratio, 3.07:1	4,879	22
KC4	Engine Oil Cooler	7,394	110
K34	Cruise Control	34,197	185
MM4	4-Speed Manual Transmission	6,835	0
NN5	California Emission Requirements	5,697	99
UL5	Radio Delete	166	-256
UM6	AM-FM Stereo Cassette	2,039	122
UU8	Stereo System, Delco-Bose	32,478	895
V01	Heavy-Duty Radiator	10,423	40
Z51	Performance Handling Package (coupe)	12,821	470
Z6A	Rear Window+Side Mirror Defog (coupe)	21,837	165
4001ZA	Malcolm Konner Special Edition (coupe)	50	500

1987 Options

RPO #	DESCRIPTION	QTY	RETAIL $
1YY07	Base Corvette Sport Coupe	20,007	$27,999
1YY67	Base Corvette Convertible	10,625	33,172

OPTIONS

CODE	DESCRIPTION	QTY	RETAIL $
AC1	Power Passenger Seat	17,123	240
AC3	Power Driver Seat	29,561	240
AQ9	Sport Seats, leather	14,119	1,025
AR9	Base Seats, leather	14,561	400
AU3	Power Door Locks	29,748	190
B2K	Callaway Twin Turbo (not GM installed)	188	19,995
B4P	Radiator Boost Fan	7,291	75
C2L	Dual Removable Roof Panels	5,017	915
24S	Removable Roof Panel, blue tint	8,883	615
64S	Removable Roof Panel, bronze tint	5,766	615
C68	Electronic Air Conditioning Control	20,875	150
DL8	Twin Remote Heated Mirrors (convertible)	6,840	35
D74	Illuminated Driver Vanity Mirror	14,992	58
D84	Two-Tone Paint (coupe)	1,361	428
FG3	Delco-Bilstein Shock Absorbers	1,957	189
G92	Performance Axle Ratio, 3.07:1	7,286	22
KC4	Engine Oil Cooler	6,679	110
K34	Cruise Control	29,594	185
MM4	4-Speed Manual Transmission	4,298	0
NN5	California Emission Requirements	5,423	99
UL5	Radio Delete	247	-256
UM6	AM-FM Stereo Cassette	2,236	132
UU8	Stereo System, Delco-Bose	27,721	905
V01	Heavy-Duty Radiator	7,871	40
Z51	Performance Handling Package (coupe)	1,596	795
Z52	Sport Handling Package	12,662	470
Z6A	Rear Window+Side Mirror Defog (coupe)	19,043	165

1988 Options

RPO #	DESCRIPTION	QTY	RETAIL $
1YY07	Base Corvette Sport Coupe	15,382	$29,489
1YY67	Base Corvette Convertible	7,407	34,820
AC1	Power Passenger Seat	18,779	240
AC3	Power Driver Seat	22,084	240
AQ9	Sport Seats, leather	12,724	1,025
AR9	Base Seats, leather	9,043	400
B2K	Callaway Twin Turbo (not GM installed)	125	25,895
B4P	Radiator Boost Fan	19,035	75
C2L	Dual Removable Roof Panels	5,091	915
24S	Removable Roof Panel, blue tint	8,332	615
64S	Removable Roof Panel, bronze tint	3,337	615
C68	Electronic Air Conditioning Control	19,372	150
DL8	Twin Remote Heated Mirrors (convertible)	6,582	35
D74	Illuminated Driver Vanity Mirror	14,249	58
FG3	Delco-Bilstein Shock Absorbers	18,437	189
G92	Performance Axle Ratio, 3.07:1	4,497	22
KC4	Engine Oil Cooler	18,877	110
MM4	4-Speed Manual Transmission	4,282	0
NN5	California Emission Requirements	3,882	99
UL5	Radio Delete	179	-297
UU8	Stereo System, Delco-Bose	20,304	773
V01	Heavy-Duty Radiator	19,271	40
Z01	35th Special Edition Package (coupe)	2,050	4,795
Z51	Performance Handling Package (coupe)	1,309	1,295
Z52	Sport Handling Package	16,017	970
Z6A	Rear Window+Side Mirror Defog (coupe)	14,648	165

1989 Options

RPO #	DESCRIPTION	QTY	RETAIL $
1YY07	Base Corvette Sport Coupe	16,663	$31,545
1YY67	Base Corvette Convertible	9,749	36,785
AC1	Power Passenger Seat	20,578	240
AC3	Power Driver Seat	25,606	240
AQ9	Sport Seats, leather	1,777	1,025
AR9	Base Seats, leather	23,364	400
B2K	Callaway Twin-Turbo (not GM installed)	67	25,895
B4P	Radiator Boost Fan	20,281	75
CC2	Auxiliary Hardtop (convertible)	1,573	1,995
C2L	Dual Removable Roof Panels	5,274	915
24S	Removable Roof Panel, blue tint	8,748	615
64S	Removable Roof Panel, bronze tint	4,042	615
C68	Electronic Air Conditioning Control	24,675	150
D74	Illuminated Driver Vanity Mirror	17,414	58
FX3	Selective Ride and Handling, electronic	1,573	1,695
G92	Performance Axle Ratio	10,211	22
K05	Engine Block Heater	2,182	20
KC4	Engine Oil Cooler	20,162	110
MN6	6-Speed Manual Transmission	4,113	0
MN5	California Emission Requirements	4,501	100
UJ6	Low Tire Pressure Warning Indicator	6,976	325
UU8	Stereo System, Delco-Bose	24,145	773
V01	Heavy-Duty Radiator	20,888	40
V56	Luggage Rack (convertible)	616	140
Z51	Performance Handling Package (coupe)	2,224	575

1990 Options

RPO #	DESCRIPTION	QTY	RETAIL $
1YY07	Base Corvette Sport Coupe	16,016	$31,979
1YY67	Base Corvette Convertible	7,630	37,264
AC1	Power Passenger Seat	20,419	270
AC3	Power Driver Seat	23,109	270
AQ9	Sport Seats, leather	11,457	1,050
AR9	Base Seats, leather	11,649	425
B2K	Callaway Twin-Turbo (not GM installed)	58	26,895
CC2	Auxiliary Hardtop (convertible)	2,371	1,995
C2L	Dual Removable Roof Panels	6,422	915
24S	Removable Roof Panel, blue tint	7,852	615
64S	Removable Roof Panel, bronze tint	4,340	615
C68	Electronic Air Conditioning Control	22,497	180
FX3	Selective Ride and Handling, electronic	7,576	1,695
G92	Performance Axle Ratio	9,362	22
K05	Engine Block Heater	1,585	20
KC4	Engine Oil Cooler	16,221	110
MN6	6-Speed Manual Transmission	8,101	0
MN5	California Emission Requirements	4,035	100
UJ6	Low Tire Pressure Warning Indicator	8,432	325
UU8	Stereo System, Delco-Bose	6,701	823
U1F	Stereo System with CD, Delco-Bose	15,716	1,219
V56	Luggage Rack (convertible)	1,284	140
Z51	Performance Handling Package (coupe)	5,446	460
ZR1	Special Performance Package (coupe)	3,049	27,016

1991 Options

RPO #	DESCRIPTION	QTY	RETAIL $
1YY07	Base Corvette Sport Coupe	14,967	$32,455
1YY67	Base Corvette Convertible	5,672	38,770
AR9	Base Seats, leather	9,505	425

333

APPENDIX

CODE	DESCRIPTION	QTY	RETAIL $
AQ9	Sport Seats, leather	10,650	1,050
AC1	Power Passenger Seat	17,267	290
AC3	Power Driver Seat	19,937	290
B2K	Callaway Twin-Turbo (not GM installed)	71	33,000
CC2	Auxiliary Hardtop (convertible)	1,230	1,995
C2L	Dual Removable Roof Panels	5,031	915
24S	Removable Roof Panel, blue tint	6,991	615
64S	Removable Roof Panel, bronze tint	3,036	615
C68	Electronic Air Conditioning Control	19,233	180
FX3	Selective Ride and Handling, electronic	6,894	1,695
G92	Performance Axle Ratio	3,453	22
KC4	Engine Oil Cooler	7,525	110
MN6	6-Speed Manual Transmission	5,875	0
MN5	California Emission Requirements	3,050	100
UJ6	Low Tire Pressure Warning Indicator	5,175	325
UU8	Stereo System, Delco-Bose	3,786	823
U1F	Stereo System with CD, Delco-Bose	15,345	1,219
V56	Luggage Rack (convertible)	886	140
Z07	Adjustable Suspension Package (coupe)	733	2,155
ZR1	Special Performance Package (coupe)	2,044	31,683

1992 Options

RPO #	DESCRIPTION	QTY	RETAIL $
1YY07	Base Corvette Sport Coupe	14,604	$33,635
1YY67	Base Corvette Convertible	5,875	40,145
AR9	Base Seats, leather	10,565	475
AR9	Base Seats, white leather	752	555
AQ9	Sport Seats, leather	7,973	1,100
AQ9	Sport Seats, white leather	709	1,180
AC1	Power Passenger Seat	16,179	305
AC3	Power Driver Seat	19,378	305
CC2	Auxiliary Hardtop (convertible)	915	1,995
C2L	Dual Removable Roof Panels	3,739	950
24S	Removable Roof Panel, blue tint	6,424	650
64S	Removable Roof Panel, bronze tint	3,005	650
C68	Electronic Air Conditioning Control	18,460	205
FX3	Selective Ride and Handling, electronic	5,840	1,695
G92	Performance Axle Ratio	2,283	50
MN6	6-Speed Manual Transmission	5,487	0
NN5	California Emission Requirements	3,092	100
UJ6	Low Tire Pressure Warning Indicator	3,416	325
UU8	Stereo System, Delco-Bose	3,241	823
U1F	Stereo System with CD, Delco-Bose	15,199	1,219
V56	Luggage Rack (for convertible)	845	140
Z07	Adjustable Suspension Package (coupe)	738	2,045
ZR1	Special Performance Package (coupe)	502	31,683

1993 Options

RPO #	DESCRIPTION	QTY	RETAIL $
1YY07	Base Corvette Sport Coupe	15,898	$34,595
1YY67	Base Corvette Convertible	5,692	41,195
AR9	Base Seats, leather	8,509	475
AR9	Base Seats, white leather	766	555
AQ9	Sport Seats, leather	11,267	1,100
AQ9	Sport Seats, white leather	622	1,180
AC1	Power Passenger Seat	18,067	305
AC3	Power Driver Seat	20,626	305
CC2	Auxiliary Hardtop (convertible)	976	1,995
C2L	Dual Removable Roof Panels	4,204	950

CODE	DESCRIPTION	QTY	RETAIL $
24S	Removable Roof Panel, blue tint	6,203	650
64S	Removable Roof Panel, bronze tint	4,288	650
C68	Electronic Air Conditioning Control	19,550	205
FX3	Selective Ride and Handling, electronic	5,740	1,695
G92	Performance Axle Ratio	2,630	50
MN6	6-Speed Manual Transmission	5,330	0
NN5	California Emission Requirements	2,101	100
UJ6	Low Tire Pressure Warning Indicator	3,353	325
UU8	Stereo System, Delco-Bose	2,685	823
U1F	Stereo System with CD, Delco-Bose	16,794	1,219
V56	Luggage Rack (for convertible)	765	140
Z07	Adjustable Suspension Package (coupe)	824	2,045
Z25	40th Anniversary Package	6,749	1,455
ZR1	Special Performance Package (coupe)	448	31,683

1994 Options

RPO #	DESCRIPTION	QTY	RETAIL $
1YY07	Base Corvette Sport Coupe	17,984	$36,185
1YY67	Base Corvette Convertible	5,346	42,960
AC1	Power Passenger Seat	17,863	305
AC3	Power Driver Seat	21,592	305
AQ9	Sport Seats	9,023	625
CC2	Auxiliary Hardtop (convertible)	682	1,995
C2L	Dual Removable Roof Panels	3,875	950
24S	Removable Roof Panel, blue tint	7,064	650
64S	Removable Roof Panel, bronze tint	3,979	650
FX3	Selective Ride and Handling, electronic	4,570	1,695
G92	Performance Axle Ratio	9,019	50
MN6	6-Speed Manual Transmission	6,012	0
NG1	New York Emission Requirements	1,363	100
UJ6	Low Tire Pressure Warning Indicator	5,097	325
U1F	Stereo System with CD, Delco-Bose	17,579	396
WY5	Tires, Extended Mobility	2,781	70
YF5	California Emission Requirements	2,372	100
Z07	Adjustable Suspension Package (coupe)	887	2,045
ZR1	Special Performance Package (coupe)	448	31,258

1995 Options

RPO #	DESCRIPTION	QTY	RETAIL $
1YY07	Base Corvette Sport Coupe	15,771	$36,785
1YY67	Base Corvette Convertible	4,971	43,665
AG1	Power Driver Seat	19,012	305
AG2	Power Passenger Seat	15,323	305
AQ9	Sport Seats	7,90	625
CC2	Auxiliary Hardtop (convertible)	459	1,995
C2L	Dual Removable Roof Panels	2,979	950
24S	Removable Roof Panel, blue tint	4,688	650
64S	Removable Roof Panel, bronze tint	2,871	650
FX3	Selective Ride and Handling, electronic	3,421	1,695
G92	Performance Axle Ratio	10,056	50
MN6	6-Speed Manual Transmission	4,784	0
NG1	New York Emission Requirements	268	100
N84	Spare Tire Delete	418	-100
UJ6	Low Tire Pressure Warning Indicator	5,300	325
U1F	Stereo System with CD, Delco-Bose	15,528	396
WY5	Tires, Extended Mobility	3,783	70
YF5	California Emission Requirements	2,026	100
Z07	Adjustable Suspension Package (coupe)	753	2,045

OPTIONS

CODE	DESCRIPTION	QTY	RETAIL $
Z4Z	Indy 500 Pace Car Replica (convertible)	527	2,816
ZR1	Special Performance Package (coupe)	448	31,258

1996 Options

RPO #	DESCRIPTION	QTY	RETAIL $
1YY07	Base Corvette Sport Coupe	17,167	$37,225
1YY67	Base Corvette Convertible	4,369	45,060
AG1	Power Driver Seat	19,798	305
AG2	Power Passenger Seat	17,060	305
AQ9	Sport Seats	12,016	625
CC2	Auxiliary Hardtop (convertible)	429	1,995
C2L	Dual Removable Roof Panels	3,983	950
24S	Removable Roof Panel, blue tint	6,626	650
64S	Removable Roof Panel, bronze tint	2,492	650
F45	Selective Real Time Damping, electronic	2,896	1,695
G92	Performance Axle Ratio	9,801	50
LT4	350ci, 330hp Engine	6,359	1,450
MN6	6-Speed Manual Transmission	6,359	0
N84	Spare Tire Delete	986	-100
UJ6	Low Tire Pressure Warning Indicator	6,865	325
U1F	Compact Disc Delco-Bose (reqs PEG 1*)	17,037	396
	* Preferred Equipment Group		
WY5	Tires, Extended Mobility	4,945	70
Z15	Collector Edition	5,412	1,250
Z16	Grand Sport Package ($2,880 w/convertible)	1,000	3,250
Z51	Performance Handling Package	1,869	350

1997 Options

RPO #	DESCRIPTION	QTY	RETAIL $
1YY07	Base Corvette Sport Coupe	9,752	$37,495
AAB	Memory Package	6,186	150
AG2	Power Passenger Seat	8,951	305
AQ9	Sport Seats	6,711	625
B34	Floor Mats	9,371	25
B84	Body Side Moldings	4,366	75
CC3	Removable Roof Panel, blue tint	7,213	650
C2L	Dual Removable Roof Panels	416	950
CJ2	Electronic Dual Zone Air Conditioning	7,999	365
D42	Luggage Shade and Parcel Net	8,315	50
F45	Selective Real Time Damping, electronic	3,094	1,695
G92	Performance Axle Ratio (Automatic only)	2,739	100
MN6	6-Speed Manual Transmission	2,809	815
NG1	Massachusetts/New York Emissions	677	170
T96	Fog Lamps	8,829	69
UN0	Delco Stereo System with CD	6,282	100
U1S	Remote Compact 12-Disc Changer	4,496	600
V49	Front License Plate Frame	2,258	15
YF5	California Emissions	885	170
Z51	Performance Handling Package	1,077	350

1998 Options

RPO #	DESCRIPTION	QTY	RETAIL $
1YY07	Base Corvette Sport Coupe	19,235	$37,495
1YY67	Base Corvette Convertible	11,849	44,425
AAB	Memory Package	24,234	150
AG2	Power Passenger Seat	28,575	305
AQ9	Sport Seats	22,675	625
B34	Floor Mats	30,592	25
B84	Body Side Moldings	17,070	75
C2L	Dual Removable Roof Panels	5,640	950
CC3	Removable Roof Panel, blue tint	6,957	650
CJ2	Dual Zone Air Conditioning	26,572	365
D42	Luggage Shade and Parcel Net (coupe)	16,549	50
F45	Selective Real Time Damping, electronic	8,374	1,695
G92	Performance Axle (3.15 ratio for automatic)	13,331	100
JL4	Active Handling System	5,356	500
MN6	6-Speed Manual Transmission	7,106	815
NG1	Massachusetts/New York Emissions	2,701	170
N73	Magnesium Wheels	1,425	3,000
T96	Fog Lamps	29,310	69
UN0	Delco Stereo System with CD	18,213	100
U1S	Remote Compact 12-Disc Changer	16,513	600
V49	Front License Plate Frame	6,087	15
YF5	California Emissions	3,111	170
Z4Z	Indy Pace Car Replica ($5,804 w/manual)	1,163	5,039
Z51	Performance Handling Package	4,249	350

1999 Options

RPO #	DESCRIPTION	QTY	RETAIL $
1YY07	Base Corvette Sport Coupe	18,078	$39,171
1YY37	Base Corvette Hardtop	4,031	38,777
1YY67	Base Corvette Convertible	11,161	45,579
AAB	Memory Package (coupe & conv)	23,829	150
AG1	Power Driver Seat (hardtop)	3,716	305
AG2	Power Passenger Seat (coupe & conv)	27,089	305
AQ9	Sport Seats (coupe & conv)	24,573	625
AP9	Parcel Net (hardtop)	2,738	15
B34	Floor Mats	32,706	25
B84	Body Side Moldings	19,348	75
C2L	Dual Removable Roof Panels	6,307	950
CC3	Removable Roof Panel, blue tint	5,235	650
CJ2	Dual Zone Air Conditioning	25,672	365
D42	Luggage Shade and Parcel Net (coupe)	18,058	50
F45	Selective Real Time Damping (coupe & conv)	7,515	1,695
G92	Performance Axle (3.15 ratio for automatic	14,525	100
JL4	Active Handling System	20,174	500
MN6	6-Speed Manual Trans (coupe & conv)	13,729	825
N37	Telescopic Steering, Power (coupe & conv)	16,847	350
N73	Magnesium Wheels	2,029	3,000
T82	Twilight Sentinel (coupe & conv)	18,895	60
T96	Fog Lamps	28,546	69
TR9	Lighting Package (hardtop only)	3,037	95
UN0	Delco Stereo System with CD	20,442	100
UV6	Head Up Instrument Display	19,034	375
UZ6	Bose Speaker Package (hardtop)	3,348	820
U1S	Remote Compact 12-Disc Changer	16,997	600
V49	Front License Plate Frame	17,742	15
YF5	California Emissions	3,336	170
Z51	Performance Handling Pkg (coupe & conv)	10,244	350
86U	Magnetic Red Metallic Paint (coupe & conv)	2,733	500

2000 Options

RPO #	DESCRIPTION	QTY	RETAIL $
1YY07	Base Corvette Sport Coupe	18,113	$39,475
1YY37	Base Corvette Hardtop	2,090	38,900
1YY67	Base Corvette Convertible	13,479	45,900
AAB	Memory Package (coupe & conv)	26,595	150

335

APPENDIX

CODE	DESCRIPTION	QTY	RETAIL $
AG1	Power Driver Seat (hardtop)	1,841	305
AG2	Power Passenger Seat (coupe & conv)	29,462	305
AQ9	Sport Seats (coupe & conv)	27,103	700
AP9	Parcel Net (hardtop)	938	15
B34	Floor Mats	33,188	25
B84	Body Side Moldings	18,773	75
C2L	Dual Removable Roof Panels	6,280	1,100
CC3	Removable Roof Panel, blue tint	5,605	650
CJ2	Dual Zone Air Conditioning	29,428	365
D42	Luggage Shade and Parcel Net (coupe)	15,689	50
F45	Selective Real Time Damping (cpe & conv)	6,724	1,695
G92	Performance Axle (3.15 ratio for automatic)	14,090	100
JL4	Active Handling System	22,668	500
MN6	6-Speed Manual Trans (cpe & conv)	13,320	815
N37	Telescopic Steering, Power (cpe & conv)	22,182	350
N73	Magnesium Wheels	2,652	2,000
QF5	Polished Aluminum Wheels	15,204	895
T82	Twilight Sentinel (coupe & conv)	23,508	60
T96	Fog Lamps	31,992	69
TR9	Lighting Package (hardtop)	1,527	95
UN0	Delco Stereo System with CD	24,696	100
UV6	Head Up Instrument Display	26,482	375
UZ6	Bose Speaker Package (hardtop)	1,766	820
U1S	Remote Compact 12-Disc Changer	15,809	600
V49	Front License Plate Frame	17,380	15
YF5	California Emissions	3,628	0
Z51	Performance Handling (std w/hardtop)	7,775	350
79U	Millennium Yellow (coupe & conv)	3,578	500
86U	Magnetic Red Metallic Paint (cpe & conv)	2,941	500

2001 Options

RPO #	DESCRIPTION	QTY	RETAIL $
1YY07	Base Corvette Sport Coupe	15,681	$40,475
1YY37	Base Corvette Z06 Hardtop	5,773	47,500
1YY67	Base Corvette Convertible	14,173	47,000
1SB	Preferred Equipment Group, Sport Coupe	2,514	1,639
1SB	Preferred Equipment Group, Convertible	1,710	1,769
1SC	Preferred Equipment Group, Sport Coupe	11,558	2,544
1SC	Preferred Equipment Group, Convertible	11,881	2,494
AAB	Memory Package (Z06)	4,780	150
B34	Floor Mats	34,907	25
B84	Body Side Moldings	20,457	75
C2L	Dual Removable Roof Panels	5,099	1,100
CC3	Removable Roof Panel, blue tint	4,769	650
DD0	Electrochromic Mirrors (Z06)	4,576	120
F45	Selective Real Time Damping (cpe & conv)	5,620	1,695
G92	3.15:1 Perf. Axle (automatic cpe & conv)	12,882	300
MN6	6-Speed Manual Trans (cpe & conv)	16,019	815
N73	Magnesium Wheels (cpe & conv)	1,022	2,000
QF5	Polished Aluminum Wheels (cpe & conv)	22,980	895
R8C	Corvette Museum Delivery	457	490
UL0	Delco Stereo Cassette (replaces std radio)	6,844	-100
UN0	Delco Stereo System with CD	28,783	100
U1S	Remote 12-Disc Changer (cpe & conv)	14,198	600
V49	Front License Plate Frame	18,935	15
Z51	Performance Handling Pkg (cpe & conv)	7,817	350
79U	Millennium Yellow w/tint coat	3,887	600
86U	Magnetic Red Metallic Paint (cpe & conv)	3,322	600

2002 Options

RPO #	DESCRIPTION	QTY	RETAIL $
1YY07	Base Corvette Sport Coupe	14,760	$41,450
1YY37	Base Corvette Z06 Hardtop	8,297	50,150
1YY67	Base Corvette Convertible	12,710	47,975
1SB	Preferred Equipment Group, Sport Coupe	1,359	1,700
1SB	Preferred Equipment Group, Convertible	1,379	1,800
1SC	Preferred Equipment Group, Sport Coupe	11,136	2,700
1SC	Preferred Equipment Group, Convertible	10,964	2,600
AAB	Memory Package (Z06)	7,794	150
B84	Body Side Moldings	21,422	75
C2L	Dual Removable Roof Panels	5,079	1,200
CC3	Removable Roof Panel, blue tint	4,208	750
DD0	Electrochromic Mirrors (Z06)	7,394	120
F45	Selective Real Time Damping (cpe & conv)	4,773	1,695
G92	3.15:1 Perf. Axle (automatic, cpe & conv)	9,646	300
MN6	6-Speed Manual Transmission (cpe & conv)	8,553	815
N73	Magnesium Wheels (cpe & conv)	114	1,500
QF5	Polished Aluminum Wheels (cpe & conv)	22,597	1,200
R8C	Corvette Museum Delivery	371	490
UL0	Delco Stereo Cassette (cpe & conv)	4,210	-100
U1S	Remote 12-Disc Changer (cpe & conv)	13,725	600
V49	Front License Plate Frame	19,948	15
Z51	Performance Handling Pkg (cpe & conv)	6,106	350
79U	Millennium Yellow w/tint coat	4,040	600
86U	Magnetic Red Metallic Paint (cpe & conv)	3,298	600

2003 Options

RPO #	DESCRIPTION	QTY	RETAIL $
1YY07	Base Corvette Sport Coupe	12,812	$43,895
1YY37	Base Corvette Z06 Hardtop	8,635	51,155
1YY67	Base Corvette Convertible	14,022	50,370
1SB	Preferred Equipment Group, Sport Group	7,310	1,200
1SB	Preferred Equipment Group, Convertible	6,643	1,200
1SC	50th Anniversary Edition Sport Coupe	4,085	5,000
1SC	50th Anniversary Edition Convertible	7,547	5,000
AAB	Memory Package (Z06)	8,241	175
B84	Body Side Moldings	22,243	150
C2L	Dual Removable Roof Panels	5,184	1,200
CC3	Removable Roof Panel, blue tint	3,150	750
DD0	Electrochromic Mirrors (Z06)	8,227	120
F55	Magnetic Selective Ride Control (cpe/conv)	14,992	1,695
G92	3.15:1 Perf. Axle (automatic cpe & conv)	9,785	395
MN6	6-Speed Manual Transmission (cpe/conv)	8,590	915
N73	Magnesium Wheels (cpe & conv)	293	1,500
QF5	Polished Aluminum Wheels (cpe & conv)	10,290	1,295
R8C	Corvette Museum Delivery	787	490
UL0	Delco Stereo Cassette (cpe & conv)	4,664	0
U1S	Remote 12-Disc Changer (cpe & conv)	14,979	600
V49	Front License Plate Frame	20,605	15
Z51	Performance Handling Package (cpe/conv)	2,592	395
79U	Millennium Yellow w/tint coat	3,900	750

2004 Options

RPO #	DESCRIPTION	QTY	RETAIL $
1YY07	Base Corvette Sport Coupe	16,165	$44,535
1YY37	Base Corvette Z06 Hardtop	5,683	52,385
1YY67	Base Corvette Convertible	12,216	51,535
1SB	Preferred Equipment Group, Sport Coupe	11,446	1,200

OPTIONS

CODE	DESCRIPTION	QTY	RETAIL $
1SB	Preferred Equipment Group, Convertible	9,334	1,200
1SB	Commemorative Edition (Z06)	2,025	4,335
1SC	Commemorative Edition Sport Coupe	2,215	3,700
1SC	Commemorative Edition Convertible	2,659	3,700
AAB	Memory Package (Z06)	5,446	175
B84	Body Side Moldings	20,626	150
C2L	Dual Removable Roof Panels	5,079	1,400
CC3	Removable Roof Panel, blue tint	4,356	750
DD0	Auto-dimming Mirrors (Z06)	5,446	160
F55	Magnetic Selective Ride Control (cpe/conv)	5,843	1,695
G92	3.15:1 Perf. Axle (automatic cpe & conv)	10,367	395
MN6	6-Speed Manual Transmission (cpe/conv)	6,928	915
N73	Magnesium Wheels (cpe & conv)	1,110	995
QF5	Polished Aluminum Wheels (cpe & conv)	22,487	1,295
R8C	Corvette Museum Delivery	142	490
UL0	Delco Stereo Cassette (cpe & conv)	3,860	0
U1S	Remote 12-Disc Changer (cpe & conv)	14,668	600
V49	Front License Plate Frame	19,520	15
Z51	Performance Handling Package (cpe/conv)	3,672	395
79U	Millennium Yellow w/tint coat	2,641	750
86U	Magnetic Red II	3,596	750

2005 Options

RPO #	DESCRIPTION	QTY	RETAIL $
1YY07	Base Corvette Coupe	26,728	$44,245
1YY67	Base Corvette Convertible	10,644	52,245
1SA	Preferred Equipment Group, Coupe	3,763	1,405
1SB	Preferred Equipment Group, Coupe	22,319	4,360
1SB	Preferred Equipment Group, Convertible	10,306	2,955
C2L	Dual Removable Roof Panels	2,585	1,400
CC3	Removable Roof Panel, transparent	8,469	750
CM7	Power Convertible Top	7,541	1,995
F55	Magnetic Selective Ride Control	9,041	1,695
G90	3.15:1 Perf. Axle (w/automatic trans)	15,112	395
MX0	4-Speed Automatic Transmission	22,380	0
QG7	Polished Aluminum Wheels	27,080	1,295
R8C	Corvette Museum Delivery	831	490
UE1	OnStar System	19,634	695
U2K	XM Satellite Radio	21,896	325
U3U	Bose Premium AM/FM CD w/Navigation	4,676	1,400
Z51	Performance Package	15,345	1,495
45U	Velocity Yellow exterior Paint (late)	760	750
19U	LeMans Blue exterior paint	3,759	300
79U	Millennium Yellow exterior paint	2,002	750
80U	Monterey Red exterior paint (late)	717	750
86U	Magnetic Red exterior paint	3,409	750

2006 Options

RPO #	DESCRIPTION	QTY	RETAIL $
1YY07	Base Corvette Coupe	16,598	$44,600
1YY67	Base Corvette Convertible	11,151	52,335
1YY87	Z06 Coupe	6,272	65,800
2LT	Equipment Group, Coupe	1,904	1,495
2LZ	Preferred Equipment Group (Z06)	4,854	2,900
3LT	Preferred Equipment Group, Coupe	10,953	4,795
3LT	Preferred Equipment Group, Convertible	9,972	3,395
C2L	Dual Removable Roof Panels	3,726	1,400
CC3	Removable Roof Panel, transparent	3,561	750
CM7	Power Convertible Top	8,537	1,995
F55	Magnetic Selective Ride Control (cpe/conv)	5,709	1,695
MX0	6-Spd Paddle Shift Auto Trans (cpe/conv)	19,094	1,250
QG7	Polished Aluminum Wheels (cpe/conv)	16,133	1,295
QL9	Polished Z06 Aluminum Wheels	3,449	1,295
QX1	Competition Gray Alum. Wheels (cpe/conv)	279	295
QX3	Chrome Aluminum Wheels (cpe/conv)	2,803	1,995
R8C	Corvette Museum Delivery	1,172	490
UE1	OnStar System	12,869	695
U3U	AM/FM CD w/Navigation ($1,600 w/2LZ)	17,474	3,340
US9	AM/FM 6-CD, Bose Speakers, XM Radio	10,690	1,740
Z51	Performance Package (cpe/conv)	10,338	1,895
19U	LeMans Blue exterior paint	3,459	300
45U	Velocity Yellow exterior paint	4,122	750
80U	Monterey Red exterior paint	5,052	750

2007 Options

RPO #	DESCRIPTION	QTY	RETAIL $
1YY07	Base Corvette Coupe	21,484	$44,995
1YY67	Base Corvette Convertible	10,418	52,910
1YY67	Indy 500 Pace Car Convertible (Z4Z)	500	66,995
1YY87	Z06 Coupe (increased to $70,000 7-26-06)	7,760	66,465
1YY87	Ron Fellows Z06 Special Edition (Z33)	399	77,500
2LT	Equipment Group, Coupe	2,606	1,495
2LZ	Equipment Group (Z06)	6,487	3,485
3LT	Equipment Group, Coupe	11,934	4,945
3LT	Equipment Group, Convertible	9,533	5,540
C2L	Dual Removable Roof Panels	3,558	1,400
CC3	Removable Roof Panel, transparent	4,370	750
D30	Non-recommended color/trim/top combo	126	590
F55	Magnetic Selective Ride Control (cpe/conv)	5,619	1,995
MX0	6-Spd Paddle Shift Auto Trans (cpe/conv)	24,422	1,250
Q44	Competition Gray Aluminum Wheels (Z06)	545	395
QG7	Polished Aluminum Wheels (cpe/conv)	3,461	1,295
QL9	Polished Aluminum Wheels (Z06)	3,459	1,495
QX1	Competition Gray Alum. Wheels (cpe/conv)	1,091	395
QX3	Chrome Aluminum Wheels (cpe/conv)	19,850	1,850
R8C	Corvette Museum Delivery	1,104	490
UE1	OnStar System (3LT or 2LZ required)	18,074	695
U3U	AM/FM, CD Nav, XM, Bose ($3,640 1LZ)	20,653	1,750
US9	AM/FM 6-CD, XM, Bose (w/1LZ)	9,970	1,890
Z51	Performance Package (coupe/conv)	13,696	1,695
19U	LeMans Blue exterior paint	3,854	300
45U	Velocity Yellow exterior paint	3,755	750
80U	Monterey Red exterior paint	5,023	750
83U	Atomic Orange exterior paint	3,790	750
**6	Two-Tone Seats w/Embroidery (cpe/conv)	—	695

2008 Options

RPO #	DESCRIPTION	QTY	RETAIL $
1YY07	Base Corvette Coupe	19,796	$45,995
1YY67	Base Corvette Convertible	7,283	54,335
1YY07	Indy 500 Pace Car Coupe (Z4Z)	234	59,090
1YY67	Indy 500 Pace Car Convertible (Z4Z)	266	68,160
1YY87	Z06 Coupe	7,226	71,000
1YY87	427 Limited Edition Z06 (Z44)	505	84,195
2LT	Equipment Group, Coupe	3,086	1,495
2LZ	Equipment Group (Z06)	4,929	3,045
3LT	Equipment Group, Coupe	9,201	4,505
3LT	Equipment Group, Convertible	5,879	5,100

APPENDIX

CODE	DESCRIPTION	QTY	RETAIL $
3LZ	Equipment Group (Z06)	1,460	6,545
4LT	Equipment Group, Convertible	528	8,600
4LT	Equipment Group, Coupe	834	8,005
C2L	Dual Removable Roof Panels	2,773	1,400
CC3	Removable Roof Panel, transparent	3,251	750
D30	Non-recommended color/trim/top combo	189	590
F55	Magnetic Selective Ride Control (cpe/conv)	4,666	1,995
GU2	2.73:1 Axle Ratio (w/automatic trans)	8,839	395
MX0	6-Spd Paddle Shift Auto Trans (cpe/conv)	19,136	1,250
NPP	Dual Mode Exhaust System (cpe/conv)	13,454	1,195
Q44	Competition Gray Aluminum Wheels (Z06)	395	395
Q76	Chrome Aluminum Wheels (Z06)	5,101	1,995
Q9V	Chrome Forged-Alum. Wheels (cpe/conv)	2,932	1,850
QL9	Polished Forged-Alum. Wheels (cpe/conv)	5,412	1,295
QX1	Competition Gray Alum. Wheels (cpe/conv)	1,781	395
QX3	Original Design Chrome Wheels (cpe/conv)	9,626	1,850
R8C	Corvette Museum Delivery	954	490
U3U	AM/FM, CD, Navigation, Bose	16,807	1,750
Z51	Performance Package (coupe/conv)	10,706	1,695
U3U	AM/FM, CD, Navigation, Bose	16,807	1,750
Z51	Performance Package (coupe/conv)	10,706	1,695
45U	Velocity Yellow exterior paint	3,264	750
83U	Atomic Orange exterior paint	2,246	300
89U	Crystal Red exterior paint	5,420	750
**6	Modified Two-Tone Seats (2LT & 3LT)	5,807	695

2009 Options

RPO #	DESCRIPTION	QTY	RETAIL $
1YY07	Base Corvette Coupe	8,632	$47,895
1YY67	Base Corvette Convertible	3,326	52,550
1YY07	Competition Sport Special Edition	52	55,765
1YY07	GT1 Championship Edition Coupe	53	65,410
1YY67	GT1 Championship Edition Convertible	17	71,915
1YY87	Z06 Coupe	3,386	73,255
1YY87	Competition Sport Special Edition Z06	20	77,600
1YY87	GT1 Championship Special Edition Z06	55	86,486
1YY87	ZR1 Coupe	1,415	103,300
2LT	Equipment Group, Coupe	1,193	1,545
2LT	Equipment Group, Convertible	686	3,540
2LZ	Equipment Group (Z06)	1,654	3,015
3LT	Equipment Group, Coupe	3,256	4,555
3LT	Equipment Group, Convertible	1,974	6,550
3LZ	Equipment Group (Z06)	1,163	6,515
3ZR	Equipment Group (ZR1)	1,202	10,000
4LT	Equipment Group, Convertible	449	10,050
4LT	Equipment Group, Coupe	307	8,055
C2L	Dual Removable Roof Panels	1,051	1,400
CC3	Removable Roof Panel, transparent	1,417	750
D30	Non-recommended color/trim/top combo	71	590
F55	Magnetic Selective Ride Control (cpe/conv)	2,105	1,995
GU2	Rear Axle 2.73 Ratio (w/auto trans)	3,971	395
MX0	6-Spd Paddle Shift Auto Trans (cpe/conv)	8,560	1,250
NPP	Dual Mode Exhaust System (coupe/conv)	6,238	1,195
Q44	Competition Gray Aluminum Wheels (Z06)	372	395
Q6B	Chrome 20-spoke Aluminum Wheels (ZR1)	1,132	2,000
Q76	Chrome Aluminum Wheels (Z06)	795	1,995
Q8A	Spider Chrome Aluminum Wheels (Z06)	1,903	1,995
Q9V	Chrome Forged-Alum. Wheels (cpe/conv)	5,649	1,850
QG7	Polished Forged-Alum. Wheels (cpe/conv)	1,989	1,295
QL9	Polished Aluminum Wheels (Z06)	13	1,495

CODE	DESCRIPTION	QTY	RETAIL $
QX1	Competition Gray Alum. Wheels (cpe/conv)	1,028	395
R8C	Corvette Museum Delivery	392	490
R8E	Gas Guzzler Tax (ZR1)	1,415	1,700
U3U	AM/FM, CD, Navigation, Bose	7,903	1,750
VPK	Exterior Appearance Package	67	2,095
VPL	Exterior Appearance Package w/wheels	433	5,475
Z51	Performance Package (coupe/conv)	4,515	1,695
45U	Velocity Yellow exterior paint	1,370	750
83U	Atomic Orange exterior paint	726	300
85U	Jetstream Blue exterior paint	1,367	750
89U	Crystal Red exterior paint	1,987	750
**6	Modified Two-Tone Seats w/embroidery	1,543	695

2010 Options

RPO #	DESCRIPTION	QTY	RETAIL $
1YY07	Base Corvette Coupe	3,054	$49,880
1YY67	Base Corvette Convertible	1,003	54,530
1YG07	Grand Sport Coupe	3,707	55,720
1YG67	Grand Sport Convertible	2,335	59,530
1YY87	Z06 Coupe	518	75,235
1YY87	ZR1 Coupe	1,577	107,830
2LT	Equipment Group, Coupe	831	1,195
2LT	Equipment Group, Convertible	214	3,190
2LZ	Equipment Group (Z06)	193	2,665
3LT	Equipment Group, Coupe	3,458	4,205
3LT	Equipment Group, Convertible	2,304	6,200
3LZ	Equipment Group (Z06)	193 7,	170
3ZR	Equipment Group (ZR1)	1,426	10,000
4LT	Equipment Group, Coupe	431	7,705
4LT	Equipment Group, Convertible	541	9,700
C2L	Dual Removable Roof Panels	1,050	1,400
CC3	Removable Roof Panel, transparent	1,336	750
D30	Non-recommended color/trim/top combo	86	590
F55	Magnetic Selective Ride Control (cpe/conv)	2,334	1,995
MX0	6-Spd Paddle Shift Auto Trans (cpe/conv)	6,913	1,250
NPP	Dual Mode Exhaust System (cpe/conv)	6,268	1,195
OB1	Crossed Flag Headrest Embroidery	1,679	300
PYD	Competition Gray Alum. Wheels (GS)	534	395
PYE	Chrome Aluminum Wheels (GS)	4,693	1,995
Q44	Competition Gray Aluminum Wheels (Z06)	84	395
Q6B	Chrome 20-spoke Aluminum Wheels (ZR1)	1,347	2,000
Q6J	Competition Gray Aluminum Wheels (ZR1)	123	395
Q76	Chrome Aluminum Wheels (Z06)	82	1,995
Q8A	Spider Chrome Aluminum Wheels (Z06)	286	1,995
Q9V	Chrome Forged-Alum. Wheels (cpe/conv)	2,617	1,850
QX1	Competition Gray Alum. Wheels (cpe/conv)	183	395
R8C	Corvette Museum Delivery	307	490
R8E	Gas Guzzler Tax (ZR1)	1,557	1,300
U3U	AM/FM, CD, Navigation, Bose	6,626	1,750
US9	AM/FM 6-CD	1,905	395
VK3	Front License Plate Bracket	7,952	15.00
VPK	Exterior Appearance Package	10	2,110
VPL	Exterior Appearance Package w/wheels	11	1,195
Z15	Grand Sport Heritage Package	1,531	1,195
45U	Velocity Yellow exterior paint	694	850
85U	Jetstream Blue exterior paint	969	850
89U	Crystal Red exterior paint	1,714	850
**6	Modified Two-Tone Seats w/embroidery	—	695

OPTIONS

2011 Options

RPO #	DESCRIPTION	QTY	RETAIL $
1YY07	Base Corvette Coupe	3,112	$49,900
1YY67	Base Corvette Convertible	780	54,550
1YG07	Grand Sport Coupe	5,212	55,740
1YG67	Grand Sport Convertible	2,782	59,950
1YY87	Z06 Coupe	904	75,255
1YY87	Z06 Carbon Ltd. Edition ($98,130 w/3LZ)	252	90,960
1YY87	ZR1 Coupe	806	110,750
2LT	Equipment Group, Coupe	1,060	1,195
2LT	Equipment Group, Convertible	251	3,190
2LZ	Equipment Group (Z06)	191	2,665
3LT	Equipment Group, Coupe	4,351	4,205
3LT	Equipment Group, Convertible	2,491	6,200
3LZ	Equipment Group (Z06)	523	7,170
3ZR	Equipment Group (ZR1)	666	10,000
4LT	Equipment Group, Coupe	385	7,705
4LT	Equipment Group, Convertible	513	9,700
C2L	Dual Removable Roof Panels (cpe)	1,116	1,400
CC3	Removable Roof Panel, transparent (cpe)	1,509	750
CFZ	Carbon Fiber Package (Z06)	535	3,995
D30	Non-recommended color/trim/top combo	58	590
ER1	Battery Protection Package	1,487	100
F55	Magnetic Selective Ride ($1,695 w/GS)	3,892	1,995
H33	Gray Headlamps (H34=Silver, H35=Black)	842	590
J55	Cross Drilled Brake Rotors (coupe/conv)	815	500
MX0	6-Spd Paddle Shift Auto Trans (cpe/conv)	8,516	1,250
NPP	Dual Mode Exhaust System (coupe/conv)	7,365	1,195
OB1	Crossed Flag Headrest Embroidery	1,837	300
PBC	Engine Plant Build Experience (Z06 & ZR1)	16	5,800
PIN	Customer-selectable VIN	4	5,000
PYD	Competition Gray Aluminum Wheels (GS)	855	395
PYE	Chrome Aluminum Wheels (Grand Sport)	6,036	1,995
Q44	Competition Gray Aluminum Wheels (Z06)	76	395
Q6B	Chrome 20-spoke Aluminum Wheels (ZR1)	649	2,000
Q6J	Competition Gray Aluminum Wheels (ZR1)	390	395
Q8A	Spider Chrome Aluminum Wheels (Z06)	217	1,995
QX1	Competition Gray Alum. Wheels (cpe/conv)	156	395
QX3	Chrome Aluminum Wheels (cpe/conv)	2,113	1,850
RQ1	Machine Faced Alum. Wheels (cpe/conv)	277	895
R8C	Corvette Museum Delivery	271	490
R8E	Gas Guzzler Tax (ZR1)	698	1,300
U3U	AM/FM, CD, Navigation, Bose	6,115	1,795
US9	AM/FM 6-CD	1,503	395
VK3	Front License Plate Bracket	8,362	15
Z07	Z06 Ultimate Performance Package	540	9,495
Z15	Grand Sport Heritage Package	2,059	1,195
36S	Custom Stitch, Yellow (37S=Blue, 38S=Red)	267	395
28U	Inferno Orange exterior paint	790	300
45U	Velocity Yellow exterior paint	867	850
85U	Jetstream Blue exterior paint	581	850
89U	Crystal Red exterior paint	1,419	850
GLB	Supersonic Blue exterior paint	1,417	300
**6	Modified Two-Tone Seats w/embroidery	2,274	695

2012 Options

RPO #	DESCRIPTION	QTY	RETAIL $
1YY07	Corvette Coupe	2,820	$50,500
1YY67	Corvette Convertible	621	55,500
1YG07	Grand Sport Coupe	5,056	56,900
1YG67	Grand Sport Convertible	2,268	60,500
1YY87	Z06 Coupe	478	76,500
1YY87	ZR1 Coupe	404	112,500
2LT	Equipment Group, Coupe	1,378	2,095
2LT	Equipment Group, Convertible	313	2,095
2LZ	Equipment Group (Z06)	100	4,310
3LT	Equipment Group, Coupe	3,692	5,995
3LT	Equipment Group, Convertible	1,818	7,995
3LZ	Equipment Group (Z06)	274	8,815
3ZR	Equipment Group (ZR1)	1,313	10,000
4LT	Equipment Group, Coupe	414	9,495
4LT	Equipment Group, Convertible	382	11,495
BA5	Black Grand Sport Fender Badges	2,293	100
B2K	Callaway 25th Anniversary Edition	26*	52,980
	* 15 coupes, 11 convertibles		
B92	Carbon Fiber Hood (Z06)	119	2,495
C2L	Dual Removable Roof Panels	1,050	1,400
CC3	Removable Roof Panel, transparent	1,048	750
CFZ	Z06 Carbon Fiber Package	196	3,995
D30	Color Combination Override	58	590
ER1	Battery Protection Package	3,763	100
F55	Magnetic Selective Ride Control (cpe/conv)	4,478	1,995*
	* priced at $1,695 for Grand Sport, $2,495 for Z06		
H33	Cyber Gray Metallic Headlamps*	548	590
	* H34=Blade Silver Metallic, H35=Black		
J55	Cross Drilled Brake Rotors	981	500
J6C	Silver Brake Calipers*	8,769	595
	* J6D=Dark Gray, J6E=Yellow, J6F=Red		
MYC	6-Spd Paddle Shift Automatic Trans	7,586	1,250
NPP	Dual Mode Exhaust (coupe/convertible)	6,664	1,195
PBC	Engine Plant Build Experience (Z06/ZR1)	12	5,800
PDE	ZR1 Performance Package	184	1,495
PIN	Customer Selectable VIN	6	5,000
PYD	Competition Gray Alum. Wheels (GS)	709	395
PYE	Chrome Aluminum Wheels (GS)	4,577	1,995
Q44	Competition Gray Wheels (Z06)	42	395
Q5V	Machine Faced Cup Style Wheels (Z06)	54	995
Q6B	Chrome Aluminum Wheels (ZR1)	101	2,000
Q6J	Competition Gray Alum. Wheels (ZR1)	14	395
Q8A	Chrome Aluminum Wheels (Z06)	103	1,995
QX1	Competition Gray Alum. Wheels (cpe/conv)	221	395
QX3	Chrome Aluminum Wheels (cpe/conv)	1,441	1,850
RQ1	Machine Faced Alum. Wheels (cpe/conv)	321	895
R8C	Museum Delivery	181	490
R8E	Gas Guzzler Tax (ZR1)	311	1000
T43	ZR1-Style Spoiler, Body Color	1,542	350
VK3	Front License Plate Bracket	7,140	15
ZLC	Chevrolet Centennial Special Edition*	2.201	4,950
	* required 3LT, 4LT, 3LZ or 3ZR		
Z07	Z06 Ultimate Performance Package	117	7,500
Z15	Grand Sport Heritage Package	1,034	1,195
**6	Modified Two-Tone Seats w/embroidery	1,277	695
36S	Custom Leather Stitching* (Yellow)	2,571	395
	* 37S=Blue, 38S=Red		
0B1	Cross Flag Embroidery on Headrests	1,081	300
28U	Inferno Orange Metallic exterior paint	471	300
45U	Velocity Yellow Premium Tintcoat paint	663	850
89U	Crystal Red Premium Tintcoat paint	1,117	850
GLB	Supersonic Blue exterior paint	893	300

339

APPENDIX

2013 Options

RPO #	DESCRIPTION	QTY	RETAIL $
1YY07	Corvette Coupe	2,597	$50,575
1YY67	Corvette Convertible	720	55,575
1YG07	Grand Sport Coupe	4,908	56,975
1YG67	Grand Sport Convertible	1,736	60,575
1YG67	427 Collector Edition Convertible	2,552	76,575
1YY87	Z06 Coupe	471	76,575
1YY87	ZR1 Coupe	482	112,575
2LT*	Equipment Group, Cpe/Convertible	2,569	2,145
	* Included w/427 Convertible, listed as 1SA		
2LZ	Equipment Group (Z06)	120	4,660
3LT	Equipment Group, Coupe	2,723	5,995
3LT*	Equipment Group, Convertible	1,789	7,995
	* Listed as 1SB for 427 Convertible, priced at $6,000		
3LZ	Equipment Group (Z06)	241	8,860
3ZR	Equipment Group (ZR1)	390	10,000
4LT	Equipment Group, Coupe	648	9,495
4LT*	Equipment Group, Convertible	2,119	11,495
	* Listed as 1SC for 427 Convertible, priced at $9,500		
BA5	Black Grand Sport Fender Badges	989	100
B92	Carbon Fiber Hood (Z06)	125	2,495
C2L	Dual Removable Roof Panels	978	1,400
CC3	Removable Roof Panel, transparent	834	750
CFZ	Carbon Fiber Package* (Z06)	1,605	3,995
	* $2,995 w/427 Convertible		
D30	Non-recommended Color/Trim/Top Combo	21	590
ERI	Battery Protection Package	5,641	100
F55	Magnetic Selective Ride Control*	5,605	1,995
	* $1,695 w/Grand Sport, $2,495 w/Z06		
H3-	Headlamp Color*	2,477	590
	* H33=Gray, H34=Silver, H35=Black		
J55	Cross-Drilled Brakes	804	500
J6-	Brake Calipers*	10,380	595
	* J6C=Silver, J6D=Gray, J6E=Yellow, J6F=Red		
MYC	6-Spd Paddle Shift Auto Trans (cpe/conv)	7,229	1,250
NPP	Dual Mode Perf. Exhaust (cpe/conv)	9,922	1,195
PBC	Engine Build Experience (427, Z06, ZR1)	23	5,800
PDE	Performance Package (ZR1)	191	1,495
PIN	Customer-selectable VIN	12	5,000
PYE	Chrome Aluminum Wheels (GS)	4,567	1,995
Q5V	Machine Face Cup Alum. Wheels (Z06, ZR1)	57	995
Q6B	Chrome 20-spoke Alum. Wheels (427, ZR1)	1,029	2,000
Q6U	Competition Black Aluminum Wheels (ZR1)	65	495
Q7Z	Competition Black Aluminum Wheels (Z06)	66	495
Q8A	Spider Chrome Aluminum Wheels (Z06)	131	1,995
Q8B	Competition Black Aluminum Wheels	202	495
Q8K	Machine Face Comp. Gray Cup Wheels	1,396	995
QX3	Chrome Aluminum Wheels (cpe/conv)	1,270	1,850
QZW	Competition Black Aluminum Wheels (GS)	1,253	495
RQ1	Machine Face Alum.Wheels (cpe/conv)	1,396	895
RUV	Black Cup Alum. Wheels (427, Z06, ZR1)	681	495
R8C	Museum Delivery	265	490
R8E	Gas Guzzler Tax (ZR1)	414	1,000
T43	ZR1-style Spoiler (Included w/427 Conv)	5,926	350
VK3	Front License Plate Bracket	8,504	15
Z07	Z06 Ultimate Perf. Pck (required F55)	83	7,500
Z15	Grand Sport Heritage Package	878	1,195
Z25	60th Anniversary Package*	2,059	1,075
	* Limited to cpe/conv/GS w/4LT, Z06 w/3LZ, & ZR1 w/3ZR		

CODE	DESCRIPTION	QTY	RETAIL $
Z27	427 Heritage Package (427)	691	1,195
Z30	60th Anniversary Graphics (required Z25)	1,753	750*
	* Price w/Z25 coupe; $850 w/Z25 convertible		
**6	Modified Two-Tone Seats w/embroidery	1,317	695
0B1	Crossed flag headrest embroidery	1,951	300
3-S	Custom Stitching*	780	395
	* 36S=Yellow, 37S=Blue, 38S=Red		
28U	Inferno Orange exterior paint	156	300
45U	Velocity Yellow exterior paint	605	850
89U	Crystal Red exterior paint	1,260	850
GLB	Supersonic Blue exterior paint	302	300

2014 Options

RPO #	DESCRIPTION	QTY	RETAIL $
1YY07	Stingray Coupe	11,134	$51,995
1YX07	Stingray Coupe w/Z51	14,931	54,775
1YY67	Stingray Convertible	4,493	56,995
1YX67	Stingray Convertible w/Z51	5,680	59,795
1YX07	Stingray Coupe Premiere Edition	500	75,760
1YX67	Stingray Convertible Premiere Edition	550	77,450
2LT	Equipment Group, Coupe & Convertible	12,529	4,210
3LT	Equipment Group, Coupe & Convertible	19,725	8,005
AE4	Competition Sport Seat w/o suede	2,573	1,995*
	* $2,495 w/suede		
BV4	Personalized Dash Plaque	1,827	200
C2M	Carbon Fiber Dual Roof Panels	969	2,995
C2Q	Body-Color Dual Roof Panels	1,529	1,995
C2Z	Visible Carbon Fiber Roof*	3,393	1,995
	* w/Body Color Surround		
CC3	Removable Roof Panel, transparent	6,973	995
D30	Non-recommended Color/Trim/Top Combo	261	590
DT-	Full Racing Stripe*	820	995
	* DTH=Carbon Flash, DTN=Cyber Gray, DTP=Blade Silver, DTQ=Orange, DTR=Crystal Red		
DT-	Hood Stinger Stripe*	2,536	500
	* DTW=Carbon Flash, DTZ=Cyber Gray		
ERI	Battery Protection Package	15,773	100
FAY	Carbon Fiber Interior Package	7,495	995
FE4	Magnetic Selective Ride Control (Z51 only)	13,386	1,995
G7H	Laguna Blue Tincoat exterior paint	3,281	995
G7J	Lime Rock green Metallic exterior paint	1,566	495
G8A	Velocity YellowTincoat exterior paint	2,125	995
GBE	Crystal Red Tintcoat exterior paint	3,020	995
IL4/IL6	Suede insert seats for 2LT/3LT	6,003	395
IWE	Suede Microfiber Wrapped Interior Trim	5,108	995
J6-	Brake calipers (J6E=Yellow, J6F=Red)	18,682	595
MYC	6-Spd Paddle Shift Automatic Trans	24,088	1,250
NPP	Performance Exhaust (added 5hp)	31,173	1,195
PIN	Customer-selectable VIN	3	5,000
Q7E	Chrome Z51 Aluminum Wheels	7,197	1,995
Q7T	Black Z51 Wheels	10,143	495
QX3	Chrome Aluminum Wheels	10,291	1,995
R8C	Museum Delivery (Included interior plaque)	1,231	990
RQ1	Machine Face Silver Wheels (n/a Z51)	1,156	1,495
TTV	Carbon Flash spoiler & black mirrors (Z51)	5,034	100
UY4	MyLink Navigation System	30,427	795
VK3	Front License Plate Bracket	23,790	15
ZF1	Appearance Package (Stingray w/o Z51)	142	1,995

340

OPTIONS

2015 Options

RPO #	DESCRIPTION	QTY	RETAIL $
1YY07	Stingray Coupe	9,667	$54,995
1YX07	Stingray Coupe w/Z51	11,090	59,995
1YY67	Stingray Convertible	2,397	59,995
1YX67	Stingray Convertible w/Z51	2,433	64,995
1YZ07	Z06 Coupe	6,980	78,995
1YZ67	Z06 Convertible	1,673	83,995
2LT	Equipment Group	10,625	4,160
2LZ	Equipment Group (Z06)	2,202	3,270
3LT	Equipment Group	9,899	9,450
3LZ	Equipment Group (Z06)	5,932	8,650
ATI	Atlantic Conv Design Pck (Z51 & 2LT or 3LT required)	17	6,350
AE4	Competition Sport Seats ($2,495 w/suede inserts)	4,884	1,995
BV4	Personalized Dash Plaque	1,109	200
C2M	Carbon Fiber Dual Roof Package	1,187	2,995
C2Q	Body-Colored Dual Roof Package	616	1,995
C2Z	Visible Carbon Fiber Roof	4,044	1,995
CC3	Removable Roof Panel, Transparent	8,506	995
CFV	Carbon Fiber Ground Effects	1,562	3,995
CFZ	Carbon Fiber Ground Effects (splitter/spoiler)	3,029	2,995
D30	Non-recommended Color/Trim/Top Combo	283	590
DT-	Hood Stinger (--W, --Z, --M, depending on color)	4,351	500
DT-	Race Stripes (--H, --N, --P, --Q, --R, depending on color)	1,352	950
EFX	Shark Gray Exterior Vents	659	595
EFY	Body Color Exterior Vents	173	999
ERI	Battery Protection Package	15,063	100
EYK	Chrome Badge Package (Z06)	21,572	100
EYT	Carbon Flash Badge Package (incl. w/Z06)	12,668	100
FAY	Carbon Fiber Interior Appearance Pck	7,944	995
FE4	Magnetic Selective Ride w/Z51 (std Z06)	7,365	1,795
G1H	Daytona Sunrise Orange Exterior Paint	1,417	495
G7H	Laguna Blue Tintcoat Exterior Paint	3,015	995
G8A	Velocity Yellow Tintcoat Exterior Paint	2,065	995
GBE	Crystal Red Tintcoat Exterior Paint	2,191	995
IL4/6	Suede Microfiber Seat Inserts	4,544	395
J6A	Black Calipers (J6D = Gray Calipers)	13,339	595
J6E	Yellow Calipers (J6F = Red, incl. w/Z06)	20,901	595
M5U	8-Spd Paddle Shift Automatic Trans	23,232	1,725
NPP	Perf. Exhaust (added 5 HP, incl. w/Z51 & Z06)	30,896	1,195
PFI	Pacific Coupe Design Pck (required Z51 & 2LT or 3LT)	97	7,090
PIN	Customer-selectable VIN	9	5,000
Q6B	Chrome Aluminum Wheels (Z06)	2,289	1,995
Q6J	Spectra Gray Alum. Wheels w/Groove (Z06)	782	595
Q6U	Black Aluminum Wheels (Z06)	4,656	495
Q7E	Chrome Z51 Alum. Wheels (non-Z06 w/ZF1)	5,164	1,995
Q7T	Black Painted Z51 Alum. Wheels (ZF1/Z51)	9,178	495
QCC	Black Machine Alum. Wheels (ZF1/Z51)	1,838	1,495
QX1	5-Spoke Black Alum. Wheels (cpe/conv)	1,148	495
QX3	Chrome Alum. Wheels (cpe/conv)	3,124	1,995
RPK	Black Machined Alum. Wheels (cpe/conv)	297	1,495
R8C	Corvette Museum Delivery w/Plaque	1,049	990
TTV	Carbon Flash Painted Spoiler & Outside Mirrors	7,887	100
UQT	Performance Data Video Recorder (incl. Navigation)	25,803	1,795
VK3	Front License Plate Bracket	25,050	15
ZF1	Appearance Package (Stingray w/o Z51)	5,387	1,995
Z07	Z07 Performance Package (Z06 only)	3,378	7,995

2016 Options

RPO #	DESCRIPTION	QTY	RETAIL $
1YY07	Stingray Coupe	10,972	$56,395
1YX07	Stingray Coupe w/Z51	10,415	61,395
1YY67	Stingray Convertible	2,705	60,395
1YX67	Stingray Convertible w/Z51	2,322	65,395
1YZ07	Z06 Coupe	11,543	80,395
1YZ67	Z06 Convertible	2,732	84,395
2LT	Equipment Group	11,048	4,455
2LZ	Equipment Group (Z06)	4,210	3,565
3LT	Equipment Group	7,308	9,745
3LZ	Equipment Group (Z06)	8,464	8,945
36S	Yellow Stitch Option (38S = Red Stitches)	2,744	395
AE4	Competition Sport Seats ($2,495 w/suede inserts)	5,587	1,995
B92	Carbon Fiber Hood Section (no stripes)	3,434	1,995
BV4	Personalized Dash Plaque (incl. w/R8C)	696	200
C2M	Carbon Fiber Dual Roof Package	1,143	2,995
C2Q	Body-Colored Dual Roof Package	1,837	1,995
C2Z	Visible Carbon Fiber Roof	3,955	1,995
CC3	Removable Roof Panel, transparent	5,698	995
CFV	Carbon Fiber Ground Effects	2,646	3,995
CFZ	Carbon Fiber Ground Effects (splitter/spoiler)	5,040	2,995
D30	Non-recommended Color/Trim/Top Combo	281	590
DT-	Hood Stinger (--W, --Z, --M, depending on color)	3,713	500
DT-	Full Stripe (--H, --N, --P, --Q, --R, DUP, DWA, depending on color)	1,338	950
E57	Carbon Fiber Tonneau Inserts (conv)	587	995
EFX	Shark Gray Exterior Vents	1,335	595
EFY	Body Color Exterior Vents	1,094	999
ERI	Battery Protection Package	19,212	100
EYK	Chrome Badge Package (Z06)	21,329	100
EYT	Carbon Flash Badge Pck (incl. w/Z06)	18,710	100
FAY	Carbon Fiber Interior Appearance Pck	8,404	995
FE2	Magnetic Selective Ride w/non-Z51	4,014	3,495
FE4	Magnetic Selective Ride w/Z51 (std Z06)	5,644	1,795
G1E	Long Beach Red Metallic Tintcoat Paint	3,224	995
G1H	Daytona Sunrise Orange Metallic Paint	965	495
G7H	Laguna Blue Tintcoat Paint	3,585	995
GC6	Corvette Racing Yellow Tintcoat Paint	2,715	995
IL4/6	Suede Microfiber Seat Inserts	5,806	395
J57	Ceramic Brake Rotors (Z06 w/o Z07)	120	7,495
J6A	Black Calipers (J6D = Gray Calipers)	14,950	595
J6E	Yellow Calipers (J6F = Red, incl. w/Z06)	25,528	595
M5U	8-Spd Paddle Shift Automatic Trans	31,440	1,725
NPP	Perf. Exhaust (added 5hp, incl. w/Z51 & Z06)	35,983	1,195
PIN	Customer-selectable VIN	8	5,000
PBC	Engine Plant Build Experience	73	5,000
Q6B	Chrome Aluminum Wheels (Z06)	3,687	1,995
Q6J	Spectra Gray Alum. Wheels w/Groove (Z06)	1,052	595
Q6U	Black Aluminum Wheels (Z06)	6,852	495
Q7E	Chrome Z51 Aluminum Wheels	5,081	1,995
Q7T	Black Painted Z51 Alum. Wheels	7,761	495
Q9B	Satin Black Alum. Wheels w/Red Stripe (Z06)	193	1,495
QCC	Black Machine Alum. Wheels (std Z51, FE2)	1,311	1,495
QX1	5-Spoke Black Alum. Wheels (cpe/conv)	2,924	495
QX3	Chrome Alum. Wheels (cpe/conv)	2,967	1,995
RPK	Black Machined Alum. Wheels (cpe/conv)	496	1,495
R87	Z51 Style Satin Black Alum. Wheels w/Red Stripes	261	1,495
R8C	Corvette Museum Delivery	675	990
R8E	Gas Guzzler Tax (Z06 w/M5U)	10,119	1,300
TTV	Carbon Flash Painted Spoiler & Outside Mirrors	8,522	100

APPENDIX

CODE	DESCRIPTION	QTY	RETAIL $
TU7	Two-Tone Seats (Napa/Suede)	1,474	395
UQT	Performance Data Video Recorder (incl. Navigation)	27,782	1,795
VK3	Front License Plate Bracket	25,754	15
ZCR	Z06 C7R Edition Coupe (Conv $24,150)	650	23,055
ZLD	Twilight Blue Design Pck	371	3,500
ZLE	Spice Red Design Pck	535	3,000
ZLG	Black Suede Design Pck	454	3,995
Z07	Z07 Performance Pck (Z06 only)	4,955	7,995

2017 Options

RPO #	DESCRIPTION	QTY	RETAIL $
1YY07	Stingray Coupe	7,548	$56,445
1YX07	Stingray Coupe w/Z51	3,705	61,445
1YY67	Stingray Convertible	1,571	60,445
1YX67	Stingray Convertible w/Z51	722	65,445
1YW07	Grand Sport Coupe	9,912	66,445
1YW67	Grand Sport Convertible	2,046	70,445
1YZ07	Z06 Coupe	6,197	80,445
1YZ67	Z06 Convertible	1,076	84,445
2LT	Equipment Group	10,366	4,455
2LZ	Equipment Group (Z06)	2,552	3,565
3LT	Equipment Group	6,659	9,745
3LZ	Equipment Group (Z06)	3,337	8,945
3F9	Safety Belt Color, Red	2,010	395
36S	Yellow Stitch Option (38S = Red Stitches)	1,584	395
AE4	Competition Sport Seats ($2,495 w/suede inserts)	3,642	1,995
B92	Carbon Fiber Hood Section (no stripes)	1,586	1,995
BV4	Personalized Dash Plaque	309	200
C2M	Carbon Fiber Dual Roof Pck	459	2,995
C2Q	Body-Colored Dual Roof Pck	872	1,995
C2Z	Visible Carbon Fiber Roof	1,459	1,995
CC3	Removable Roof Panel, transparent	2,251	995
CFV	Carbon Fiber Ground Effects	2,512	3,995
CFZ	Carbon Fiber Ground Effects (splitter/spoiler)	3,981	2,995
D30	Non-recommended Color/Trim/Top Combo	163	590
D--	GS Full Stripe (-NE, -XP, -XT, -XU, -XX, depending on color)	1,500	995
D--	Hood Stinger (-TW, -TZ, -VQ, -XM, depending on color)	1,834	500
D--	Full Stripe (-TH, -TN, -TP, -TQ, -TR, -UP, -WA, depending on color)	755	950
E57	Carbon Fiber Tonneau Inserts (conv)	235	995
EFX	Shark Gray Exterior Vents	512	595
EFY	Body Color Exterior Vents	1,162	999
ERI	Battery Protection Package	13,588	100
EYK	Chrome Badge Package	17,939	100
EYT	Carbon Flash Badge Pck (incl. w/Z06)	14,843	100
FAY	Carbon Fiber Interior Appearance Pck	5,240	995
FE2	Magnetic Selective Ride w/non-Z51	1,578	3,495
FE4	Magnetic Selective Ride w/Z51 (std Z06, GS)	1,532	1,795
G1E	Long Beach Red Metallic Tintcoat Paint	2,004	995
GC6	Corvette Racing Yellow Tintcoat Paint	1,371	995
GGA	Black Rose Metallic Paint	1,333	495
J57	Ceramic Brake Rotors (Z06 w/o Z07, GS)	1,301	7,495
J6A	Black Calipers (J6D = Gray Calipers)	14,393	595
J6E	Yellow Calipers (J6F = Red, incl. w/Z06)	18,147	595
M5U	8-Spd Paddle Shift Automatic Trans	25,556	1,725
N26	Suede Steering Wheel, Seats, Shift Knob, Boot	3,600	695
NPP	Performance Exhaust (added 5hp, incl. w/Z51, Z06 & GS)	28,793	1,195

CODE	DESCRIPTION	QTY	RETAIL $
PIN	Customer-selectable VIN	7	5,000
PBC	Engine Plant Build Experience	32	5,000
PD5	Buyer's Plant Tour (required PBC)	20	2,500
Q6B	Chrome Aluminum Wheels (Z06)	1,620	1,995
Q6J	Spectra Gray Alum. Wheels w/Groove (Z06)	381	595
Q6U	Black Aluminum Wheels (Z06)	3,490	495
Q7E	Chrome Z51 Alum. Wheels	1,943	1,995
Q7T	Black Painted Z51 Alum. Wheels	2,319	495
Q8K	GS Machined Face Alum. Wheels	804	1,495
Q8X	GS Black Aluminum Wheels	6,093	495
Q8Z	GS Chrome Alum. Wheels	2,571	1,995
Q9B	Z06 Satin Black Alum. Wheels w/Red Stripe	658	1,495
QCC	Black Machine Alum. Wheels	495	1,495
QX1	5-Spoke Black Alum. Wheels (cpe/conv)	2,437	495
QX3	Chrome Alum. Wheels (cpe/conv)	2,098	1,995
RPK	Black Machined Alum. Wheels (cpe/conv)	225	1,495
R87	Z51 Style Satin Black Alum. Wheels w/Red Stripes	394	1,495
R8C	Corvette Museum Delivery	372	990
R8E	Gas Guzzler Tax (Z06 w/M5U)	4,462	1,300
RUR	GS Satin Black Alum. Wheels w/Red Stripe	960	1,495
TTV	Carbon Flash Painted Spoiler & Outside Mirrors	4,875	100
TU7	Two-Tone Seats (Napa/Suede)	1,527	395
UQT	Performance Data Video Recorder (incl. Navigation)	20,364	1,795
VK3	Front License Plate Bracket	19,956	15
ZLD	Twilight Blue Design Pck	231	3,500
ZLE	Spice Red Design Package	307	3,000
ZLG	Black Suede Design Pck	270	3,995
Z07	Z07 Performance Pck (Z06 only)	3,788	7,995
Z15	Grand Sport Heritage Pck	3,863	795
Z25	Grand Sport Collector Edition	935	4,995

2018 Options

RPO #	DESCRIPTION	QTY	RETAIL $
1YY07	Stingray Coupe	2,352	$56,490
1YX07	Stingray Coupe w/Z51	716	61,490
1YY67	Stingray Convertible	537	60,490
1YX67	Stingray Convertible w/Z51	198	65,490
1YW07	Grand Sport Coupe	2,569	66,490
1YW67	Grand Sport Convertible	512	70,490
1YZ07	Z06 Coupe	2,353	80,490
1YZ67	Z06 Convertible	449	84,490
2LT	Equipment Group	2,688	4,455
2LZ	Equipment Group (Z06)	1,002	3,565
3LT	Equipment Group	1,758	9,745
3LZ	Equipment Group (Z06)	1,239	8,945
3F9	Safety Belt Color, Red	780	395
36S	Yellow Stitch Option (37S = Blue, 38S = Red)	1,042	395
AE4	Competition Sport Seats ($2,495 w/suede inserts)	1,567	1,995
B92	Carbon Fiber Hood Section (no stripes)	1,122	1,995
BV4	Personalized Dash Plaque	94	200
C2M	Carbon Fiber Dual Roof Pck	84	2,995
C2Q	Body-Colored Dual Roof Pck	182	1,995
C2Z	Visible Carbon Fiber Roof	919	1,995
CC3	Removable Roof Panel, transparent	672	995
CFV	Carbon Fiber Ground Effects	1,276	3,995
CFZ	Carbon Fiber Ground Effects (splitter/spoiler)	829	2,995
D30	Non-recommended Color/Trim/Top Combo	25	590
D--	GS Full Stripe (-NE, -UR, -XP, -XT, -XU, -XX, depending on color)	259	995

OPTIONS

CODE	DESCRIPTION	QTY	RETAIL $
D--	Hood Stinger (-TW, -VQ, -XM, -YV, depending on color)	723	500
D--	Full Stripe (-TH, -TN, -TP, -TQ, -TR, -UO, -UP, -WA, depending on color)	272	950
E57	Carbon Fiber Tonneau Inserts (conv)	156	995
EFX	Shark Gray Exterior Vents	90	595
EFY	Body Color Exterior Vents	272	999
ERI	Battery Protection Pck	3,831	100
EYK	Chrome Badge Package	4,845	100
EYT	Carbon Flash Badge Pck (incl. w/Z06)	4,841	100
FAY	Carbon Fiber Interior Appearance Pck	1,259	995
FE2	Magnetic Selective Ride w/non-Z51	158	1,795
FE4	Magnetic Selective Ride w/Z51 (std Z06, GS)	358	1,795
G1E	Long Beach Red Metallic Tintcoat Paint	599	995
G26	Sebring Orange Metallic Tintcoat Paint	85	995
GC6	Corvette Racing Yellow Tintcoat Paint	399	995
GGA	Black Rose Metallic Paint	219	495
J57	Ceramic Brake Rotors (Z06 w/o Z07, GS)	305	7,495
J6A	Black Calipers (J6K = Gray, J6E = Yellow, J6F = Red, incl. w/Z06)	9,686	595
M5U	8-Spd Paddle Shift Automatic Trans	7,563	1,725
N26	Suede Steering Wheel, Seats, Shift Knob, Boot	1,677	695
NPP	Performance Exhaust (added 5hp, incl. w/Z51, Z06 & GS	8,429	1,195
PIN	Customer-selectable VIN	2	5,000
Q6B	Chrome Aluminum Wheels (Z06)	134	1,995
Q6J	Spectra Gray Alum. Wheels w/Groove (Z06)	1,102	595
Q6U	Black Aluminum Wheels (Z06)	541	495
Q7E	Chrome Z51 Aluminum Wheels	988	1,995
Q7T	Black Painted Z51 Alum. Wheels	1,235	495
Q8K	GS Machined Face Alum. Wheels	135	1,495
Q8U	Z06 10-Spoke Pearl Nickel Painted Wheels	44	1,495
Q8X	GS Black Aluminum Wheels	1,232	495
Q8Z	GS Chrome Alum. Wheels	727	1,995
Q9B	Z06 Satin Black Alum. Wheels w/Red Stripe	223	1,495
Q9C	Z06 Black Alum. Wheels w/Yellow Stripe	79	1,495
QCC	Black Machine Aluminum Wheels	121	1,495
RNM	Torque Directional Painted Alum. Wheels	9	1,495
RNN	Motorsports Black Painted Alum. Wheels	84	1,495
RNP	Z51 Type Black Painted Wheels w/Yellow Stripe	32	1,495
R2Q	Torque Directional Chrome Alum. Wheels	70	2,495
R86	Motorsports Polished Alum. Wheels	36	1,995
R87	Z51 Style Satin Black Alum. Wheels w/Red Stripes	185	1,495
R8C	Corvette Museum Delivery	88	990
R8E	Gas Guzzler Tax (Z06 w/M5U)	2,019	1,300
RUR	GS Satin Black Alum. Wheels w/Red Stripe	68	1,495
TTV	Carbon Flash Painted Spoiler & Outside Mirrors	1,885	100
TU7	Two-Tone Seats (Napa/Suede)	398	395
UQT	Performance Data Video Recorder (incl. Navigation)	6,125	1,795
VK3	Front License Plate Bracket	6,034	40
Z07	Z07 Performance Package	823	7,995
Z15	Grand Sport Heritage Pck	760	795
Z30	Carbon 65 Edition	650	15,000

2019 Options

RPO #	DESCRIPTION	QTY	RETAIL $
1YY07	Stingray Coupe	9,771	$56,590
1YX07	Stingray Coupe w/Z51	1,728	61,590
1YY67	Stingray Convertible	1,868	60,590
1YX67	Stingray Convertible w/Z51	324	65,590
1YW07	Grand Sport Coupe	9,496	66,590
1YW67	Grand Sport Convertible	1,745	70,590
1YZ07	Z06 Coupe	5,965	80,590
1YZ67	Z06 Convertible	972	84,590
1YV07	ZR1 Coupe	2,441	119,995
1YV67	ZR1 Convertible	512	123,995
2LT	Equipment Group	10,604	4,455
2LZ	Equipment Group (Z06)	3,109	3,565
3LT	Equipment Group	3,502	9,745
3LZ	Equipment Group (Z06)	2,232	8,945
3ZR	Equipment Group (ZR1)	2,761	10,000
3F9	Safety Belt Color, Red	2,949	395
36S	Yellow Stitch Option (37S = Blue, 38S = Red)	2,105	395
AE4	Competition Sport Seats ($2,495 w/suede inserts)	4,783	1,995
B92	Carbon Fiber Hood Section (no stripes)	4,349	1,995
BV4	Personalized Dash Plaque	680	200
C2M	Carbon Fiber Dual Roof Pck	706	2,995
C2Q	Body-Colored Dual Roof Pck	440	1,995
C2Z	Visible Carbon Fiber Roof	2,981	1,995
CC3	Removable Roof Panel, transparent	1,902	995
CFV	Carbon Fiber Ground Effects	4,714	3,995
CFZ	Carbon Fiber Ground Effects (splitter/spoiler)	2,722	2,995
D30	Non-recommended Color/Trim/Top Combo	134	590
D--	GS Center Stripe (-NE, -UR, -XP, -XT, -XU, -XX, depending on color)	624	995
D--	Hood Stinger (-TW, -VQ, -XM, -YV, depending on color)	2,703	500
D--	Full Stripe (-TH, -TN, -TP, -TQ, -TR, -UO, -UP, -WA, depending on color)	1,068	950
DFZ	Satin Black Stripe w/Jake Logo (ZR1)	228	500
E57	Carbon Fiber Tonneau Inserts (conv)	669	995
EFX	Shark Gray Exterior Vents	200	595
EFY	Body Color Exterior Vents	567	999
ERI	Battery Protection Package	13,227	100
EYK	Chrome Badge Package	16,693	100
EYT	Carbon Flash Badge Pck (incl. w/Z06)	15,575	100
FAY	Carbon Fiber Interior Appearance Pck	3,487	995
FE2	Magnetic Selective Ride w/non-Z51	569	1,795
FE4	Magnetic Selective Ride w/Z51 (std Z06, GS, ZR1)	622	1,795
G1E	Long Beach Red Metallic Tintcoat Paint	2,397	995
G26	Sebring Orange Metallic Tintcoat Paint	2,980	995
GC6	Corvette Racing Yellow Tintcoat Paint	1,291	995
J57	Ceramic Brake Rotors (Z06 w/o Z07, GS)	3,163	7,495
J6A	Black Calipers (J6K = Gray, J6E = Yellow, J6F = Red, incl. w/Z06)	33,177	595
M5U	8-Spd Paddle Shift Automatic Trans	27,017	1,725
N26	Suede Steering Wheel, Seats, Shift Knob, Boot	2,618	695
NPP	Performance Exhaust (added 5hp, incl. w/Z51, Z06, GS & ZR1	24,871	1,195
PBC	Engine Plant Build Experience	54	5,000
PDA	Driver Series Pck, GS Coupe, Antonio Garcia	14	4,995*
PDJ	Driver Series Pck, GS Coupe, Jan Magnussen	25	4,995*
PDO	Driver Series Pck, GS Coupe, Oliver Gavin	35	4,995*
PDT	Driver Series Pck, GS Coupe, Tommy Milner	21	4,995*
	* $5,995 w/3LT		
PIN	Customer-selectable VIN	19	5,000
Q6B	Chrome Alum. Wheels (Z06)	1,230	1,995
Q6J	Spectra Gray Alum. Wheels w/Groove (Z06)	498	595
Q6U	Black Aluminum Wheels (Z06)	3,892	495
Q7E	Chrome Z51 Alum. Wheels	3,028	1,995

343

APPENDIX

CODE	DESCRIPTION	QTY	RETAIL $
Q7T	Black Painted Z51 Alum. Wheels	5,591	495
Q8K	GS Machined Face Alum. Wheels	371	1,495
Q8U	Z06 10-Spoke Pearl Nickel Painted Alum. Wheels	151	1,495
Q8X	GS Black Aluminum Wheels	6,035	495
Q8Z	GS Chrome Alum. Wheels	2,757	1,995
Q9B	Z06 Satin Black Alum. Wheels w/Red Stripe	441	1,495
Q9C	Z06 Black Alum. Wheels w/Yellow Stripe	92	1,495
Q9H	ZR1 Carbon Flash Painted Alum. Wheels	1,849	495
Q9J	ZR1 Satin Graphite Painted Alum. Wheels	425	595
Q9K	ZR1 Chrome Alum. Wheels	496	1,995
QCC	Black Machine Alum. Wheels	393	1,495
RNM	Torque Directional Painted Alum. Wheels	35	1,495
RNN	Motorsports Black Painted Alum. Wheels	457	1,495
RNP	Z51 Type Black Painted Wheels w/Yellow Stripe	77	1,495
RUR	GS Satin Black Alum. Wheels w/Red Stripes	721	1,495
R2Q	Torque Directional Chrome Alum. Wheels	121	2,495
R86	Motorsports Polished Alum. Wheels	202	1,995
R87	Z51 Style Satin Black Wheels w/Red Stripes	425	1,495
R8C	Corvette Museum Delivery	587	990
R8E	Gas Guzzler Tax, Z06 w/M5U (ZR1 $2,100)	6,830	1,300
TTV	Carbon Flash Painted Spoiler & Outside Mirrors	5,329	100
TU7	Two-Tone Seats (Napa/Suede)	1,507	395
UQT	Performance Data Video Recorder (incl. Navigation)	18,004	1,795
VK3	Front License Plate Bracket	20,811	40
ZLZ	Sebring Orange Design Pck (ZR1)	758	6,995
ZTK	Track Performance Pck (ZR1)	2,077	2,995
Z07	Z07 Performance Pck	1,350	7,995
Z15	Grand Sport Heritage Pck	2,178	795

2020 Options

RPO #	DESCRIPTION	QTY	RETAIL $
1YC07	Stingray Coupe 1LT	2,663	$59,995
1YC07	Stingray Coupe 2LT	6,685	67,295
1YC07	Stingray Coupe 3LT	7,439	71,945
1YC67	Stingray Convertible 1LT	283	67,495
1YC67	Stingray Convertible 2LT	1,281	74,295
1YC67	Stingray Convertible 3LT	2,017	78,945
379	Orange Seat Belts	701	395
3A9	Tension Blue Seat Belts	1,133	395
3F9	Torch Red Seat Belts	3,621	395
3M9	Yellow Seat Belts	749	395
3N9	Tan Seat Belts	1,291	395
36S	Yellow Custom Leather Stitch	518	395
38S	Adrenaline Red Custom Leather Stitch	2,339	395
AE4	Competition Sport Seats ($500 w/3LT)	1,891	1,995
AH2	GT2 Seats (incl. w/3LT)	12,548	1,495
BV4	Personalized Dash Plaque (w/frame & VIN)	1,009	200
C2M	Carbon Fiber Dual Roof Package	560	3,495
C2Q	Body-Colored Dual Roof Package	711	1,995
C2Z	Visible Carbon Fiber Removable Roof	1,407	2,495
CC3	Removable Roof Panel, Transparent	2,086	995
D30	Non-Recommended Color Combination Override	459	590
D84	Carbon Flash Painted Nacelles & Roof (conv)	1,162	1,295
D86	Carbon Flash Painted Nacelles & Body Color Roof (conv)	350	1,295
DTH	Full Length Racing Stripes, Carbon Flash	1,497	995
DUB	Full Length Racing Stripes, Sterling Silver	326	995
DXO	Full Length Racing Stripes, Midnight Gray	165	995
E60	Front Lift Adjustable Height w/Memory	11,899	1,495

RPO #	DESCRIPTION	QTY	RETAIL $
EFA	Exterior Trim Accents, Shadow Gray	171	995
EFY	Body Color Exterior Trim Accents	1,237	995
ERI	Battery Protection Package	7,300	100
EYK	Chrome Exterior Badge Package	1,716	100
FA5	Carbon Fiber Interior Trim	4,822	1,500
FE4	Magnetic Selective Ride Control (required Z51)	10,991	1,895
G1E	Long Beach Red Metallic Tintcoat	1,068	995
G26	Sebring Orange Tintcoat	1,377	995
GD0	Accelerate Yellow Metallic	688	500
GMO	Rapid Blue	1,243	500
IOT	Info System (w/Navigation & Performance Data Recorder)	17,740	1,795
J6E	Yellow Brake Calipers	1,698	595
J6F	Bright Red Brake Calipers	8,638	595
J6N	Edge Red Brake Calipers	2,548	595
N26	Sueded Microfiber-Wrapped Steering Wheel	3,251	595
NPP	Performance Exhaust (incl. w/Z51)	18,174	1,195
Q8Q	Carbon Flash Aluminum Wheels	8,500	995
Q8S	Sterling Silver Trident Spoke Wheels	1,924	1,495
Q8T	Spectra Gray Machined Trident Spoke Wheels	2,420	1,495
R6X	Custom Interior Trim/Seat Override (w/3LT)	1,056	590
R8C	Corvette Museum Delivery w/Personalized Plaque	943	995
TU7	Two-Tone Seats	2,767	395
VK3	Front License Plate Bracket	11,712	15
Z51	Performance Package	15,476	5,000
ZYC	Carbon Flash Metallic Painted Outside Mirrors	6,289	100
ZZ3	Engine Appearance Package (Coupe)	7,468	995

2021 Options

RPO #	DESCRIPTION	QTY	RETAIL $
1YC07	Stingray Coupe 1LT	3,083	$60,995
1YC07	Stingray Coupe 2LT	7,110	68,295
1YC07	Stingray Coupe 3LT	4,919	71,945
1YC67	Stingray Convertible 1LT	848	68,495
1YC67	Stingray Convertible 2LT	4,455	75,295
1YC67	Stingray Convertible 3LT	5,801	79,945
379	Orange Seat Belts	688	395
3A9	Tension Blue Seat Belts	1,766	395
3F9	Torch Red Seat Belts	4,100	395
3M9	Yellow Seat Belts	1,054	395
3N9	Tan Seat Belts	1,282	395
36S	Yellow Custom Leather Stitch	431	395
38S	Red Custom Leather Stitch (3LT only)	2,283	395
AE4	Competition Sport Seats, 2LT ($995 w/1LT, $500 w/3LT)	1,770	1,995
AH2	GT2 Seats, 2LT (std w/3LT)	15,766	1,495
BV4	Personalized Dash Plaque (w/Frame & VIN)	1,111	200
C2M	Carbon Fiber Dual Roof Package	352	3,495
C2Q	Body-Colored Dual Roof Package	680	1,995
C2Z	Visible Carbon Fiber Removable Roof	753	2,495
CC3	Removable Roof Panel, Transparent	1,665	995
D30	Non-Recommended Color Combination Override	901	590
D84	Carbon Flash Painted Nacelles & Roof (conv)	3,919	1,295
D86	Carbon Flash Painted Nacelles & Body Colored Roof (conv)	975	1,295
DSY	Orange Full Length Dual Racing Stripes	39	995
DSZ	Red Full Length Dual Racing Stripes	115	995
DT0	Yellow Full Length Dual Racing Stripes	76	995
DTH	Full Length Racing Stripes, Carbon Flash	2,006	995

OPTIONS

RPO #	DESCRIPTION	QTY	RETAIL $
DUB	Full Length Racing Stripes, Sterling Silver	400	995
DUW	Full Length Racing Stripes, Blue	48	995
DXO	Full Length Racing Stripes, Midnight Gray	183	995
DZU	Carbon Flash/Edge Yellow Stinger Stripe	97	500
DZV	Carbon Flash/Midnight Silver Stinger Stripe	330	500
DZX	Carbon Flash/Edge Red Stinger Stripe	247	500
E60	Front Lift Adjustable Height w/Memory	16,026	1,995
EFA	Exterior Trim Accents, Shadow Gray	205	995
EFY	Body Color Exterior Trim Accents	1,797	995
ERI	Battery Protection Package	10,178	100
EYK	Chrome Exterior Badge Package	1,992	100
FA5	Carbon Fiber Interior Trim	5,178	1,500
FE2	Magnetic Selective Ride Control (w/o Z51)	2,022	1,895
FE4	Magnetic Selective Ride Control (w/Z51)	12,037	1,895
G26	Sebring Orange Tintcoat	1,255	995
GD0	Accelerate Yellow Metallic	839	500
GMO	Rapid Blue	2,384	500
GPH	Red Mist Metallic Tintcoat	3,476	995
IOT	Info System (w/Navigation & Performance Data Recorder)	22,595	1,795
J6E	Yellow Brake Calipers	2,267	595
J6F	Bright Red Brake Calipers	10,076	595
J6N	Edge Red Brake Calipers	4,770	595
N26	Sueded Microfiber-Wrapped Steering Wheel	4,573	595
NPP	Performance Exhaust (incl. w/Z51)	22,872	1,195
PIN	Customer Selectable VIN	5	5,000
Q8Q	Carbon Flash Aluminum Wheels	10,114	995
Q8S	Sterling Silver Trident Spoke Wheels	1,832	1,495
Q8T	Spectra Gray Machined Trident Spoke Wheels	4,327	1,495
R6X	Custom Interior Trim/Seat Override (w/3LT)	1,220	590
R8C	Corvette Museum Delivery w/Personalized Plaque	1,274	995
TU7	Two-Tone Seats (2LT & 3LT)	4,202	395
VK3	Front License Plate Bracket	14,381	15
Z51	Performance Package	18,223	5,995
ZYC	Carbon Flash Metallic Painted Outside Mirrors	8,856	100
ZZ3	Engine Appearance Package (coupe)	5,592	995

2022 Options

RPO #	DESCRIPTION	QTY	RETAIL $
1YC07	Stingray Coupe 1LT	2,991	$62,195
1YC07	Stingray Coupe 2LT	6,108	69,495
1YC07	Stingray Coupe 3LT	4,352	74,145
1YC67	Stingray Convertible 1LT	991	69,695
1YC67	Stingray Convertible 2LT	4,952	76,495
1YC67	Stingray Convertible 3LT	6,437	81,145
379	Orange Seat Belts (3LT only)	748	395
3A9	Tension Blue Seat Belts	1,644	395
3F9	Torch Red Seat Belts	3,859	395
3M9	Yellow Seat Belts	2,278	395
3N9	Tan Seat Belts	1,266	395
36S	Yellow Custom Leather Stitch (3LT only)	474	395
38S	Adrenaline Red Custom Leather Stitch (3LT only)	1,923	395
AE4	Competition Sport Seats, 2LT ($995 w/1LT, $500 w/3LT)	1,791	1,995
AH2	GT2 Seats, 2LT (std w/3LT)	15,477	1,495
BV4	Personalized Dash Plaque (w/frame & VIN)	744	200
C2M	Carbon Fiber Dual Roof Package	215	3,495
C2Q	Body-Colored Dual Roof Package	653	1,995
C2Z	Visible Carbon Fiber Removable Roof	544	2,495

RPO #	DESCRIPTION	QTY	RETAIL $
CC3	Removable Roof Panel, Transparent	1,347	995
D30	Non-Recommended Color Combination Override	814	590
D84	Carbon Flash Painted Nacelles & Roof (conv)	4,637	1,295
D86	Carbon Flash Painted Nacelles & Body Colored Roof (conv)	1,054	1,295
DSY	Orange Full Length Dual Racing Stripes	27	995
DSZ	Red Full Length Dual Racing Stripes	116	995
DT0	Yellow Full Length Dual Racing Stripes	801	995
DTH	Full Length Racing Stripes, Carbon Flash	1,542	995
DUB	Full Length Racing Stripes, Sterling Silver	286	995
DUW	Full Length Racing Stripes, Blue	65	995
DZU	Carbon Flash/Edge Yellow Stinger Stripe	74	500
DZV	Carbon Flash/Midnight Silver Stinger Stripe	279	500
DZX	Carbon Flash/Edge Red Stinger Stripe	309	500
E60	Front Lift Adjustable Height w/Memory	15,196	2,660
EFA	Exterior Trim Accents, Shadow Gray	148	995
EFY	Body Color Exterior Trim Accents	1,583	995
ERI	Battery Protection Package	9,997	100
EYK	Chrome Exterior Badge Package	1,368	100
FA5	Carbon Fiber Interior Trim	5,193	1,500
FE2	Magnetic Selective Ride Control (w/o Z51)	1,945	1,895
FE4	Magnetic Selective Ride Control (w/Z51)	11,757	1,895
GC5	Amplify Orange Tintcoat	1,375	995
GD0	Accelerate Yellow Metallic	1,193	500
GMO	Rapid Blue	2,261	500
GPH	Red Mist Metallic Tintcoat	3,274	995
IOT	Info System (w/Navigation & Performance Data Recorder)	22,021	1,795
J6E	Yellow Brake Calipers	3,319	595
J6F	Bright Red Brake Calipers	8,895	595
J6N	Edge Red Brake Calipers	4,803	595
N26	Sueded Microfiber-Wrapped Steering Wheel	6,027	595
NPP	Performance Exhaust (incl. w/Z51)	22,441	1,195
PIN	Customer Selectable VIN	10	5,000
Q8Q	Carbon Flash Aluminum Wheels	9,789	995
Q8S	Sterling Silver Trident Spoke Wheels	1,823	1,495
Q8T	Spectra Gray Machined Trident Spoke Wheels	4,087	1,495
R6X	Custom Interior Trim/Seat Override (w/3LT)	908	590
R8C	Corvette Museum Delivery w/Personalized Plaque	1,146	995
TU7	Two-Tone Seats (2LT & 3LT)	3,958	395
TVS	Low Rear Spoiler & Front Splitter	3,413	595
VK3	Front License Plate Bracket	14,299	15
ZCR	IMSA-GTLM Championship C8.R Edition	1,007	6,595
ZF1	Aero Delete for Z51 (deletes front splitter & rear spoiler)	749	0
Z51	Performance Package	17,727	6,345
ZYC	Carbon Flash Metallic Painted Outside Mirrors	10,330	100
ZZ3	Engine Appearance Package (coupe)	2,803	995

2023 Options

RPO #	DESCRIPTION	QTY	RETAIL $
1YC07	Stingray Coupe 1LT	6,732	$64,200
1YC07	Stingray Coupe 2LT	10,450	71,500
1YC07	Stingray Coupe 3LT	7,652	76,150
1YC67	Stingray Convertible 1LT	1,865	71,700
1YC67	Stingray Convertible 2LT	8,149	78,500
1YC67	Stingray Convertible 3LT	12,524	83,150
1YH07	Z06 Coupe 1LZ	204	105,000
1YH07	Z06 Coupe 2LZ	562	114,200

APPENDIX

RPO #	DESCRIPTION	QTY	RETAIL $
1YH07	Z06 Coupe 3LZ	2,343	118,850
1YH67	Z06 Convertible 1LZ	45	112,500
1YH67	Z06 Convertible 2LZ	350	121,200
1YH67	Z06 Convertible 3LZ	2,909	125,850
379	Orange Seat Belts (3LT only)	1,609	495
3A9	Tension Blue Seat Belts	2,760	495
3F9	Torch Red Seat Belts	14,625	495
3M9	Yellow Seat Belts	2,300	495
3N9	Tan Seat Belts	2,276	495
36S	Yellow Custom Leather Stitch (3LT black interior only)	1,191	495
38S	Adrenaline Red Custom Leather Stitch (3LT black interior only)	3,437	495
AE4	Competition Sport Seats, 2LT/2LZ ($995 w/1LT & 1LZ, $500 w/3LT & 3 LZ)	4,230	1,995
AH2	GT2 Seats, 2LT/2LZ (std w/3LT & 3LZ)	33,434	1,695
BAZ	Stealth Interior Trim (2LT/2LZ & 3LT/3LZ)	9,154	595
BV4	Personalized Dash Plaque (w/frame & VIN)	1,052	295
C2M	Carbon Fiber Dual Roof Package	294	3,495
C2Q	Body-Colored Dual Roof Package	1,070	1,995
C2Z	Visible Carbon Fiber Removable Roof	1,504	2,495
CC3	Removable Roof Panel, Transparent	2,748	995
CFV	Carbon Fiber Visible Ground Effects (Z06)	1,734	3,995
CFZ	Carbon Fiber Painted Ground Effects (Z06)	1,537	2,995
D30	Non-Recommended Color Combination Override	1,442	695
D84	Carbon Flash Painted Nacelles & Roof (conv)	9,974	1,295
D86	Carbon Flash Painted Nacelles & Body Colored Roof (conv)	1,800	1,295
DSY	Orange Full Length Dual Racing Stripes	51	995
DSZ	Red Full Length Dual Racing Stripes	313	995
DT0	Yellow Full Length Dual Racing Stripes	163	995
DTF	70th Anniversary Full Length Gray Stripes	1,214	995
DTH	Full Length Racing Stripes, Carbon Flash	2,363	995
DTK	70th Anniversary Full Length Black Stripes	637	995
DUB	Full Length Racing Stripes, Sterling Silver	442	995
DUW	Full Length Racing Stripes, Blue	136	995
DXO	Full Length Racing Stripes, Midnight Gray	249	995
DZU	Carbon Flash/Edge Yellow Stinger Stripe	131	500
DZV	Carbon Flash/Midnight Silver Stinger Stripe	429	500
DZX	Carbon Flash/Edge Red Stinger Stripe	436	500
E60	Front Lift Adjustable Height w/Memory	30,215	2,595
EFA	Exterior Trim Accents, Dark Shadow Metallic (Stingray)	417	995
EFY	Body Color Exterior Trim Accents	2,650	995
ERI	Battery Protection Package	19,167	100
EYK	Chrome Exterior Badge Package	2,040	295
FA5	Carbon Fiber Interior Trim (2LT/2LZ & 3LT/3LZ)	7,606	1,500
FA6	Carbon Fiber Interior Trim Level 2 (Z06 3LZ only)	1,926	4,995
FE2	Magnetic Selective Ride Control (Stingray w/o Z51)	4,097	1,895
FE4	Magnetic Selective Ride Control (Stingray w/Z51)	17,829	1,895
GC5	Amplify Orange Tintcoat	3,030	995
GD0	Accelerate Yellow Metallic	1,898	500
GMO	Rapid Blue	3,078	500
GPH	Red Mist Metallic Tintcoat	5,785	995
IOT	Info System (w/Navigation & Performance Data Recorder, 1LT/1LZ)	45,014	1,795
J6E	Yellow Brake Calipers	4,231	695
J6F	Bright Red Brake Calipers	19,305	695
J6L	Orange Brake Calipers (available w/J57 carbon-ceramic brakes)	387	695
J6N	Edge Red Brake Calipers	11,013	695
J57	Carbon Ceramic Brakes (Z06, incl. w/Z07)	n/a	8,495
N26	Sueded Microfiber-Wrapped Steering Wheel	11,816	695
NPP	Performance Exhaust (incl. w/Z51)	47,239	1,195
PIN	Customer Selectable VIN	52	5,000
Q8Q	Carbon Flash Open Spoke Wheels (Stingray)	10,169	995
Q8S	Sterling Silver Trident Spoke Wheels (Stingray)	2,113	1,495
Q8T	Spectra Gray Machined Trident Spoke Wheels (Stingray)	3,928	1,495
Q99	20-Spoke Machine-Face Wheels (Stingray)	2,901	1,995
Q9A	20-Spoke Carbon Paint Wheels (Stingray)	8,164	1,495
Q9I	20-Spoke Black Paint Wheels (Stingray)	9,779	995
R8C	Corvette Museum Delivery w/Personalized Plaque	2,504	995
ROY	Carbon Flash Painted Carbon Fiber Wheels (Z06, J57 Brakes required)	285	9,995
ROZ	Visible Carbon Fiber Wheels (Z06, J57 Brakes required)	871	11,995
SOA	Spider Black Wheels (Z06)	2,314	495
SOC	Spider Satin Graphite Wheels (Z06)	584	495
SOD	Spider Machined Wheels (Z06)	421	1,495
T0F	Carbon Aero Package (Z06, incl. CFZ)	964	8,495
T0G	Visible Carbon Aero Package (Z06, incl. CFV)	1,131	10,495
TU7	Two-Tone Seats (2LT/2LZ & 3LT/3LZ)	6,592	595
TVS	Low Rear Spoiler & Front Splitter (Stingray)	8,300	595
VK3	Front License Plate Bracket	30,363	15
Y70	70th Anniversary Edition (3LT/3LZ)	5,601	5,995
Z07	Performance Package (Z06)	1,845	8,995
Z51	Performance Package (Stingray)	30,927	6,345
ZF1	Aero Delete for Z51 (deletes front splitter & rear spoiler)	1,841	0
ZYC	Carbon Flash Metallic Painted Outside Mirrors	19,045	195
ZZ3	Engine Appearance Package (coupe)	1,521	995

2024 Options

RPO #	DESCRIPTION	RETAIL $
1YC07	Stingray Coupe 1LT	$67,895
1YC07	Stingray Coupe 2LT	74,995
1YC07	Stingray Coupe 3LT	86,745
1YC67	Stingray Convertible 1LT	74,895
1YC67	Stingray Convertible 2LT	81,995
1YC67	Stingray Convertible 3LT	93,745
1YH07	Z06 Coupe 1LZ	109,695
1YH07	Z06 Coupe 2LZ	118,595
1YH07	Z06 Coupe 3LZ	132,145
1YH67	Z06 Convertible 1LZ	116,695
1YH67	Z06 Convertible 2LZ	125,595
1YH67	Z06 Convertible 3LZ	139,145
1YG07	E-Ray Coupe 1LZ	104,495
1YG07	E-Ray Coupe 2LZ	109,995
1YG07	E-Ray Coupe 3LZ	120,945
1YG67	E-Ray Convertible 1LZ	111,495
1YG67	E-Ray Convertible 2LZ	116,995
1YG67	E-Ray Convertible 3LZ	127,945
379	Orange Seat Belts	495
3A9	Tension Blue Seat Belts	495
3F9	Torch Red Seat Belts	495
3M9	Yellow Seat Belts	495
3N9	Tan Seat Belts	495
5V5	Visible Carbon Fiber Low Spoiler (Z06)	5,995
36S	Yellow Custom Leather Stitch (3LT black interior only)	495
38S	Red Custom Leather Stitch (3LT black interior only)	495
AE4	Competition Sport Seats, 2LT ($995 w/1LT, $500 w/3LT)	1,995

OPTIONS

RPO #	DESCRIPTION	RETAIL $
AH2	GT2 Seats, 2LT (std w/3LT)	1,695
B6P	Engine Appearance Package w/Carbon Fiber Trim (coupe)	895
BAZ	Stealth Interior Trim (2LT/2LZ or 3LT/3LZ)	595
BCP	LT2 Engine Cover in Edge Red (Stingray & E-Ray)	495
BCS	LT2 Engine Cover in Sterling Silver (Stingray & E-Ray)	495
BV4	Personalized Dash Plaque	295
C2M	Carbon Fiber Dual Roof Package	3,495
C2Q	Body-Colored Dual Roof Package	1,995
C2Z	Visible Carbon Fiber Removable Roof	2,495
CC3	Removable Roof Panel, Transparent	995
CFV	Carbon Fiber Visible Ground Effects (Z06 & E-Ray)	3,995
CFZ	Carbon Fiber Painted Ground Effects (Z06 & E-Ray)	2,995
D3V	Engine Bay Lighting (Coupe)	100
D30	Non-Recommended Color Combination Override	695
D84	Carbon Flash Painted Nacelles & Roof (conv)	1,295
D86	Carbon Flash Painted Nacelles & Body Colored Roof (conv)	1,295
DPB	Carbon Flash Dual Full Length Racing Stripes w/Blue Accent	995
DPC	Carbon Flash Dual Full Length Racing Stripes w/Yellow Accent	995
DPG	Carbon Flash Dual Full Length Racing Stripes w/Orange Accent	995
DPL	Carbon Flash Dual Full Length Racing Stripes w/Red Accent	995
DPT	Carbon Flash Dual Full Length Racing Stripes w/Silver Accent	995
DSY	Edge Orange Full Length Dual Racing Stripes	995
DSZ	Edge Red Full Length Dual Racing Stripes	995
DT0	Edge Yellow Full Length Dual Racing Stripes	995
DTB	Full Length Racing Stripes, Electric Blue (E-Ray Only)	995
DTH	Full Length Racing Stripes, Carbon Flash	995
DUB	Full Length Racing Stripes, Sterling Silver	995
DUW	Full Length Racing Stripes, Edge Blue	995
DZU	Carbon Flash/Edge Yellow Stinger Stripe	500
DZV	Carbon Flash/Midnight Silver Stinger Stripe	500
DZX	Carbon Flash/Edge Red Stinger Stripe	500
E60	Front Lift Adjustable Height w/Memory (2LT & 3LT)	2,595
EFA	Exterior Trim Accents, Shadow Gray (Stingray)	995
EFY	Body Color Exterior Trim Accents (incl. E-Ray)	995
ERI	Battery Protection Package	100
EYK	Chrome Exterior Badge Package	295
FA5	Carbon Fiber Interior Trim (2LT & 3LT)	1,500
FA6	Carbon Fiber Interior Trim Level 2 (Z06 & E-Ray, 3LZ only)	4,995
FE2	Magnetic Selective Ride Control (Stingray w/o Z51)	1,895
FE4	Magnetic Selective Ride Control (Stingray w/Z51)	1,895
GC5	Amplify Orange Tintcoat	995
GD0	Accelerate Yellow Metallic	500
GMO	Rapid Blue	500
GPH	Red Mist Metallic Tintcoat	995
GXO	Sea Wolf Gray Tintcoat	995
J6A	Black Brake Calipers (std Z06, n/a w/J57 Brakes)	n/c
J6D	Dark Gray Brake Calipers (std w/Stingray & E-Ray, incl. w/J57 Brakes	n/c
J6E	Edge Yellow Brake Calipers	695
J6F	Bright Red Brake Calipers (Gloss Finish)	695
J6L	Orange Brake Calipers (Z06 w/J57 brakes & E-Ray)	695
J6N	Edge Red Brake Calipers (Darker Shade, Satin Finish)	695
J57	Carbon Ceramic Brakes (Z06, incl. w/Z07)	8,495
N26	Sueded Microfiber-Wrapped Steering Wheel (required AE4)	695
NGA	Black Exhaust Tips (NPP required w/LT2)	395
NPP	Performance Exhaust (incl. w/Z51)	1,195
PBC	LT6 Custom Engine Build Program (Z06)	n/a
PIN	Customer Selectable VIN	5,000
Q8P	Bright Silver 5-Open-Spoke Aluminum Wheels, std Stingray	n/c
Q8Q	Carbon Flash Open Spoke Wheels	995

RPO #	DESCRIPTION	RETAIL $
Q99	20-Spoke Bright Machine-Face Wheels (Stingray)	1,995
Q9A	20-Spoke Midnight Gray Wheels w/Red Stripe (Stingray)	1,495
Q9I	20-Spoke Black Paint Wheels (Stingray)	995
Q9O	Satin Graphite 5-split-spoke Wheels w/machined edge (Stingray)	1,495
Q9Y	Sterling Silver Machine-Face 5-split-spoke Wheels (Stingray)	1,995
R6X	Custom Interior Trim & Seat Combination (w/3LT, $1,390 w/Sueded Microfiber Inserts)	695
R8C	Corvette Museum Delivery w/Personalized Plaque	1,495
ROU	Pearl Nickel 5-Spoke Aluminum Wheels, std E-Ray	n/c
ROX	Carbon Flash Painted Aluminum Wheels w/Machined Edge (E-Ray)	995
ROY	Carbon Flash Painted Carbon Fiber Wheels (Z06 w/J57 Brakes & E-Ray)	9,995
ROZ	Visible Carbon Fiber Wheels (Z06 w/J57 Brakes & E-Ray)	11,995
RYQ	Visible Carbon Fiber Door Intake Trim (Stingray)	3,395
RZ9	Visible Carbon Fiber Grille Insert (Stingray)	2,495
SB7	Corvette Racing Themed Graphics w/Jake & Stingray R Logos (Stingray)	495
SHT	Jake Hood Graphics w/Tech Bronze Accent	495
SLN	Visible Carbon Fiber Engine Cross Brace w/Jake Logo	2,895
SOA	Spider Black Aluminum Wheels (Z06)	495
SOC	Spider Satin Graphite Aluminum Wheels (Z06)	495
SOD	Spider Machine-Face Aluminum Wheels (Z06)	1,495
SOE	Spider Titanium Satin Aluminum Wheels, std Z06	n/c
SOM	Bright Polished Aluminum Wheels (E-Ray)	1,495
SON	Gloss Black Aluminum Wheels (E-Ray)	495
STZ	Visible Carbon Fiber Wheels w/Red Stripe (Z06 w/J57 Brakes & E-Ray)	13,500
T0F	Carbon Aero Package (Z06, incl. CFZ)	8,495
T0G	Visible Carbon Aero Package (Z06, incl. CFV)	10,495
TU7	Two-Tone Seats (2LT & 3LT)	595
TVS	Low Rear Spoiler & Front Splitter (Stingray)	595
UQT	Performance Data & Video Recorder (1LT Only)	1,300
V8X	Exposed Carbon Fiber Sill Plates (Stingray)	1,995
VK3	Front License Plate Bracket	15
Z07	Performance Package (Z06)	8,995
Z51	Z51 Performance Package (Stingray)	6,345
ZER	Performance Package (E-Ray)	500
ZF1	Aero Delete for Z51 (deletes front splitter & rear spoiler)	n/c
ZYC	Carbon Flash Metallic Painted Outside Mirrors	195
ZZ3	Engine Appearance Package (conv)	995

Index

A

Advanced 4 studio, 208
Advanced Concepts Center (ACC), 171, 172, 208
Advanced Engineering, 143, 151–152, 194, 289
Advanced Vehicle engineering (AVE) group, 162
Agner, Bud, 190
Albert, Jon, 211
Allen, Chris, 189
Almond, Dick, 216
ALMS Manufacturer's Championship, 218, 233
Amend, James, 308
American Sunroof Company (ASC), 194
Andretti, Michael, 151
Android Auto, 298
Anthony, Joe, 189, 190
Apple CarPlay, 298
AP Racing, 164
Argosy magazine, 18
Arkus-Duntov, Zora. *See* Duntov, Zora
Auto Age magazine, 29
Automobile magazine, 171, 194, 210, 261, 263
Automobile Manufacturers Association (AMA), 40, 45, 68, 70, 232
Automotive News, 261
AutoWeek magazine, 112, 114, 205, 243, 259, 260, 261, 262, 281

B

Bailey, Shaun, 244
Baker, Buck, 33
Baker, Ken, 162
Baldick, Rick, 217, 219, 222, 223, 229
Balsley, Dick, 162, 164
Barr, Harry, 61
Barra, Mary, 264, 322
Beamon, Bob, 318
Belskus, Jeff, 254
Bennion, Kirk, 263, 288–289, 291
Biden, Joe, 322
Bond, John, 23, 47
Bondurant, Bob, 215
Brown, John, 135
Brunswick Corporation, 189
Burden, Melissa, 290
Bush, George W., 260
Buttner, Dave, 298

C

Cafaro, John, 144–145, 171, 172, 191, 207, 208
Campbell, Jim, 220, 254
Campbell-Ewald ad agency, 31
Car and Driver, 80, 103, 109, 110, 114, 125, 128, 149, 194, 195, 210, 213, 229, 240, 261, 264, 285 285, 290, 299, 304, 307, 309, 312
Caray, Harry, 303
Car Craft magazine, 110
Car Life magazine, 89, 92, 101, 103
Catsburg, Nicky, 296, 313
Ceppos, Rich, 149–150, 171, 194, 304, 305, 309, 312
Charles, Harlan, 226, 241, 267, 288–290, 309
Cherry, Wayne, 207, 208
Chevrolet, Louis, 250–251
Chevrolet-Pontiac-Canada (CPC), 161, 187, 189
Chevy 3 studio, 143, 144, 162, 171, 191, 194, 207, 208
Chitwood, Joie, 238
Claudio, Gary, 260
Cole, Ed, 15, 17, 19, 20, 27, 34, 36, 37, 112, 114, 120, 142, 171, 212
Cooksey, Wil, 203, 241
Cooper, Bill, 161
Corporate Average Fuel Economy (CAFE) standard, 259–260
Corvette Challenge, 160–161
Coventry Climax, 189
Covid-19 pandemic, 285, 297, 299
Crosley Corporation, 18
Csere, Csaba, 194, 210, 213
Curtice, Harlow, 19–20, 33

D

Dahlquist, Eric, 89, 99–100
Danahy, Jim, 309
Davis, David E., Jr., 210
Davis, Denny, 86
Davis, Grady, 65
Daytona 500, 86
Daytona Speed Week, 30, 31, 33, 35
Delco Moraine, 42, 106
DeLorean, John, 95, 101, 111, 112, 114, 141
DeMatio, Joe, 261
Demere, Mac, 199
Detroit International Auto Show, 152, 162–163, 172, 210, 231, 261, 290
Dolza, John, 34
Donnelly, Dick, 189
Duff, Mike, 299
Duke University, 244
Dunne, Jim, 194
Duntov, Zora, 10, 12–13, 15, 17–18, 28, 29–30, 31, 34, 35, 36, 37, 39, 40, 46, 49, 50, 55, 59, 60, 61, 62, 63, 68, 70, 73, 74, 83, 89, 90, 91, 92, 94, 95, 97, 100, 107, 108, 109, 112, 114, 116, 120–121, 141, 142, 143, 144, 154, 171, 184, 217, 219, 220, 232, 246, 259, 264, 279, 286–287, 288, 289, 307, 309
Dura Convertible Systems, Inc., 214
Dustin, Mike, 318
Dyer, Ezra, 263

348

INDEX

E

Earl, Harley, 14, 15, 17, 19, 31, 36, 45, 285
Earl, Jerry, 31, 33
Engle, Barry, 296
Environmental Protection Agency (EPA), 134

F

Falconer, Ryan, 158
Fangio, Juan, 40
Fedders Corporation, 134–135
Fehan, Doug, 233
Fellows, Ron, 218, 237
Fermoyle, Ken, 27, 28
Fishel, Herb, 218
Fitch, John, 30, 31, 36, 40, 286
Fittipaldi, Emerson, 238
Frincke, Fred, 97, 100
Froling, Tom, 230

G

Gamble, Hank, 23
Garcia, Antonio, 296
Gardner, Dustin, 318
Garlits, Don, 294
Gee, Russ, 187, 189
Glasspar company, 19
Goodrich, Bill, 268
Goodyear, Scott, 158
Gordon, Jeff, 273

H

Hagerty magazine, 287
Hall, Jim, 96
Hansgen, Walt, 31
Hayner, Stu, 161
Healey, James, 262
Heinricy, John, 197, 260
Henderson, Fritz, 261

Hendrick Motorsports, 158
Hertz company, 241
Hicks, Jim, 234
Hill, David C., 15, 141, 171, 207, 208, 209, 211, 212, 213, 214, 219, 220, 224, 229, 243, 287, 288, 289–290, 307
Holder, Josh, 30
Hot Rod magazine, 15, 89, 101
Hufstader, Gib, 112, 121
Huntington, Roger, 29, 59

I

Ilmore, 162–163
Indianapolis 500, 89, 125, 141, 151, 175, 207, 229, 237, 238, 252–253, 261, 273, 283
Indianapolis Motor Speedway Hall of Fame museum, 40
International Motor Sports Association (IMSA), 157, 313

J

Jackson, Billy, 203
Jeffords, Jim, 33
Jenkins, Ab, 197
John Deere, 189
Jordan, Charles, 171, 207–208
Juechter, Tadge, 13, 15, 229, 230, 243, 244, 245, 248, 251, 260, 261, 262, 263, 264, 272, 273, 285, 288, 289, 290–291, 292, 294, 296, 299, 304, 307, 308–309, 310, 314, 321
Juriga, John, 212
Jurnecka, Rory, 253

K

Kaiser, Henry, 19
Kansas City Star, 260
Keating, Ben, 313
Keating, Thomas, 19
Kelly, Edward, 133

Kelly, Steve, 101
Kelsey-Hayes company, 123
Kent, Mark, 233
Kimble, David, 189, 191
Koch, Chuck, 101, 103–104
Koerner, Ed, 212
Kott, Douglas, 158
Krambeck, Fernando, 229–230
Krieger, Harold, 49
Kulkarni, Anil, 169, 170, 212
Kurtis, Frank, 18

L

Lamm, John, 165, 172
LaReau, Jamie, 309
Lawrence, Floyd, 23–24
Lee, Jordan, 270
Lee Racing, 158
LeMans. *See* 24 Hours of LeMans
LeMay, Curtis, 18
Leone, Dave, 309
Life magazine, 19
Lind, Kim, 289
Lotus, 151–152, 154, 158, 162, 163, 164, 171, 187, 189–190, 194, 200, 203
Luft, Alex, 291
Ludvigsen, Karl, 31
Lund, Robert, 124, 129
Lutz, Bob, 259, 261, 290

M

MacDonald, Don, 24, 28
Magill, Robert, 133
Majoros, Steve, 303
Massachusetts Institute of Technology, 121
McFarland, Jim, 15
McLaren, Bruce, 96–97
McLean, Robert, 19, 20, 60
McLellan, Dave, 15, 51, 121, 125, 131, 136, 141, 143, 144, 145, 169, 171, 187, 188, 191, 203, 207, 287, 288, 307

349

INDEX

Mears, Rick, 125, 151
Mecom, John, 70
Meegan, Ron, 260
MerCruiser, 189–190
Mercury Marine, 189, 202
Michaelson, Rod, 210
Michigan Technological University, 171
Midgley, Roy, 187, 189–190, 203
Mitchell, Bill, 33, 45–46, 47, 51, 53, 62–63, 69, 71, 77, 89, 101, 241
Models
    Acura NSX, 171
    American Motors AMX/3, 95
    Aston Martin DBS Volante, 255
    Audi R8 RSI Spyder, 255
    Chevrolet
        Camaro, 95, 101, 136, 143, 171
        Chevelle, 108–109, 184
        Vega, 122
    Chrysler 300B, 35
    Corvette
        25th Anniversary, 99, 125, 127
        35th Anniversary, 141, 173
        40th Anniversary, 141, 172, 197, 200, 202, 207
        50th Anniversary, 207, 223, 224, 226, 229, 231, 248, 254
        60th Anniversary, 229, 254, 255
        70th Anniversary, 285, 302, 303, 304, 309
        283, 35–36
        427 Special Edition Z06, 240–241
        Aerovette, 114, 142, 287
        Astro II, 95, 287
        C2, 58–87
        C3, 98–139
        C4, 13, 121, 136, 140–185, 205, 207, 209–211, 213, 214
        C5, 13, 47, 56, 146, 154, 171, 206–227, 288
        C5-R racer, 217–219
        C6, 13, 171, 228–257, 259, 288, 318
        C6.R racer, 232–233, 273
        C7, 13, 258–283, 288, 289, 290, 291, 292, 294, 295, 296, 297, 300, 304, 309
        C7.R racer, 273, 275, 313
        C8, 10, 13, 261, 284–296, 297, 298, 299, 300, 303, 304, 309, 314, 322
        "C8.R Edition" Stingray, 301–302
        C8.R racer, 296, 301–302, 304, 313
        Carbon Limited Edition Z06, 248–249, 255
        Centennial Edition (2012), 249–251, 255
        CERV I, 49–50, 59, 94, 287
        CERV II, 73–74, 94, 143, 152, 287
        CERV III, 162–165
        CERV IV, 209
        Championship Edition Z06, 243
        Collector Edition (1982), 125, 136, 138
        Collector Edition (1996), 178, 181, 184
        Collector Edition 427 Convertible (2013), 254
        Commemorative Edition (2004), 226
        Competition Sport Package (2009), 242
        Corvette Challenge, 160–161
        Corvette Indy, 143, 151–152, 154, 163
        E-Ray, 285, 309–312, 314
        EVs, future of, 322
        Four-Rotor Corvette, 114
        Grand Sport, 68–71, 181, 184–185, 246, 248, 250, 277, 279–280
        GS 3, 94
        GT1 Championship Edition (2009), 242–243
        GT3.R racer, 313
        GTP, 157–158
        Launch Edition Stingrays, 300
        Limited Edition Corvette (1978), 125, 127, 129
        LT1, 122, 169–171, 173, 178, 199, 203, 212, 264, 267, 277, 294
        LT4, 178, 181, 270
        LT5, 122, 152, 163, 164, 187–191, 194–195, 197, 199, 200, 203
        Mako Shark I, 53–54
        Mako Shark II, 77–79, 89, 91, 92–93, 101
        Manta Ray, 101
        "Opel," 20
        Q-Corvette, 45, 47, 49, 60–62, 288
        Ron Fellows ALMS GT1 Champion Edition, 229, 237
        SR-2, 31, 33
        SS racer, 36–40, 49, 286
        Stingray, 45–47, 53, 241–243, 261–262, 263–264, 267, 269, 272, 273, 275–277, 279–280, 291, 292, 294, 295, 296, 298, 301, 303, 304, 309, 314, 322
        Sting Ray, 10, 13, 50, 51, 53, 59–60, 63, 65, 68, 69–70, 71, 80, 82, 89, 90, 91, 93, 213, 241, 272, 321, 322
        Stingray (C7), 262, 290, 291
        Sting Ray III, 171–172, 208
        Two-Rotor Corvette, 112
        XP-64, 37
        XP-700, 51, 53, 62
        XP-720, 62
        XP-755, 53–54, 62, 77
        XP-819, 94

INDEX

XP-880, 94–95
XP-882, 94–95, 111, 112, 142, 287
XP-895, 110–111
XP-897GT, 112, 114
Z01, 156–157, 172–173
Z4Z, 175, 178, 237–238
Z06, 64–65, 68, 99, 220–222, 222–223, 226, 233–234, 236–237, 238, 240–241, 242–243, 246, 248, 251, 253, 259, 267, 269–273, 275–277, 288, 295, 296, 303, 304–307, 309, 313, 314, 315, 318, 321
Z07, 116, 167, 253, 273, 305, 307, 321
Z15, 184
Z16, 184
Z25, 172–173, 202, 255, 279
Z44, 240–241
Z51, 147, 149, 150, 151, 156, 158, 160, 167, 184–185, 214, 231–232, 236, 242, 267, 269, 276, 279, 292, 295, 298, 300, 301, 304, 312, 315
Z52, 156, 157
ZHZ, 241
ZLC, 250–251
ZR-1, 104, 106, 109, 172–173, 184–185, 186–205, 213
ZR1, 191, 197, 199, 202, 242, 243–245, 246, 248–249, 253, 259, 281, 283, 288, 292, 295, 309, 314–321
ZR-1 Spyder, 194, 255
Crosley Hot Shot, 18
Dodge Viper, 12, 200
Ferrari California, 255
Ford
    GT, 12
    Model T, 59
    Mustang, 298
    Mustang Mach-E, 322
    Pantera, 95
    Thunderbird, 17, 28
Jaguar D-type, 33, 36, 38
Kaiser Darrin, 19
Kurtis Kraft, 18
Lola T-600, 157–158
Lotus Super 7, 171
MG
    TC, 171
    TD, 17
Mormon Meteor III, 197
Porsche
    904, 90
    911E, 103–104
    911 Turbo S Cabriolet, 255
    914, 144
    928, 144
Shelby Cobra, 70
Woodill Wildfire, 19
Molded Fiber Glass Company, 20
Morrison, Mac, 260–261
Morrison, Tommy, 197
Morrison Motorsports, 197
Mosher, William, 24
Mosport, 70
Moss, Stirling, 40
Motorama auto show, 19–20, 23, 120
Motor Life magazine, 23, 27, 101
Motor Trend magazine, 23, 24, 28, 33–34, 59, 99, 100, 103–104, 106–107, 146, 195, 199, 213, 216, 243–244, 253, 260, 296, 322
Muscaro, Dave, 234

N

Nash, Doug, 143, 158
Nash-Healey, 18
Nassau Speed Weeks, 70
National Corvette Museum, 13, 40, 145, 203, 261
Negri, Joe, 217
New York International Auto Show, 77, 79, 95
Nichols, Bill, 229
Nickey Chevrolet, 33

O

O'Blenes, Russ, 313
Okulski, Travis, 291
Olley, Maurice, 19, 20, 23
Ongais, Danny, 125
Opszynski, Ron, 189
Owens Corning, 19

P

Palmer, Jerry, 143–145, 158, 162, 191, 207
Paris Automobile Salon, 79
Payne, Henry, 291
Peerless Automotive, 158
Penfound, Barry, 194
Peper, Ed, 238, 240
Performance Build Center, 291
Perkins, Jim, 175, 203, 208–209
Perry, Chris, 254
Peters, Tom, 208, 230, 261, 263
Piatek, Ed, 11, 299, 303, 307, 309, 310, 312, 322
Piggins, Vince, 96–97, 101, 158
Powell, John, 161
Powell Development America, 161
Powertrain Engineering, 187
Premo, Jim, 20
Preve, Fran, 100
Protofab, 161
Pund, Daniel, 229
Purdy, Ken, 18

R

Rahal, Bobby, 151
Reuss, Lloyd, 187–188, 191, 194, 203, 207
Reuss, Mark, 13, 263, 281, 307, 309, 322
Reynolds Metals Company, 111
Road America, 31, 53, 158
Road & Track magazine, 23, 28, 29, 34, 35, 41, 47, 51, 82, 93, 95, 103, 114, 142, 150, 154, 158, 162, 163, 165, 172, 194, 210, 244, 253, 263, 270, 290, 291, 318
Road Atlanta, 158
Robinson, Doug, 191
Rochester Products, 34
Rolex 24 endurance race, 218, 295
Rolls-Royce, 19
Roma, Tony, 309, 321
Rose, Mauri, 26–27
Rudd, Tony, 162, 187, 190
Runkle, Don, 162, 165, 188, 194
Rusz, Joe, 154, 163

S

Schaafsma, Fred, 143
Schag, Timothy, 236
Scheidt, Rick, 251
Schinella, John, 171, 208
Schulke, Terri, 294
Schwartz, Randy, 309
Sebring. *See* 12 Hours of Sebring
Sessions, Ron, 194
Shaw, Bill, 309
Shelby, Carroll, 70
Sherman, Don, 261, 285–287, 290, 291, 292
Shinoda, Larry, 45, 53–54, 59, 62, 77, 89–90
Simcoe, Michael, 13
Skelton, Betty, 30–31
Sloan School of Management (MIT), 121
Smith, John "Jack," 203, 205, 207, 209
Smith, Kevin, 149
Smith, Roger, 187, 207
Smith, Sam, 270
Sneva, Tom, 125
Society of Automotive Engineers (SAE), 17–18, 20, 36, 111–112, 188–189
Specialized Vehicles Inc. (SVI), 161
Specialty Equipment Manufacturers Association (SEMA), 194
Sperry, Ron, 212, 218
Sports Car Club of America (SCCA), 18, 33, 46, 47, 65, 160, 161, 171, 232
Sports Car Graphic magazine, 93, 107
Sports Car Illustrated magazine, 29, 31, 51
Sports Car International, 215
St. Antoine, Arthur, 243–244
Stefanyshyn, Gene, 14, 309
Stein, Ralph, 18
Stempel, Robert, 141, 163, 203, 207
Stielow, Mark, 313
Stinson, Terry, 189–190
Stone, Matt, 260
Stracke, Karl-Friedrich, 261
Studio 3. *See* Chevy 3 studio
Sullivan, Danny, 151

T

Taruffi, Piero, 40
Taylor, Jordan, 296
Thompson, Dick "The Flying Dentist," 46
Transformers: Revenge of the Fallen, 241
True magazine, 18
12 Hours of Sebring, 30–31, 33, 34, 35, 36, 37, 38, 39, 40, 73, 232–233, 286
24 Hours of LeMans, 37, 40, 73, 83, 218, 232, 234, 242, 248

U

University of Michigan, 171
Unser, Al, Sr., 125
USA Today, 262, 291

V

Van Valkenburgh, Paul, 93, 107
Varrone, Nico, 313
Villeneuve, Jacques, 158

W

Wagoner, Rick, 243, 244, 290
Wakefield, Ron, 95
Wallace, Tom, 15, 245, 260–261, 288, 290, 307, 309
Wall Street Journal, 126–127
Ward's Auto World, 308
Webster, Larry, 240
Welburn, Ed, 261–262
Wesoloski, Steve, 233
Winchell, Frank, 94, 101
Winegarden, Sam, 231
Winters Foundry, 97
Woron, Walt, 33–34

Y

Yeager, Chuck, 151

Z

Zak, Phil, 321
Zeyte, Walt, 49
Zimmerman, Dave, 229–230